Methods in Molecular Biology

Series Editor
John M. Walker
School of Life and Medical Sciences
University of Hertfordshire
Hatfield, Hertfordshire, UK

For further volumes:
http://www.springer.com/series/7651

For over 35 years, biological scientists have come to rely on the research protocols and methodologies in the critically acclaimed *Methods in Molecular Biology* series. The series was the first to introduce the step-by-step protocols approach that has become the standard in all biomedical protocol publishing. Each protocol is provided in readily-reproducible step-by-step fashion, opening with an introductory overview, a list of the materials and reagents needed to complete the experiment, and followed by a detailed procedure that is supported with a helpful notes section offering tips and tricks of the trade as well as troubleshooting advice. These hallmark features were introduced by series editor Dr. John Walker and constitute the key ingredient in each and every volume of the *Methods in Molecular Biology* series. Tested and trusted, comprehensive and reliable, all protocols from the series are indexed in PubMed.

Phytoplasma

Methods and Protocols

Edited by

Katrin Janik and Mattia Tabarelli

Functional Genomics, Laimburg Research Centre, Auer/Ora, Italy

Humana Press

Editors
Katrin Janik
Functional Genomics
Laimburg Research Centre
Auer/Ora, Italy

Mattia Tabarelli
Functional Genomics
Laimburg Research Centre
Auer/Ora, Italy

ISSN 1064-3745 ISSN 1940-6029 (electronic)
Methods in Molecular Biology
ISBN 978-1-0716-5103-2 ISBN 978-1-0716-5104-9 (eBook)
https://doi.org/10.1007/978-1-0716-5104-9

This Humana imprint is published by the registered company Springer Science+Business Media, LLC, part of Springer Nature.
The registered company address is: 1 New York Plaza, New York, NY 10004, U.S.A.

Preface

This edition of *Phytoplasma: Methods and Protocols* builds upon a rich legacy of methodological development, offering an expanded and updated set of tools for detection, classification, vector biology, functional studies, and genomic analysis of these elusive plant pathogens.

The chapters gathered here reflect both the breadth of the field and its rapid technological progress: portable field diagnostics, advanced molecular assays, refined insect–pathogen interaction studies, and powerful sequencing workflows in combination with established laboratory protocols.

We thank the contributing authors for sharing their expertise and the readers for carrying forward the work of understanding and controlling phytoplasma-associated diseases. We hope this volume serves as both a practical laboratory companion and an inspiration for research and future innovations in the field.

Auer/Ora, Italy

Katrin Janik
Mattia Tabarelli

Contents

Contributors

MAKO AKAHORI • *Department of Agricultural and Environmental Biology, Graduate School of Agricultural and Life Sciences, The University of Tokyo, Tokyo, Japan*
ALBERTO ALMA • *Department of Agricultural, Forest and Food Sciences, University of Turin, Turin, Italy*
ELISA ANGELINI • *CREA Research Centre for Viticulture and Enology, Conegliano, Italy*
HARVINDER BENNYPAUL • *Canadian Food Inspection Agency, Sidney Laboratory, Centre for Plant Health, North Saanich, BC, Canada*
ASSUNTA BERTACCINI • *Alma Mater Studiorum, University of Bologna, Bologna, Italy*
EDOARDO BERTINI • *Department of Biotechnology, University of Verona, Verona, Italy*
KAJOHN BOONROD • *Plant Pathogen Diagnostic Laboratory and Plant Protection, Dienstleitungszentrum Ländlicher Raum, DLR, Rheinhessen-Nahe-Hunsrück, Bad Kreuznach, Germany*
MATTEO BRILLI • *Department of Biosciences, University of Milan, Milan, Italy*
YANGHUI CAO • *Illinois Natural History Survey, Prairie Research Institute, University of Illinois at Urbana-Champaign, Champaign, IL, USA; Key Laboratory of Plant Protection Resources and Pest Management of Ministry of Education, College of Plant Protection, Northwest A&F University, Yangling, Shaanxi, China*
YI-CHING CHIU • *Institute of Biochemistry, National Chung Hsing University, Taichung, Taiwan; Doctoral Program in Microbial Genomics, National Chung Hsing University and Academia Sinica, Taichung, Taiwan*
GRAZIANA DA ROLD • *U.O. Ecopathology SCT2-Belluno, Istituto Zooprofilattico Sperimentale delle Venezie, Belluno, Italy*
MARINA DERMASTIA • *Department of Biotechnology and Systems Biology, National institute of Biology, Ljubljana, Slovenia*
CHRISTOPHER H. DIETRICH • *Illinois Natural History Survey, Prairie Research Institute, University of Illinois at Urbana-Champaign, Champaign, IL, USA*
BOJAN DUDUK • *Institute of Pesticides and Environmental Protection, Laboratory of Phytopathology, Belgrade, Serbia*
TIM J. DUMONCEAUX • *Canadian Food Inspection Agency, Sidney Laboratory, Centre for Plant Health, North Saanich, BC, Canada; Agriculture and Agri-Food Canada Saskatoon Research and Development Centre, Saskatoon, SK, Canada*
LUISA FILIPPIN • *CREA Research Centre for Viticulture and Enology, Conegliano, Italy*
AMIR FINE • *Functional Genomics, Laimburg Research Centre, Auer/Ora, BZ, Italy*
LUCA FONTANESI • *Functional Genomics, Laimburg Research Centre, Auer/Ora, Italy*
ANDREAS GALLMETZER • *Virology and Diagnostics, Laimburg Research Centre, Auer/Ora, Italy*
LUCA GALTAROSSA • *Functional Genomics, Laimburg Research Centre, Auer/Ora, BZ, Italy*
ELENA GONELLA • *Department of Agricultural, Forest and Food Sciences, University of Turin, Turin, Italy*
CHRISTINE HAMMOND • *Agriculture and Agri-Food Canada Saskatoon Research and Development Centre, Saskatoon, SK, Canada*
ZAHRA GOLESTANI HOTKANI • *Department of Biosciences, University of Milan, Milan, Italy; Research and Innovation Centre, Edmund Mach Foundation, San Michele, Italy*

Nozomu Iwabuchi • *Department of Agricultural and Environmental Biology, Graduate School of Agricultural and Life Sciences, The University of Tokyo, Tokyo, Japan*
Katrin Janik • *Functional Genomics, Laimburg Research Centre, Auer/Ora, BZ, Italy*
Hiroaki Koinuma • *Department of Agriculture, School of Agriculture, Tokai University, Kumamoto, Japan*
Chih-Horng Kuo • *Doctoral Program in Microbial Genomics, National Chung Hsing University and Academia Sinica, Taichung, Taiwan; Institute of Biotechnology, National Chung Hsing University, Taichung, Taiwan; Institute of Plant and Microbial Biology, Academia Sinica, Taipei, Taiwan*
Kensaku Maejima • *Department of Agricultural and Environmental Biology, Graduate School of Agricultural and Life Sciences, The University of Tokyo, Tokyo, Japan*
Helen Mae Mejia • *Institute of Biochemistry, National Chung Hsing University, Taichung, Taiwan*
Akio Miyazaki • *Department of Agricultural and Environmental Biology, Graduate School of Agricultural and Life Sciences, The University of Tokyo, Tokyo, Japan*
Mirko Moser • *Research and Innovation Centre, Edmund Mach Foundation, San Michele, Italy*
Kevin Muirhead • *Agriculture and Agri-Food Canada Saskatoon Research and Development Centre, Saskatoon, SK, Canada*
Rita Musetti • *Department of Land, Environment, Agriculture and Forestry, TESAF, University of Padova, Legnaro, PD, Italy*
Shigetou Namba • *Department of Agricultural and Environmental Biology, Graduate School of Agricultural and Life Sciences, The University of Tokyo, Tokyo, Japan*
Maruša Pompe Novak • *Department of Biotechnology and Systems Biology, National institute of Biology, Ljubljana, Slovenia; University of Nova Gorica, Nova Gorica, Slovenia*
Anna Pedroncelli • *Functional Genomics, Laimburg Research Centre, Auer/Ora, Italy*
Mario Pezzotti • *Department of Biotechnology, University of Verona, Verona, Italy*
Margot Raffeiner • *Faculty of Biology and Biotechnology, Ruhr University Bochum, Bochum, Germany*
Daniel Sanderson • *Canadian Food Inspection Agency, Sidney Laboratory, Centre for Plant Health, North Saanich, BC, Canada*
Justin Shih • *Molecular Plant Pathology Laboratory, Beltsville Agricultural Research Center, Agricultural Research Service, United States Department of Agriculture, Beltsville, MD, USA*
Rebeka Strah • *Department of Biotechnology and Systems Biology, National institute of Biology, Ljubljana, Slovenia*
Mattia Tabarelli • *Functional Genomics, Laimburg Research Centre, Auer/Ora, Italy*
Rosemarie Tedeschi • *Department of Agricultural, Forest and Food Sciences (DISAFA), University of Turin, Turin, Italy*
Valeria Trivellone • *Illinois Natural History Survey, Prairie Research Institute, University of Illinois at Urbana-Champaign, Champaign, IL, USA*
Wei Wei • *Molecular Plant Pathology Laboratory, Beltsville Agricultural Research Center, Agricultural Research Service, United States Department of Agriculture, Beltsville, MD, USA*
Yasuyuki Yamaji • *Department of Agricultural and Environmental Biology, Graduate School of Agricultural and Life Sciences, The University of Tokyo, Tokyo, Japan*
Jun-Yi Yang • *Institute of Biochemistry, National Chung Hsing University, Taichung, Taiwan; Doctoral Program in Microbial Genomics, National Chung Hsing University and*

Academia Sinica, Taichung, Taiwan; Institute of Biotechnology, National Chung Hsing University, Taichung, Taiwan; Advanced Plant and Food Crop Biotechnology Center, National Chung Hsing University, Taichung, Taiwan

YINONG YANG • *Department of Plant Pathology and Environmental Microbiology and the Huck Institute of the Life Sciences, The Pennsylvania State University, University Park, PA, USA*

SARA ZENONI • *Department of Biotechnology, University of Verona, Verona, Italy*

Chapter 1

Phytoplasmas: An Introduction

Mattia Tabarelli and Katrin Janik

Abstract

In this introductory chapter, we provide an overview of phytoplasma biology and outline the historical milestones that shaped the field, from their first discovery to their current taxonomic status. We also highlight how this third edition of "Phytoplasmas: Methods and Protocols" differs from previous volumes, reflecting both the adoption of new molecular and sequencing technologies and the revival of classical approaches. Taken together, these developments illustrate the evolving trajectory of phytoplasma research and set the stage for the detailed protocols presented in the following chapters.

Key words Detection, Molecular diagnostics, Vector–pathogen interactions, Functional characterization, Comparative genomics

Invisible to the naked eye, elusive, and devastating: phytoplasmas are wall-less bacterial bacteria that colonize the phloem of plants, triggering a cascade of symptoms in their host plants. With no cure in sight and hundreds of plant species affected, the urgency to understand and manage phytoplasma-associated diseases has never been greater. This book is a sequel to the molecular toolbox designed to help detect, identify, and decode phytoplasmas and seeks to offer researchers and diagnosticians a comprehensive guide to the latest methods and protocols in the field.

Phytoplasmas are confined to the phloem tissue of plants and are associated to diseases affecting diverse plant species of agricultural value [1, 2]. Their economic impact is considerable, and currently, no effective curative treatments exist. The development of rapid and sensitive detection methods, along with the ability to investigate phytoplasma–host interactions, is essential for managing diseases associated with these pathogens.

In 1967, Doi and colleagues identified mycoplasma-like organisms in sieve tubes of plants with yellowing symptoms and witches' broom, and in 1994 these bacteria were designated with the trivial name "Phytoplasma" [3]. Ten years later, they were officially grouped in the genus '*Candidatus* Phytoplasma'

Katrin Janik and Mattia Tabarelli (eds.), *Phytoplasma: Methods and Protocols*, Methods in Molecular Biology, vol. 3008,
https://doi.org/10.1007/978-1-0716-5104-9_1,

[4]. The provisional status "*Candidatus*" refers to microorganisms that cannot be cultured in pure culture and are incompletely characterized according to the International Committee on Systematics of Prokaryotes [5, 6]. Due to their unavailability in pure culture, the Koch postulates cannot be fulfilled to confirm their role as pathogens [7, 8]. Phytoplasmas belong to the class *Mollicutes*, they colonize the plant phloem and insect–vector tissues and are characterized by very small genomes with a low G+C content [3, 9]. Transmission occurs via insect feeding from infected plants, i.e., phloem-feeding insects can acquire phytoplasmas, which then replicate within their bodies and are newly released via plant feeding [10, 11].

Detection and identification of phytoplasmas are essential for understanding their epidemiology and developing effective management strategies across different crops and environments.

Molecular techniques have improved identification, but challenges persist due to low pathogen concentrations, variable symptoms, seasonal fluctuations, and environmental factors affecting detection reliability. An updated overview of current phytoplasma identification and classification strategies can be found in Chapter 2.

Being the third edition of the book series *Phytoplasmas: Methods and Protocols*, a neat basis and a solid practical background for researchers is now paved. Considering the book *Plant and Insect Mycoplasma Techniques* from 1982 [12] as the precursor of the series, major shifts in techniques can be observed.

1 Detection and Diagnostics

Over four decades, detection has shifted from time-consuming, expert-driven microscopy and serology toward ultra-sensitive, DNA-based diagnostic technologies (Chapters 3, 5, 6, and 7), with PCR remaining the gold standard, while isothermal methods such as LAMP and CRISPR-based assays have further advanced rapid and field-applicable diagnostics. The widely adopted iPhyClassifier tool, initially presented in an earlier edition of this book series, has become a cornerstone for the classification of phytoplasmas [13]. Additionally, the development of anchored hybrid enrichment (AHE) as described in Chapter 4 represents a novel, highly sensitive approach for both their detection and genotyping. The inclusion of a universal endogenous internal control in multiplexed detection protocols (Chapter 5) ensures maximum flexibility and quality control across diverse eukaryotic samples. These advancements, integrated with robust bioinformatic pipelines, emphasize the continued centrality of accurate diagnostics as a frontline tool for phytoplasma disease management.

2 Vector Biology and Interaction Studies

The study of insect vectors has evolved from primarily morphological and ecological observations in 1982 to advanced transmission and cell culture–based functional studies (Chapters 8 and 10), along with integrated bioinformatic workflows for phytoplasma discovery and characterization in insect vectors (Chapter 9), reflecting a shift from descriptive approaches toward mechanistic and systems-level understanding of vector–phytoplasma interactions.

3 Functional Characterization and Experimental Systems

Functional characterization of phytoplasmas and their interactions has progressed from basic microscopy and biochemical methods used in earlier decades to sophisticated experimental systems, including protocols for plant transformation and protoplast cultivation (Chapter 11), recombinant protein expression in *Escherichia coli* (Chapter 12), and transient expression in planta using agroinfiltration (Chapter 13). This is complemented by detailed molecular assays such as in vitro ubiquitination to study effector functions (Chapter 14), advanced immunolabeling techniques for electron microscopy (Chapter 15), and the use of model organisms like *Spiroplasma melliferum* to explore potential antimicrobials (Chapter 16), illustrating a shift from observational to molecular and functional experimental approaches.

4 Genomics and Sequencing

Genomics and sequencing of phytoplasmas have evolved dramatically from the early reliance on limited molecular markers toward advanced, high-resolution techniques. This evolution is exemplified by the use of multiplexed multilocus sequence typing with hybridization probe–based enrichment for precise phytoplasma identification (Chapter 17), comprehensive genome reconstruction workflows from DNA enrichment to assembly (Chapter 18), and the application of cutting-edge long-read nanopore sequencing on high-molecular-weight DNA to obtain complete phytoplasma genome sequences (Chapter 19). This progression reflects the integration of increasingly sophisticated sequencing technologies and bioinformatic tools, enabling in-depth genomic insights that were previously unattainable.

5 Future Perspectives

The past decades of phytoplasma research have been characterized by a succession of technological leaps and the transformation from labor-intensive, low-throughput approaches to highly sensitive, integrated, and data-rich methodologies. The different areas covered in this book reflect this trajectory: diagnostics have embraced DNA-based analyses and portable, field-ready assays; vector biology has moved from morphological observations to in vitro cellular systems; functional studies now employ molecular, structural, and biochemical toolkits in tandem; and genomics has advanced into high-resolution, long-read sequencing combined with bioinformatic pipelines. These developments have not only increased the speed and accuracy of diagnostics but have also shifted the scope of phytoplasma research from descriptive and confirmatory studies toward predictive, mechanistic, and systems-level understanding.

References

1. Bertaccini A, Duduk B, Paltrinieri S et al (2014) Phytoplasmas and phytoplasma diseases: a severe threat to agriculture. AJPS 5: 1763–1788. https://doi.org/10.4236/ajps.2014.512191
2. Wang R, Bai B, Li D et al (2024) Phytoplasma: a plant pathogen that cannot be ignored in agricultural production—research progress and outlook. Mol Plant Pathol 25. https://doi.org/10.1111/mpp.13437
3. Gundersen DE, Lee IM, Rehner S-A et al (1994) Phylogeny of mycoplasmalike organisms (phytoplasmas): a basis for their classification. J Bacteriol 176:5244–5254. https://doi.org/10.1128/jb.176.17.5244-5254.1994
4. IRPCM Phytoplasma/Spiroplasma Working Team - Phytoplasma taxonomy group (2004) *Candidatus* Phytoplasma', a taxon for the wall-less, non-helical prokaryotes that colonize plant phloem and insects. Int J Syst Evol Microbiol 54:1243–1255. https://doi.org/10.1099/ijs.0.02854-0
5. Oren A, Arahal DR, Göker M et al (2023) International code of nomenclature of prokaryotes. Prokaryotic Code (2022 revision). Int J Syst Evol Microbiol 73:005585. https://doi.org/10.1099/ijsem.0.005585
6. Murray RG, Schleifer KH (1994) Taxonomic notes: a proposal for recording the properties of putative taxa of procaryotes. Int J Syst Bacteriol 44:174–176. https://doi.org/10.1099/00207713-44-1-174
7. Bertaccini A, Arocha-Rosete Y, Contaldo N et al (2022) Revision of the '*Candidatus* Phytoplasma' species description guidelines. Int J Syst Evol Microbiol 72:005353. https://doi.org/10.1099/ijsem.0.005353
8. Koch R (2018) Die Ätiologie der Tuberkulose (1884). In: Gradmann C (ed) Robert Koch: Zentrale Texte. Springer Berlin Heidelberg, Berlin, Heidelberg, pp 133–137
9. Weisburg WG, Tully JG, Rose DL et al (1989) A phylogenetic analysis of the mycoplasmas: basis for their classification. J Bacteriol 171: 6455–6467. https://doi.org/10.1128/jb.171.12.6455-6467.1989
10. Bourtzis K, Miller TA (2006) Insect symbiosis, vol 2. CRC Press. https://doi.org/10.1201/9781420005936
11. Hogenhout SA, Oshima K, Ammar E-D et al (2008) Phytoplasmas: bacteria that manipulate plants and insects. Mol Plant Pathol 9:403–423. https://doi.org/10.1111/j.1364-3703.2008.00472.x
12. Daniels MJ, Markham PG (1982) Plant and insect mycoplasma techniques. Springer Netherlands, Dordrecht
13. Zhao Y, Wei W, Lee I-M et al (2013) The iPhyClassifier, an interactive online tool for phytoplasma classification and taxonomic assignment. Methods Mol Biol 938:329–338. https://doi.org/10.1007/978-1-62703-089-2_28

Part I

Detection and Diagnostics: Molecular Methods for Detection and Identification

Chapter 2

Phytoplasma Detection and Identification: Tools for Their Classification

Assunta Bertaccini

Abstract

A summary of the situation for detecting and naming phytoplasmas associated with different diseases in the last 30 years is presented in this chapter. In the last decades, a great amount of research has been devoted to classifying and differentiating these prokaryotes by molecular and bioinformatic tools applied to the analysis of specific phytoplasma genes. The availability of a robust classification system has greatly facilitated phytoplasma identification, leading to increased knowledge of their associated plant diseases worldwide. Scientific knowledge on phytoplasma biology still needs to be improved to allow better and more focused management solutions with minimal impact on the environment to mitigate the severity of these diseases in both agricultural and natural environments, also contributing to reduce the loss of biodiversity.

Key words Molecular methods, '*Candidatus* Phytoplasma' species, Provisional taxonomy, Bacteria, Plant disease

1 Introduction

Phytoplasma-associated diseases are a major limiting factor for the quality and productivity of many ornamental, horticultural, and other economically important agricultural crops worldwide. Annual losses due to phytoplasma diseases in many crops vary, but under pathogen-favorable conditions, they always lead to disastrous consequences for the farming community. Phytoplasmas are unique, cell wall-less bacteria that inhabit the phloem tissue of plants and possess a unique lifestyle that allows them to live in hosts from both the plant and insect kingdoms. They are transmitted by insect vectors such as leafhoppers and planthoppers. Additional transmission routes for these bacteria occur through dodder, grafting, micropropagation, and seeds [1]. The latter has been recently demonstrated for phytoplasmas infecting several agriculturally relevant crops such as tomato, eggplant, carrot, corn, alfalfa, and coconut [2–5]. Phytoplasmas are associated with

Katrin Janik and Mattia Tabarelli (eds.), *Phytoplasma: Methods and Protocols*, Methods in Molecular Biology, vol. 3008, https://doi.org/10.1007/978-1-0716-5104-9_2,

economically significant plant diseases, leading to economic losses in agriculture and horticulture worldwide [6].

Detection and identification of phytoplasmas are of basic importance for studying their epidemiology and proposing appropriate management strategies for phytoplasma-associated diseases in various environments and in diverse crops or landscapes. Their detection and identification have been achieved from decades; however, their classification still poses challenges due to difficulties in cultivating them in artificial media to obtain pure isolates [7]. In particular, the elimination of plant bacterial endophytes from the colonies containing phytoplasmas is still under study [8, 9]. Phytoplasmas are commonly associated with symptoms such as yellowing, stunting, witches' broom, and abnormal flower development [10]. Their early detection is crucial for managing and mitigating their disease impact in diverse agricultural and environmental situations. Accurate identification is necessary to detect their presence, tracking their epidemiology, and devising control strategies. Despite the advancements and availability of several molecular techniques, the phytoplasma identification still faces several old challenges due to their low concentrations within host tissues, host-specific symptom variability, seasonal variations in concentration, and presence of environmental stress that can influence their identification.

2 Tools for Phytoplasma Detection

Phytoplasmas have variable sizes and shapes (Fig. 1) and survive and multiply in isotonic environments provided by plant phloem and insect hemolymph.

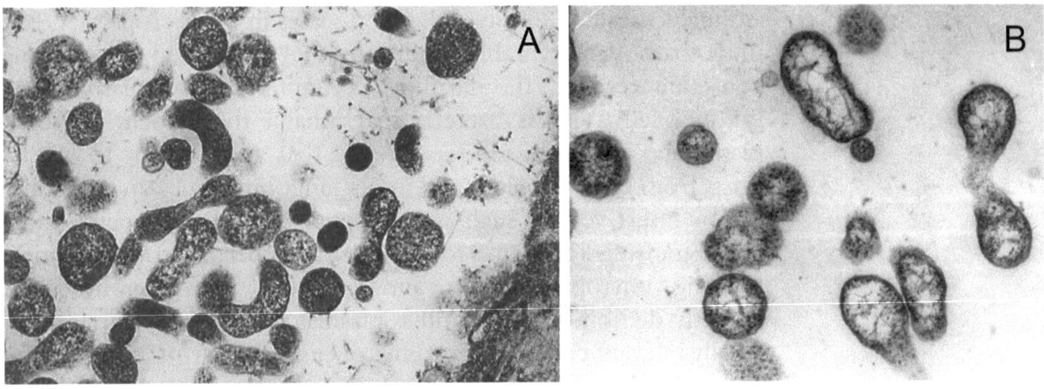

Fig. 1 Electron microscopy pictures of '*Candidatus* Phytoplasma asteris' in sieve tubes of infected gladiolus plants. In **a**, it is visible their pleomorphic shape in cells containing cytoplasm and nuclear material, while in **b**, the phytoplasmas show partial degeneration of internal structures, suggesting an initial degradation (magnifications: **a**, 8000×; **b**, 12,000×)

Fig. 2 (**a**) Sample collection from a coconut palm infected with lethal yellowing ('*Ca*. P. palmicoa') in Equatorial Guinea using the trunk boring system (courtesy of L. Rovesti). (**b**) Apple tree cultivar Renetta del Canada, about 60 years old, in Val di Sole (Italy), infected with apple proliferation phytoplasmas ('*Ca*. P. mali'). (Courtesy of B. Duduk)

They possess one of the smallest genomes among living organisms and encode metabolic pathways that allow them to interact with both their plant and insect hosts. The full genome sequence has been completed for a number of phytoplasma strains, together with a growing number of draft genome sequences [11]. Phytoplasmas are able to increase the metabolic activity of their hosts, modify insect fitness, and modify the behavior of infected plants. However, it is not uncommon to detect phytoplasmas in asymptomatic plants, and this leaves open questions related to their pathogenicity and biology as relevant characteristics that must be addressed together with phytoplasma identification when describing a new phytoplasma-associated disease.

Proper collection and storage of samples are also critical for a successful phytoplasma detection. Samples should be collected from symptomatic or asymptomatic (for latent infection) plant tissues, particularly leaf midribs, stems, phloem scrapes, or trunk borings (respectively, for dicotyledonous or monocotyledonous plant species) or flowers (Fig. 2). After appropriate preparation, they should be maintained under cool conditions (4°C) or frozen (20°C) for immediate or late processing, respectively.

Microscopy Techniques: Microscopy is one of the initial tools that was used for phytoplasma detection. Fluorescent microscopy can reveal phytoplasma presence in plant tissue by staining the DNA inside the phytoplasmas localized in the phloem cells with

DNA-binding dyes. Transmission electron microscopy was key for the phytoplasma (named at the time mycoplasma-like organisms, MLO) discovery [12] and is used to visualize phytoplasma size and shape in samples. However, both microscopic methods lack the specificity required for conclusive identification unless coupled with the use of specific antisera with appropriate methodology for preparation and staining.

Serological Methods: Serological techniques such as enzyme-linked immunosorbent assay (ELISA) and immunofluorescence assays (IFAS) have been employed for phytoplasma detection and tentative differentiation. These methods rely on antibodies specific to phytoplasma antigens. Serology is less commonly used due to the difficulty in obtaining phytoplasma-specific antigens. Moreover, phytoplasmas have high genomic plasticity and show large diversity in their membrane proteins; therefore, antisera preparation must be done for the specific strain associated with the studied epidemic. Among others, antibodies have been prepared against partial sequences of the major immunodominant proteins of some phytoplasmas [13]. More recently, it has been possible to produce antisera from phytoplasma colonies [14] these were able to detect aster yellows phytoplasmas infecting grapevine (GY-AY) by using IFAS. The IFAS positive control consisted of colonies isolated from infected grapevine samples stored at -80°C and previously used for rabbit immunization and antiserum production [14, 15]. The developed protocol could be used for the direct detection of phytoplasmas in herbaceous hosts or indirectly, after phytoplasma isolation and colony production.

Molecular Diagnostics: Several molecular methods are employed to identify phytoplasmas, such as polymerase chain reaction (PCR) and nested PCR, quantitative PCR (qPCR), loop-mediated isothermal amplification (LAMP), digital droplet PCR (ddPCR), DNA Sanger sequencing, and high-throughput sequencing technologies. Not all these techniques allow direct identification of phytoplasmas; therefore, further amplicon sequencing and/or restriction fragment length polymorphism (RFLP) analyses are required to identify or differentiate the detected phytoplasmas [16]. Next-generation sequencing (NGS) and microarray analysis provide comprehensive insights into phytoplasma (draft) genomes and complex host–pathogen microbiome interactions. NGS may allow phytoplasma detection and the study of some of their genetic variability. The development of molecular techniques, especially those incorporating advanced computational analyses, has allowed researchers to improve phytoplasma detection. Additionally, the integration of high-throughput sequencing technologies has facilitated deeper insight into phytoplasma genomes, revealing significant genetic diversity that challenges their provisional taxonomy, mainly based on comparison of their whole 16S rRNA gene. A preliminary differentiation based on this gene can be achieved by using RFLP analyses on its amplicons; this supports the differentiation of phytoplasmas into ribosomal groups and subgroups [17, 18]. Although widely used, this method is not always the most reliable for distinguishing phytoplasmas, as it has often resulted in the proposal of ribosomal subgroups that do not correspond to biologically or epidemiologically relevant features. In some cases, however, the ribosomal subgroup is the only available category to distinguish phytoplasma strains in specific areas or crops and thus the only possible differentiation for epidemiological studies. In several cases, subgrouping resulted in genetic lineages or, in some cases, artifacts deriving from incorrect or problematic cloning results. The practical suggestion is to confirm ribosomal group/subgroup assignments generated by virtual RFLP analyses with experimental RFLP analyses in the laboratory to prevent the generation and use of ribosomal groups that lack epidemiological relevance or are the result of molecular artifacts.

3 Phytoplasma Classification

Phytoplasmas are not yet available as pure culture, since in the majority of cases the percentage of phytoplasma detected in colonies is lower than 100%, and endophytic bacteria are often co-cultured with them [19]. Despite the repeatable shape of the colonies (Fig. 3) obtained after phytoplasma isolation in liquid culture, it is still not possible to utilize rules for species description accepted for culturable prokaryotes.

Phytoplasma 16S rRNA genes are highly conserved; however, they possess sufficient diversity to be used for classification. Sequencing an amplified 16S rRNA gene provides the possibility to identify the phytoplasmas as '*Candidatus* Phytoplasma' species, the provisional rank used for their official identification and designation [20]. The officially described '*Ca.* Phytoplasma' species are listed in Table 1. The '*Ca.* Phytoplasma' genus was proposed to accommodate cell-wall-less bacteria that are molecularly and biochemically incompletely characterized and colonize plant phloem and insect tissues. This is a provisional classification, but it is highly relevant due to its application in epidemiological and ecological studies that are strongly needed to keep plant diseases associated with the presence of phytoplasmas under control worldwide.

Efforts to standardize and refine classification criteria have become increasingly critical to ensure uniformity and clarity in the designation of new '*Ca.* Phytoplasma' species. After the first designation of '*Ca.* Phytoplasma' as a new genus [20], recent guidelines

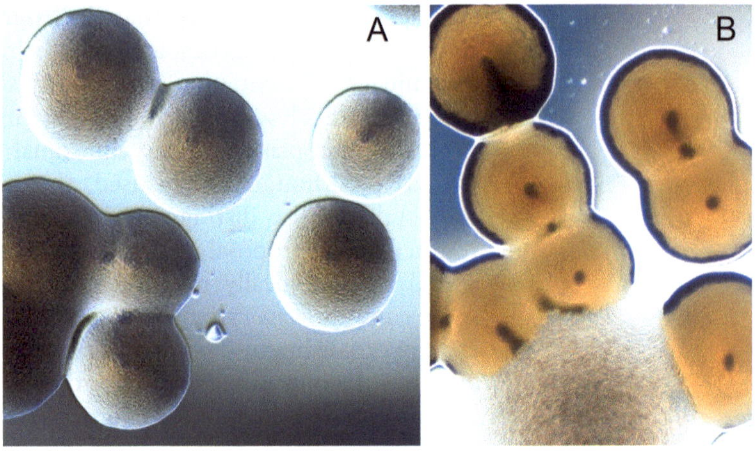

Fig. 3 (**a**) Colonies containing phytoplasmas obtained from different infected plant species after 72 h of incubation under microaerophilic conditions at 24 ± 1 °C on solid CB medium after initial isolation in CB solid medium. (**b**) A diverse colony is present, probably due to the presence of bacterial endophytes. Colonies were photographed under a binocular microscope at 40× magnification (N. Contaldo and A. Bertaccini, unpublished)

[7] updated the threshold for '*Ca.* Phytoplasma' species description based on the 16S rRNA gene from 97.50% to 98.65%, in agreement with the value used for other bacterial species. According to these guidelines, the designation of a new species must meet at least one of the following criteria: (i) the whole 16S rRNA gene sequence identity is <98.65% when compared to published '*Ca.* Phytoplasma' species; or (ii) the ANI value is of <95% when the 16S rRNA gene identity is >98.65%. If the 16S rRNA gene identity exceeds 98.65%, the description of a new species must be supported by the divergence of housekeeping genes (*tufB*, *secY*, *secA*, *rplV-rpsC*, *groEl*) at pre-established identity thresholds. The very last possibility, in case that all the above results are not sufficient for differentiation, is to use some of the biological features as reported in (IRPCM, 2004) [20] stating that "...the phytoplasmas have unique insect vectors, distinct host ranges or symptomatology in a host, and unique molecular or serological characteristics..." but this is usually not sufficient to support the proposal for a new taxon especially considering the lack of such type of information in the majority of the newly discovered phytoplasmas.

A barcode screening system was also developed in agreement with the 16S rRNA gene classification for fast phytoplasma detection and identification [21]. The *rp*, *tuf*, *groEl*, *secA*, *leu*, and *secY* genes show more variation than the 16S rRNA gene and can be very useful for epidemiological studies and for differentiation of genetically close strains within very similar '*Ca.* Phytoplasma' species [22].

Table 1
List of the published '*Ca.* Phytoplasma' species with molecular and geographical details (*phytoplasmas detected once or rarely)

'*Ca.* Phytoplasma' species	16S ribosomal subgroup (16Sr)	Geographic distribution (Continent)				GenBank accession number(s) of the reference strain/whole genome		Literature
		Europe	America	Australasia	Africa			
'*Ca.* P. allocasuarinae'*	XXXIII-A			+		AY725236		[23]
'*Ca.* P. americanum'	XVIII-A		+			DQ174122	MN227133	[24]
'*Ca.* P. asteris'	I-B	+	+	+	+	M30790	AP006628	[25]
'*Ca.* P. aurantifolia=citri'	II-B			+	+	U15442		[26]
'*Ca.* P. australasiae= australasiaticum'	II-D	+		+		Y10097	JQ868448	[27]
'*Ca.* P. australiense'	XII-B			+		L76865	AM422018	[28]
'*Ca.* P. balanitae'	V-F			+		AB689678		[29]
'*Ca.* P. bonamiae'*	nd			+			JAOSIQ000000000	[30]
'*Ca.* P. brasiliense'	XV-A	+	+			AF147708		[31]
'*Ca.* P. caricae'*	XVII-A		+			AY725234		[32]
'*Ca.* P. castaneae'*	XIX-A			+		MW264918		[33]
'*Ca.* P. cirsii'*	XI-E	+				KR869146		[34]
'*Ca.* P. cocostanzaniae'*	IV-C				+	X80117		[7]
'*Ca.* P. convolvuli'	XII-H	+				JN833705		[35]
'*Ca.* P. costaricanum'*	XXXI-A		+			HQ225630		[36]
'*Ca.* P. cynodontis'	XIV-A	+		+		AJ550984	KP019340	[37]
'*Ca.* P. dypsidis'	XXXIX-A			+		MT536195		[38]
'*Ca.* P. fabacearum'*	nd			+			JAOSIM000000000	[30]

(continued)

Table 1
(continued)

'Ca. Phytoplasma' species	16S ribosomal subgroup (16Sr)	Geographic distribution (Continent)				GenBank accession number(s) of the reference strain/whole genome		Literature
		Europe	America	Australasia	Africa			
'*Ca*. P. fragariae'	XII-E	+				HM104662	MK501641	[39]
'*Ca*. P. fraxini'	VII-A	+	+	+		AF092209		[40]
'*Ca*. P. graminis'*	XVI-A		+			AY725228		[32]
'*Ca*. P. hispanicum'	XIII-A		+			AF248960		[41]
'*Ca*. P. japonicum'	XII-D			+		AB010425		[42]
'*Ca*. P. luffae'*	VIII-A			+		AF248956	AF353090	[43]
'*Ca*. P. lycopersici'*	I-Y		+			EF199549		[44]
'*Ca*. P. malaleucae'*	nd			+			JAOSID000000000	[30]
'*Ca*. P. malaysianum'	XXXII-A			+		EU371934		[45]
'*Ca*. P. mali"	X-A	+		+		AJ542541	CU469464	[46]
'*Ca*. P. meliae'	XIII-G		+			KU850940	NZ_JAGVRH010000003	[47]
'*Ca*. P. noviguineense'	XXXVIII-A			+		LC228755		[48]
'*Ca*. P. omanense'	XXIX-A			+		EF666054		[49]
'*Ca*. P. oryzae'*	XI-A			+		AB052873		[50]
'*Ca*. P. palmae'	IV-A		+			U18747	VBRA02000009	[7]
'*Ca*. P. palmicola'	XXII-A				+	KF751387		[51]
'*Ca*. P. phoenicium'	IX-B	+		+		AF515636	KM275491	[52]
'*Ca*. P. pini'	XXI-A	+	+			AJ632155	VIAE01000001	[53]
'*Ca*. P. pruni'	III-A	+	+			JQ044397		[54]

'*Ca*. P. prunorum'	X-B	+		+		AJ542545	JQ868450	[46]
'*Ca*. P. pyri'	X-C	+	+	+		AJ542543		[46]
'*Ca*. P. rhamni'*	XX-A	+				X76431	JQ868449	[23]
'*Ca*. P. rubi'	V-E	+				AY197648		[55]
'*Ca*. P. sacchari'	XI-B			+		MN889545	VWXM01000002	[56]
'*Ca*. P. solani'	XII-A	+	+	+	+	AF248959		[57]
'*Ca*. P. spartii'	X-D	+				X92869		[23]
'*Ca*. P. stylosantis'*	XXXVII-A			+		MT431550	JQ730740	[58]
'*Ca*. P. sudamericanum'*	VI-I		+			GU292081		[59]
'*Ca*. P. tamaricis'	XXX-A			+		MW447513		[60]
'*Ca*. P. trifolii'	VI-A	+	+	+		AY390261		[61]
'*Ca*. P. tritici'	I-C	+	+	+		DQ078304	AVAO01000003	[62]
'*Ca*. P. ulmi'	V-A	+	+			AY197655	OU413475	[63]
'*Ca*. P. vignae'*	nd			+		PP575759	PP575760	[64]
'*Ca*. P. woydetiae'*	XXXVI-A			+		KY069029		[65]
'*Ca*. P. ziziphi'	V-B	+	+	+		AB052876	CP025121	[66]

However, using only genetic differentiation to characterize phytoplasmas could end in producing just lists of genotypes if knowledge of the respective phytoplasma biology is not accompanying the taxonomy. As the scientific community continues to classify '*Ca.* Phytoplasma' species, it is essential to adhere to the published standards to ensure reproducibility and maintain accuracy of the provisional taxonomy. There are currently 53 '*Ca.* Phytoplasma' species (plus the "flavescence dorée," which is not a '*Ca.* Phytoplasma' species but retains the trivial name of the disease to which it is associated, considering its relevant quarantine situation in Europe) validly published. Some other '*Ca.* Phytoplasma' species were published but not in the official journal for taxonomy (*i.e.*, the International Journal of Systematic and Evolutionary Microbiology), so those species names cannot be used by researchers in scientific publications since they were not validly published.

To correctly classify a phytoplasma, it is necessary to BLAST its 16S rRNA gene sequence against phytoplasma reference strains in the GenBank (https://blast.ncbi.nlm.nih.gov/Blast.cgi) or run it in the *i*PhyClassifier (https://plantpathology.ba.ars.usda.gov/cgi-bin/resource/iphyclassifier.cgi) or in the EPPO-QBank database classification tool (https://qbank.eppo.int/); in the latter, the majority of officially published '*Ca.* Phytoplasma' species are deposited and available.

4 Conclusion

The field of phytoplasma detection and identification is rapidly modifying and updating. The integration of artificial intelligence (AI)-powered bioinformatics tools and advanced imaging technologies holds promise for more efficient detection processes. Developing portable diagnostic kits for field use would greatly benefit agricultural practices.

Detecting and identifying phytoplasmas is integral to managing their associated plant diseases and safeguarding crop production. By employing robust tools and adhering to established guidelines, researchers and agricultural professionals can ensure accurate identification, paving the way for effective disease-control measures. Continued advancements in technology and international collaboration will be keys to overcoming the challenges posed by these enigmatic pathogens.

References

1. Bertaccini A, Weintraub PG, Rao GP et al (2019) Phytoplasmas: plant pathogenic bacteria-II transmission and management of phytoplasma associated diseases. Springer, Singapore. https://doi.org/10.1007/978-981-13-2832-9

2. Satta E, Paltrinieri S, Bertaccini A (2019) Phytoplasma transmission by seed. In: Bertaccini A, Weintraub PG, Rao GP, Mori N (eds) Phytoplasmas: plant pathogenic bacteria-II transmission and management of phytoplasma associated diseases. Springer, Singapore. https://doi.org/10.1007/978-981-13-2832-9
3. Mateeti ST, Checchi G, Messina NA et al (2022) Presence and seed transmission of phytoplasmas in tomato fields in Italy. Phytopathog Mollicutes 12(1):1–6. https://doi.org/10.5958/2249-4677.2023.00029.4
4. Darabakula M, Mateeti ST, Pacini F et al (2024) Eggplant little leaf-associated phytoplasma detection in seedlings under insect-proof conditions. Int J Plant Biol 15:217–229. https://doi.org/10.3390/ijpb15020018
5. Lu H, Wilson B, Zhang H et al (2024) Detection and identification of Bogia coconut syndrome phytoplasma from seed-associated tissues and seedlings of coconut (*Cocos nucifera*) and betel nut (*Areca catechu*). Sci Rep 14:11542. https://doi.org/10.1038/s41598-024-61916-4
6. Bertaccini A, Duduk B, Paltrinieri S et al (2014) Phytoplasmas and phytoplasma diseases: a severe threat to agriculture. Am J Plant Sci 5:1763–1788. https://doi.org/10.4236/ajps.2014.512191
7. Bertaccini A, Arocha-Rosete Y, Contaldo N et al (2022) Revision of the '*Candidatus* Phytoplasma' species description guidelines. Int J Syst Evol Microbiol 72:005353. https://doi.org/10.1099/ijsem.0.005353
8. Contaldo N, Satta E, Zambon Y et al (2016) Development and evaluation of different complex media for phytoplasma isolation and growth. J Microbiol Methods 127:105–110. https://doi.org/10.1016/j.mimet.2016.05.031
9. Zakurin A, Li J, Barbieri C, Pacini F et al (2024) Candidatus Phytoplasma ziziphi' co-cultivation from micropropagated jujube witches' broom infected shoots. Phytopathog Mollicutes 14(2):74–80. https://doi.org/10.5958/2249-4677.2024.00009.6
10. Bertaccini A (2022) Plants and phytoplasmas: when bacteria modify plants. Plants 11:1425. https://doi.org/10.3390/plants11111425
11. Kube M, Duduk B, Oshima K (2019) Genome sequencing. In: Bertaccini A, Oshima K, Kube M et al (eds) Phytoplasmas: plant pathogenic bacteria - III genomics, host pathogen interactions, and diagnosis. Springer, Singapore. https://doi.org/10.1007/978-981-13-9632-8
12. Doi YM, Teranaka M, Yora K et al (1967) Mycoplasma or PLT-group-like microorganisms found in the phloem elements of plants infected with mulberry dwarf, potato witches' broom, aster yellows, or paulownia witches' broom. Jpn J Phytopathol 33:259–266. https://doi.org/10.3186/jjphytopath.33.259
13. Bertaccini A, Fiore N, Zamorano A et al (2019) Molecular and serological approaches in detection of phytoplasmas in plants and insects. In: Bertaccini A, Oshima K, Kube M et al (eds) Phytoplasmas: plant pathogenic bacteria - III genomics, host pathogen interactions, and diagnosis. Springer, Singapore. https://doi.org/10.1007/978-981-13-9632-8
14. Contaldo N, Freo I, Lucchese C et al (2019) Preliminary evaluation of the use of an antiserum obtained from phytoplasma culture. Phytopathog Mollicutes 9(1):201–202. https://doi.org/10.5958/2249-4677.2019.00101.4
15. Contaldo N, Formica L, Bertaccini A (2022) Suitability of IFAS test to detect phytoplasmas associated with grapevine yellows. Phytopathog Mollicutes 12(1):60. https://doi.org/10.5958/2249-4677.2022.00025.1
16. Bertaccini A, Lee I-M (2018) Phytoplasmas: an update. In: Rao GP, Bertaccini A, Fiore N et al (eds) Phytoplasmas: plant pathogenic bacteria-I. Characterization and epidemiology of phytoplasma-associated diseases. Springer, Singapore. https://doi.org/10.1007/978-981-13-0119-3
17. Lee I-M, Gundersen-Rindal DE, Davis RE et al (1998) Revised classification scheme of phytoplasmas based on RFLP analysis of 16S rRNA and ribosomal protein gene sequences. Int J Syst Bacteriol 48:1153–1169. https://doi.org/10.1099/00207713-48-4-1153
18. Zhao Y, Wei W, Lee I-M et al (2009) Construction of an interactive online phytoplasma classification tool, *i*PhyClassifier, and its application in analysis of the peach X-disease phytoplasma group (16SrIII). Int J Syst Evol Microbiol 59:2582–2593. https://doi.org/10.1099/ijs.0.010249-0
19. Contaldo N, Bertaccini A, Paltrinieri S et al (2012) Axenic culture of plant pathogenic phytoplasmas. Phytopathol Mediterr 51(3):607–617. https://doi.org/10.14601/Phytopathol_Mediterr-11773
20. IRPCM (2004) '*Candidatus* Phytoplasma', a taxon for the wall-less, non-helical prokaryotes that colonise plant phloem and insects. Int J Syst Evol Microbiol 54:1243–1255. https://doi.org/10.1099/ijs.0.02854-0
21. Makarova O, Contaldo N, Paltrinieri S et al (2013) DNA barcoding for phytoplasma identification. Methods Mol Biol 938:301–317.

https://doi.org/10.1371/journal.pone.0052092

22. Martini M, Quaglino F, Bertaccini A (2019) Multilocus genetic characterization of phytoplasmas. In: Bertaccini A, Oshima K, Kube M et al (eds) Phytoplasmas: plant pathogenic bacteria - III genomics, host pathogen interactions, and diagnosis. Springer, Singapore. https://doi.org/10.1007/978-981-13-9632-8
23. Marcone C, Gibb KS, Streten C et al (2004) '*Candidatus* Phytoplasma spartii', '*Candidatus* Phytoplasma rhamni' and '*Candidatus* Phytoplasma allocasuarinae', respectively associated with spartium witches' broom, buckthorn witches' broom and allocasuarina yellows diseases. Int J Syst Evol Microbiol 54: 1025–1029. https://doi.org/10.1099/ijs.0.02838-0
24. Lee I-M, Bottner KD, Secor G et al (2006) '*Candidatus* Phytoplasma americanum', a phytoplasma associated with a potato purple top wilt disease complex. Int J Syst Evol Microbiol 56:1593–1597. https://doi.org/10.1099/ijs.0.64251-0
25. Lee I-M, Gundersen-Rindal DE, Davis RE et al (2004) Candidatus Phytoplasma asteris', a novel taxon associated with aster yellows and related diseases. Int J Syst Evol Microbiol 54: 1037–1048. https://doi.org/10.1099/ijs.0.02843-0
26. Zreik L, Carle P, Bové J-M et al (1995) Characterization of the mycoplasma-like organism associated with witches' broom disease of lime and proposition of a '*Candidatus*' taxon for the organism, '*Candidatus* Phytoplasma aurantifolia'. Int J Syst Bacteriol 45:449–453. https://doi.org/10.1099/00207713-45-3-449
27. White DT, Blackall LL, Scott PT et al (1998) Phylogenetic positions of phytoplasmas associated with dieback, yellow crinkle and mosaic diseases of papaya, and their proposed inclusion in '*Candidatus* Phytoplasma australiense' and a new taxon, '*Candidatus* Phytoplasma australasiae'. Int J Syst Bacteriol 48:941–951. https://doi.org/10.1099/00207713-48-3-941
28. Davis RE, Dally EL, Gundersen DE et al (1997) 'Candidatus Phytoplasma australiense', a new phytoplasma taxon associated with Australian grapevine yellows. Int J Syst Bacteriol 47:262–269. https://doi.org/10.1099/00207713-47-2-262
29. Win NKK, Lee S-Y, Bertaccini A et al (2013) '*Candidatus* Phytoplasma balanitae' associated with witches' broom disease of *Balanites triflora*. Int J Syst Evol Microbiol 63:636–640. https://doi.org/10.1099/ijs.0.041566-0
30. Rodrigues JB, Tran-Nguyen LTT, Gambley C et al (2023) The observation of taxonomic boundaries for the 16SrII and 16SrXXV phytoplasmas using genome-based delimitation. Int J Syst Evol Microbiol 73(7):005977. https://doi.org/10.1099/ijsem.0.005977
31. Montano HG, Davis RE, Dally EL et al (2001) '*Candidatus* Phytoplasma brasiliense', a new phytoplasma taxon associated with hibiscus witches' broom disease. Int J Syst Evol Microbiol 51:1109–1118. https://doi.org/10.1099/00207713-51-3-1109
32. Arocha Y, López M, Piñol B et al (2005) '*Candidatus* Phytoplasma graminis' and '*Candidatus* Phytoplasma caricae', two novel phytoplasmas associated with diseases of sugarcane, weeds and papaya in Cuba. Int J Syst Evol Microbiol 55:2451–2463. https://doi.org/10.1099/ijs.0.63797-0
33. Jung H-Y, Sawayanagi T, Kakizawa S et al (2002) '*Candidatus* Phytoplasma castaneae', a novel phytoplasma taxon associated with chestnut witches' broom disease. Int J Syst Evol Microbiol 52:1543–1549. https://doi.org/10.1099/00207713-52-5-1543
34. Šafářová D, Zemánek T, Válová P et al (2016) '*Candidatus* Phytoplasma cirsii', a novel taxon from creeping thistle [*Cirsium arvense* (L.) Scop]. Int J Syst Evol Microbiol 66:1745–1753. https://doi.org/10.1099/ijsem.0.000937
35. Martini M, Marcone C, Mitrović J et al (2012) '*Candidatus* Phytoplasma convolvuli', a new phytoplasma taxon associated with bindweed yellows in four European countries. Int J Syst Evol Microbiol 62:2910–2915. https://doi.org/10.1099/ijs.0.038612-0
36. Lee I-M, Bottner-Parker KD, Zhao Y et al (2011) '*Candidatus* Phytoplasma costaricanum' a novel phytoplasma associated with an emerging disease in soybean (*Glycine max*). Int J Syst Evol Microbiol 61:2822–2826. https://doi.org/10.1099/ijs.0.029041-0
37. Marcone C, Schneider B, Seemüller E (2004) '*Candidatus* Phytoplasma cynodontis', the phytoplasma associated with Bermuda grass white leaf disease. Int J Syst Evol Microbiol 54:1077–1082. https://doi.org/10.1099/ijs.0.02837-0
38. Jones LM, Pease B, Perkins SL et al (2021) '*Candidatus* Phytoplasma dypsidis', a novel taxon associated with a lethal wilt disease of palms in Australia. Int J Syst Evol Microbiol 71(5). https://doi.org/10.1099/ijsem.0.004818
39. Valiunas D, Staniulis J, Davis RE (2006) '*Candidatus* Phytoplasma fragariae', a novel phytoplasma taxon discovered in yellows diseased

strawberry, *Fragaria* x *ananassa*. Int J Syst Evol Microbiol 56:277–281. https://doi.org/10.1099/ijs.0.63935-0

40. Griffiths HM, Sinclair WA, Smart CD et al (1999) The phytoplasma associated with ash yellows and lilac witches' broom: '*Candidatus* Phytoplasma fraxini'. Int J Syst Bacteriol 49: 1605–1614. https://doi.org/10.1099/00207713-49-4-1605
41. Davis RE, Harrison NA, Zhao Y et al (2016) '*Candidatus* Phytoplasma hispanicum', a novel taxon associated with Mexican periwinkle virescence disease of *Catharanthus roseus*. Int J Syst Evol Microbiol 66:3463–3467. https://doi.org/10.1099/ijsem.0.001218
42. Sawayanagi T, Horikoshi N, Kanehira T et al (1999) '*Candidatus* Phytoplasma japonicum', a new phytoplasma taxon associated with Japanese *Hydrangea* phyllody. Int J Syst Bacteriol 49:1275–1285. https://doi.org/10.1099/00207713-49-3-1275
43. Davis RE, Zhao Y, Wei W et al (2017) '*Candidatus* Phytoplasma luffae', a novel taxon associated with witches' broom disease of loofah, *Luffa aegyptica* Mill. Int J Syst Evol Microbiol 67:3127–3133. https://doi.org/10.1099/ijsem.0.001980
44. Arocha Y, Antesana O, Montellano E et al (2007) '*Candidatus* Phytoplasma lycopersici', a phytoplasma associated with "hoja de perejil" disease in Bolivia. Int J Syst Evol Microbiol 57: 1704–1710. https://doi.org/10.1099/ijs.0.64851-0
45. Nejat N, Vadamalai G, Davis RE et al (2013) '*Candidatus* Phytoplasma malaysianum', a novel taxon associated with virescence and phyllody of Madagascar periwinkle (*Catharanthus roseus*). Int J Syst Evol Microbiol 63: 540–548. https://doi.org/10.1099/ijs.0.041467-0
46. Seemüller E, Schneider B (2004) '*Candidatus* Phytoplasma mali', '*Candidatus* Phytoplasma pyri' and '*Candidatus* Phytoplasma prunorum', the causal agents of apple proliferation, pear decline and European stone fruit yellows, respectively. Int J Syst Evol Microbiol 54:1217–1226. https://doi.org/10.1099/ijs.0.02823-0
47. Fernández FD, Galdeano E, Kornowski MV et al (2016) Description of '*Candidatus* Phytoplasma meliae', a phytoplasma associated with Chinaberry (*Melia azedarach* L.) yellowing in South America. Int J Syst Evol Microbiol 66:5244–5251. https://doi.org/10.1099/ijsem.0.001503
48. Miyazaki A, Shigaki T, Koinuma H et al (2018) '*Candidatus* Phytoplasma noviguineense', a novel taxon associated with Bogia coconut syndrome and banana wilt disease on the Island of New Guinea. Int J Syst Evol Microbiol 68: 170–175. https://doi.org/10.1099/ijsem.0.002480
49. Al-Saady NA, Khan AJ, Calari A et al (2008) '*Candidatus* Phytoplasma omanense', associated with witches' broom of *Cassia italica* (Mill.) Spreng in Oman. Int J Syst Evol Microbiol 58:461–466. https://doi.org/10.1099/ijs.0.65425-0
50. Jung H-Y, Sawayanagi T, Wongkaew P et al (2003) '*Candidatus* Phytoplasma oryzae', a novel phytoplasma taxon associated with rice yellow dwarf disease. Int J Syst Evol Microbiol 53:1925–1929. https://doi.org/10.1099/ijs.0.043315-0
51. Harrison NA, Davis RE, Oropeza C et al (2014) '*Candidatus* Phytoplasma palmicola', associated with a lethal yellowing-type disease of coconut (*Cocos nucifera* L.) in Mozambique. Int J Syst Evol Microbiol 64:1890–1899. https://doi.org/10.1099/ijs.0.060053-0
52. Verdin E, Salar P, Danet J-L et al (2003) '*Candidatus* Phytoplasma phoenicium' sp. nov., a novel phytoplasma associated with an emerging lethal disease of almond trees in Lebanon and Iran. Int J Syst Evol Microbiol 53:833–838. https://doi.org/10.1186/s12866-015-0487-4
53. Schneider B, Torres E, Martín MP et al (2005) '*Candidatus* Phytoplasma pini', a novel taxon from *Pinus silvestris* and *Pinus halepensis*. Int J Syst Evol Microbiol 55:303–307. https://doi.org/10.1099/ijs.0.63285-0
54. Davis RE, Zhao Y, Dally EL et al (2013) '*Candidatus* Phytoplasma pruni', a novel taxon associated with X-disease of stone fruits, *Prunus* spp.: multilocus characterization based on 16S rRNA, *secY*, and ribosomal protein genes. Int J Syst Evol Microbiol 63:766–776. https://doi.org/10.1099/ijs.0.041202-0
55. Malembic-Maher S, Salar P, Filippin L et al (2011) Genetic diversity of European phytoplasmas of the 16SrV taxonomic group and proposal of '*Candidatus* Phytoplasma rubi'. Int J Syst Evol Microbiol 61:2129–2134. https://doi.org/10.1099/ijs.0.025411-0
56. Kirdat K, Tiwarekar B, Thorat V et al (2020) '*Candidatus* Phytoplasma sacchari', a novel taxon - associated with sugarcane grassy shoot (SCGS) disease. Int J Syst Evol Microbiol 71: 004591. https://doi.org/10.1099/ijsem.0.004591
57. Quaglino F, Zhao Y, Casati P et al (2013) '*Candidatus* Phytoplasma solani', a novel taxon associated with "stolbur"- and "bois noir"-related diseases of plants. Int J Syst Evol

Microbiol 63:2879–2894. https://doi.org/10.1099/ijs.0.044750-0

58. Rodrigues JR, Kinoti WM, Tran-Nguyen LTT et al (2021) '*Candidatus* Phytoplasma stylosanthis', a novel taxon with a diverse host range in Australia, characterised using multilocus sequence analysis of 16S rRNA, secA, tuf, and rp genes. Int J Syst Evol Microbiol 71:004589. https://doi.org/10.1099/ijsem.0.004589
59. Davis RE, Zhao Y, Dally EL et al (2012) '*Candidatus* Phytoplasma sudamericanum', a novel taxon, and strain PassWB-Br4, a new subgroup 16SrIII-V phytoplasma, from diseased passion fruit (*Passiflora edulis* f. *flavicarpa* Deg.). Int J Syst Evol Microbiol 62:984–989. https://doi.org/10.1099/ijs.0.033423-0
60. Zhao Y, Sun Q, Wei W et al (2009) '*Candidatus* Phytoplasma tamaricis', a novel taxon discovered in witches'-broom-diseased salt cedar (*Tamarix chinensis* Lour.). Int J Syst Evol Microbiol 59:2496–2504. https://doi.org/10.1099/ijs.0.010413-0
61. Hiruki C, Wang K (2004) Clover proliferation phytoplasma: '*Candidatus* Phytoplasma trifolii'. Int J Syst Evol Microbiol 54:1349–1353. https://doi.org/10.1099/ijs.0.02842-0
62. Zhao Y, Wei W, Davis RE et al (2021) The agent associated with blue dwarf disease in wheat represents a new phytoplasma taxon, '*Candidatus* Phytoplasma tritici'. Int J Syst Evol Microbiol 71(1). https://doi.org/10.1099/ijsem.0.004604
63. Lee I-M, Martini M, Marcone C et al (2004) Classification of phytoplasma strains in the elm yellows group (16SrV) and proposal of '*Candidatus* Phytoplasma ulmi' for the phytoplasma associated with elm yellows. Int J Syst Evol Microbiol 54:337–347. https://doi.org/10.1099/ijs.0.02697-0
64. Rodrigues JB, Tran-Nguyen, Gambley C et al (2024) '*Candidatus* Phytoplasma vignae', assigning a species description to a long-known phytoplasma occurring in northern Australia. Int J Syst Evol Microbiol 74(8). https://doi.org/10.1099/ijsem.0.006502
65. Naderali N, Nejat N, Vadamalai G et al (2017) '*Candidatus* Phytoplasma wodyetiae', a new taxon associated with yellow decline disease of foxtail palm (*Wodyetia bifurcata*) in Malaysia. Int J Syst Evol Microbiol 67:3765–3772. https://doi.org/10.1099/ijsem.0.002187
66. Jung H-Y, Sawayanagi T, Kakizawa S et al (2003) '*Candidatus* Phytoplasma ziziphi', a novel phytoplasma taxon associated with jujube witches' broom disease. Int J Syst Evol Microbiol 53:1037–1041. https://doi.org/10.1099/ijs.0.02393-0

Chapter 3

End-Point PCR for Universal Detection of Phytoplasmas Based on the 16S rRNA Gene

Bojan Duduk

Abstract

End-point PCR, followed by restriction fragment length polymorphism (RFLP) or sequencing and sequence analysis of the 16S rRNA gene, is a cornerstone method for the universal detection and identification of phytoplasmas, including previously undescribed strains. This protocol describes the use of end-point PCR in direct, nested, or semi-nested systems, coupled with RFLP analysis using 17 restriction enzymes, to classify phytoplasmas into distinct 16Sr ribosomal groups and subgroups while addressing challenges such as inter-operon heterogeneity and mixed infections. Alternatively, sequencing of the PCR-amplified 16S rRNA gene, followed by sequence analysis (e.g., virtual RFLP or sequence homology), enables classification into either 16Sr groups/subgroups or '*Candidatus* Phytoplasma' species. The method's sensitivity, cost-effectiveness, and compatibility with established classification frameworks make it invaluable for epidemiological studies, quarantine measures, and the delineation of '*Candidatus* Phytoplasma' species. By providing a clear framework for the precise diagnosis of phytoplasma-associated diseases in diverse plant and insect hosts, this protocol supports rapid responses to outbreaks and helps mitigate the economic impact of phytoplasmas.

Key words Phytoplasma, End-point PCR, Detection, Identification, Ribosomal group, Ribosomal subgroup, '*Candidatus* Phytoplasma'

1 Introduction

Phytoplasmas, fastidious plant-pathogenic bacteria, are associated with severe diseases in diverse crops, resulting in significant economic losses worldwide [1]. Their low concentration, particularly in woody hosts, and uneven distribution within plant sieve tubes pose challenges for detection, requiring sensitive and reliable diagnostic tools [2]. Since phytoplasmas fundamentally cannot be cultivated in axenic culture in vitro, conventional culture-dependent methods for bacterial identification—such as morphological observation, biochemical assays, serotyping, and evaluation of antibiotic inhibition or resistance patterns—are not applicable. Detection methods such as graft transmission to indicator plants,

Katrin Janik and Mattia Tabarelli (eds.), *Phytoplasma: Methods and Protocols*, Methods in Molecular Biology, vol. 3008, https://doi.org/10.1007/978-1-0716-5104-9_3,

electron microscopy (EM), and DNA-specific staining with DAPI are labor-intensive and lack specificity [3–5]. Early identification and diagnosis of phytoplasmas and their associated diseases are critical for developing and implementing effective control strategies. This helps reduce disease spread, limits economic losses, and ensures compliance with international trade regulations. Although phytoplasma infections often manifest as distinct symptoms, asymptomatic infections have also been reported [6].

The development of polymerase chain reaction (PCR) targeting the 16S rRNA gene (16S rDNA), first sequenced in 1989 [7], marked a turning point in enabling universal detection and classification of phytoplasmas [8–11]. End-point PCR, particularly in nested or semi-nested formats, remains the gold standard for phytoplasma detection due to its sensitivity, cost-effectiveness, and compatibility with restriction fragment length polymorphism (RFLP) analysis. The 16S rRNA gene, with its conserved yet variable regions, serves as the primary target for both detection and taxonomic classification. The first comprehensive classification scheme, based on RFLP analysis of PCR-amplified 16S rRNA gene sequences, has evolved into a robust system delineating 38 ribosomal groups and more than 100 subgroups [12, 13]. Virtual RFLP tools, such as iPhyClassifier [14], have enhanced this framework by enabling rapid, automated comparison of RFLP patterns, improving subgroup delineation and the identification of novel phytoplasmas [15–17]. Complementing this group-based system, the '*Candidatus* Phytoplasma' nomenclature designates taxa defined mainly by 16S rRNA gene sequence similarity and supported by phylogenetic and ecological data [17–19]. Recent revisions to the species description guidelines have refined the criteria for delineating '*Candidatus* Phytoplasma' species, incorporating updated molecular, phylogenetic, and ecological parameters to ensure consistency and accuracy in taxonomic assignments [19]. Additionally, advancements in RFLP-based classification have expanded the taxonomic scheme, providing greater resolution for identifying phytoplasma diversity [17]. The dual classification—by ribosomal groups and "*Candidatus*" species—forms the foundation for epidemiological tracking, quarantine enforcement, and disease management.

Universal and group-specific primers designed from 16S rRNA gene sequences are used in end-point PCR to detect known and novel phytoplasmas in plants and insect vectors (Table 1). While nested PCR achieves high sensitivity, the risk of false positives, due to contamination, has encouraged the exploration of alternatives such as real-time PCR and loop-mediated isothermal amplification (LAMP) [32–34]. However, end-point PCR integrated with RFLP analysis remains unmatched for taxonomic resolution in many contexts because of its accessibility and established protocols. Challenges such as 16S rRNA gene inter-operon heterogeneity and

Table 1
Primers for amplification of 16Sr DNA of phytoplasmas

Primer[a]	Specificity 16Sr group	Sequence 5′–3′	Position[b]	Reference
P1	Universal	AAGAATTTGATCCTGGCTCAGGATT	16S rDNA	[8]
P7	Universal	CGTCCTTCATCGGCTCTT	23S rDNA	[20]
R0	Universal	GAATACCTTGTTACGACTTAACCCC	16S rDNA	[21]
P3	Universal	GGATGGATCACCTCCTT	16S rDNA	[20]
P4	Universal	GAAGTCTGCAACTCGACTTC	16Sr DNA	[20]
P5	Universal	CGGCAATGGAGGAAACT	16Sr DNA	[20]
R16F2n	Universal	GAAACGACTGCTAAGACTGG	16Sr DNA	[21]
R16R2	Universal	TGACGGGCGGTGTGTACAAACCCCG	16Sr DNA	[21]
F1	Universal	AAGACGAGGATAACAGTTGG	16Sr DNA	[9]
B6	Universal	TAGTGCCAAGGCATCCACTGTG	IS	[22]
R16nF2	Universal	CATGCAAGTCGAACGGA	16Sr DNA	[23]
R16nR2	Universal	CTTAACCCCAATCATCGA	16Sr DNA	[23]
P1A	Universal	AACGCTGGCGGCGCGCCTAATAC	16Sr DNA	[24]
16Sr-SR	Universal	GGTCTGTCAAAACTGAAGATG	IS	[25]
P7A	Universal	CCTTCATCGGCTCTTAGTGC	23Sr DNA	[24]
Pc399	Universal	AACGCCGCGTGAACGATGAA	16Sr DNA	[26]
Pc1694	Universal	ATCAGGCGTGTGCTCTAACC	IS	[26]
fU5	Universal	CGGCAATGGAGGAAACT	16Sr DNA	[27]
rU3	Universal	TTCAGCTACTCTTTGTAACA	16Sr DNA	[27]
PA2f	Universal	GCCCCGGCTAACTATGTGC	16Sr DNA	[28]
PA2r	Universal	TTGGTGGGCCTAAATGGACTC	IS	[28]
M1(758F)	Universal	GTCTTTACTGACGC	16Sr DNA	[29]
M2(1232R)	Universal	CTTCAGCTACCCTTTGTAAC	16Sr DNA	[29]
SN910601	Universal	GTTTGATCCTGGCTCAGGATT	16Sr DNA	[30]
SN910502	Universal	AACCCCGAGAACGTATTCACC	16Sr DNA	[30]
R16(I)F1	I, II, XII, XV	TAAAAGACCTAGCAATAGG	16Sr DNA	[31]
R16(I)R1	I, II, XII, XV	CAATCCGAACTAAGACTCT	16Sr DNA	[31]
R16(III)F2	III	AAGAGTGGAAAAACTCCC	16Sr DNA	[31]
R16(III)R1	III	TTCGAACTGAGATTGA	16Sr DNA	[31]
R16(V)F1	V	TTAAAAGACCTTCTTCGG	16Sr DNA	[31]
R16(V)R1	V	TTCAATCCGTACTGAGACTACC	16Sr DNA	[31]

(continued)

Table 1
(continued)

Primer[a]	Specificity 16Sr group	Sequence 5′–3′	Position[b]	Reference
R16(X)F1	X	GACCCGCAAGTATGCTGAGAGATG	16Sr DNA	[21]
R16(X)R1	X	CAATCCGAACTGAGAGTCT	16Sr DNA	[21]
fO1	X	CGGAAACTTTTAGTTTCAGT	16Sr DNA	[27]
rO1	X	AAGTGCCCAACTAAATGAT	16Sr DNA	[27]

[a]In bold: primers for which a protocol is provided here
[b]IS: spacer region between 16Sr and 23Sr DNA

mixed infections can complicate the interpretation of RFLP profiles [35–37]. Multigene analyses using single-copy genes (e.g., *secY*, *tuf*, *rpl22*, *rpS3*, *groEL*) offer finer strain differentiation, complementing 16S rRNA gene-based methods [38–40].

This chapter outlines the end-point PCR protocol (see scheme in Fig. 1) for universal phytoplasma detection, emphasizing primer selection, nested PCR, and RFLP analysis for group and subgroup identification. Alternatively, sequencing of the PCR-amplified 16S rRNA gene, followed by sequence analysis (e.g., virtual RFLP via iPhyClassifier or sequence homology), enables classification of detected phytoplasmas into either 16Sr groups/subgroups or '*Candidatus* Phytoplasma' species. By applying the updated classification framework and species description guidelines, these methods enable accurate diagnosis and effective management of phytoplasma-associated diseases in plant and insect hosts.

2 Materials

Use molecular biology–grade reagents and maintain sterile conditions to avoid contamination. Wear gloves and use filter tips for all pipetting steps.

2.1 Sample Collection and DNA Extraction

1. Fresh plant tissue (leaf midribs, shoot phloem scraps, or roots) or insect vectors (approximately 0.5–1 g).
2. Liquid nitrogen (optional for homogenization in mortars).
3. Mortar and pestle, homogenization bags, or a tissue homogenizer.
4. CTAB-based extraction buffer: 3% (w/v) cetyltrimethylammonium bromide (CTAB), 100 mM Tris-HCl, pH 8.0, 20 mM EDTA, pH 8.0, 1.4 M NaCl, 1% (w/v) polyvinylpyrrolidone (PVP-40), and 0.2% (v/v) β-mercaptoethanol (add just before use) (*see* **Note 1**).

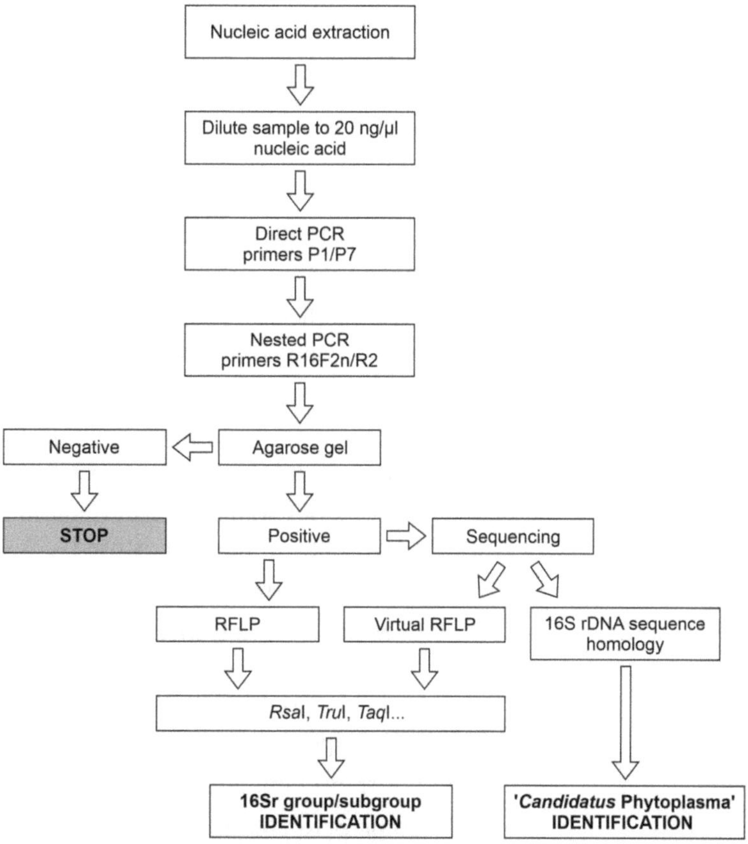

Fig. 1 Phytoplasma identification procedure (workflow)

5. Chloroform:isoamyl alcohol (24:1, v/v).
6. Isopropanol.
7. 70% v/v ethanol.
8. Sterile molecular biology–grade DNase/RNase-free distilled water.
9. TE Buffer (10 mM Tris-HCl, 1 mM EDTA, pH 8.0).
10. Microcentrifuge tubes (1.5 or 2 mL), micropipettes, and sterile filter tips (20, 200, and 1000 µL).

2.2 PCR Amplification (See Note 2)

1. Taq DNA polymerase (e.g., GoTaq, Promega) with 10× reaction buffer and 25 mM $MgCl_2$ (*see* **Note 3**).
2. dNTP mix (40 mM total; i.e., 10 mM each of dATP, dCTP, dGTP, and dTTP).
3. Universal phytoplasma primers (e.g., P1/P7, R16F2n/R16R2) and group-specific primers (*see* Table 1).
4. Template DNA (20–100 ng/µL).

5. Sterile molecular biology–grade DNase/RNase-free distilled water.
6. PCR tubes (0.2 or 0.5 mL) and micropipettes with sterile filter tips.

2.3 Agarose Gel Electrophoresis (Visualization of PCR Product)

1. Agarose (molecular biology grade).
2. 1× TAE buffer: 40 mM Tris-acetate, pH 8.0, 1 mM EDTA.
3. DNA ladder (e.g., 1 kb ladder, 100 bp ladder).
4. Loading dye 5× (30% glycerol (v/v), 0.25% bromophenol blue (w/v)), or a commercially available equivalent.
5. Gel electrophoresis apparatus and power supply.
6. DNA staining solution: ethidium bromide, 10 mg/mL stock solution, and 0.5 μg/mL working solution (*see* **Note 4**).
   - For stock solution mix ethidium bromide in distilled water, typically 10 mg/mL is used.
   - Keep the stock solution in a dark place.
   - Make a 0.5 μg/mL working solution by adding one drop (50 μL) of 10 mg/mL ethidium bromide solution to 1000 mL distilled water.
7. Micropipettes with sterile tips (20 μL).
8. UV transilluminator or gel documentation system.
9. To prepare 100 mL of 1% agarose gel:
   - Weigh out 1 g of agarose.
   - Add 100 mL of 1× TAE buffer.
   - Dissolve carefully in a water bath or microwave.
   - Cool it to about 40 °C.

2.4 RFLP Analysis and Visualization of RFLP Profiles on a Polyacrylamide Gel

1. Restriction enzymes (*see* **Note 5**).
2. 10× restriction enzyme buffers (supplied with the enzymes).
3. Sterile nuclease-free water (molecular biology grade).
4. Incubator or water bath (37 °C, 65 °C, or as specified by the enzyme manufacturer).
5. Loading dye.
6. Polyacrylamide gel electrophoresis (PAGE) system (7% polyacrylamide gel) or agarose gel (2%) for RFLP product separation (*see* **Note 6**).
7. Acrylamide/bis-acrylamide 29:1 solution.
8. Double-distilled water (ddH_2O).
9. Ammonium persulfate.

10. Tetramethylethylenediamine (TEMED).
11. Beakers (50 mL and 100 mL).
12. 1× TBE buffer: 89 mM Tris-HCl, 89 mM boric acid, 2 mM EDTA, pH ~8.
13. DNA staining solution: ethidium bromide (10 mg/mL) or a safer alternative (e.g., SYBR Green, GelRed).
14. DNA ladder (e.g., 50 bp ladder).
15. UV transilluminator or gel documentation system.

2.5 Equipment

1. PCR thermocycler.
2. Centrifuge (minimum 12000× *g*).
3. Micropipettes (20, 200, and 1000 μL).
4. UV transilluminator or gel documentation system.
5. Gel electrophoresis apparatus and power supply.
6. Incubator or water bath with adjustable temperature.
7. Instruments for quantifying DNA (spectrophotometer, NanoDrop, Qubit, or equivalent).

3 Methods

3.1 Sample Collection and DNA Extraction

1. Collect 0.5–1 g of fresh plant tissue (leaf midribs, shoot phloem scraps, or roots) from symptomatic plants or insect vectors (*see* **Note 7**).
2. Immediately freeze plant samples or store at 4 °C for a short period until processing (*see* **Note 8**). Insects can be stored in 70% ethanol until processing.
3. Homogenize the sample in liquid nitrogen using a mortar and pestle or a tissue homogenizer.
4. Extract DNA using a CTAB-based method (*see* **Note 1**).
5. Add 7 mL of CTAB buffer to 1 g of ground plant tissue (or equivalent) or 500 μL of CTAB buffer to an insect specimen. Mix and incubate at 65 °C for 20 min.
6. Add an equal volume of chloroform:isoamyl alcohol (24:1) to one of the CTAB buffer, vortex, and centrifuge at 12000× *g* for 15 min.
7. Transfer the aqueous phase to a new tube, add an equal volume of isopropanol, and incubate at −20 °C for 30 min (*see* **Note 9**).
8. Centrifuge at 12000× *g* for 15 min, discard the supernatant, and wash the pellet with 70% ethanol.

9. Air-dry the pellet and resuspend in 50–100 μL of sterile water or TE buffer.
10. Quantify DNA using a spectrophotometer or fluorometer (e.g., NanoDrop or Qubit) and adjust to 20–100 ng/μL.

3.2 PCR Amplification

1. Prepare a 25 μL PCR mix containing: 2.5 μL 10× PCR buffer, 1.5 μL 25 mM $MgCl_2$, 0.5 μL 10 mM dNTP mix, 0.5 μL of each primer (10 μM, e.g., P1/P7 for direct PCR, R16F2n/R16R2 for nested PCR), 0.2 μL Taq DNA polymerase (5 U/μ L), 1 μL template DNA (20 ng), and sterile nuclease-free water to a final volume of 25 μL.
2. Perform direct PCR using universal primers (e.g., P1/P7; *see* **Note 10**) with the following cycling protocol: initial denaturation 94 °C for 2 min; 35 cycles of 94 °C for 1 min, 50 °C for 2 min, 72 °C for 3 min; and a final extension for 10 min at 72 °C.
3. For nested PCR, dilute the PCR product 1:30 with sterile nuclease-free water and use 1 μL as the template with nested primers (e.g., R16F2n/R16R2) under the same cycling conditions as in the previous step (*see* **Note 11**).

3.3 Visualization of PCR Products

1. Run 6 μL of PCR product on a 1% agarose gel in 1× TAE buffer, stain with ethidium bromide or GelRed, and visualize under UV light to confirm amplification (expected size ~1.2–1.8 kb depending on primers; *see* **Note 12**).
2. Prepare a 1% agarose gel (*see* **Note 13**), pour the melted agarose gel into a tray, insert the comb, and remove any air bubbles.
3. Let the gel solidify at room temperature (~20 min).
4. Place the gel into the electrophoresis chamber and cover it with 1× TAE buffer.
5. Load the 6 μL of PCR product (optionally mix with loading dye; *see* **Note 14**) and 2.5 μL of DNA ladder into the wells.
6. Always load the DNA ladder in the first or last well.
7. Connect the electrodes to the power source (5 V/cm) and run the gel, ensuring the dye does not run off the edge.
8. Stain the gel by soaking it in ethidium bromide working solution (0.5 μg/mL; *see* **Note 15**) for 10–15 min, then wash in distilled water for 10–15 min.
9. Observe the gel on a transilluminator by visualizing DNA under UV light.
10. Take a picture and print and/or store in computer or in the laboratory book after proper labelling of the samples.

3.4 RFLP Analysis

3.4.1 Amplicon Restriction Digestion

1. Digest 3–8 μL of PCR product with 1 U of restriction enzyme in a 20 μL reaction. For the reaction mix, combine: 2 μL 10× appropriate enzyme buffer, 1 U restriction enzyme, 3–8 μL PCR product, and sterile nuclease-free water to make up to 20 μL (*see* **Note 16**).
2. Incubate at the enzyme-specific temperature (typically 37 °C or 65 °C) for the time specified in the restriction enzyme manufacturer's instruction or overnight (*see* **Note 17**).
3. Separate digested PCR products on a 7% polyacrylamide gel or 3% agarose gel, stain with ethidium bromide, and visualize under UV light (*see* **Note 19**).

3.4.2 Visualization of RFLP Products on an Agarose Gel

1. Prepare 100 mL of 3% agarose gel (*see* **Note 13**):
   - Weigh out 3 g of agarose.
   - Add 100 mL of 1× TAE buffer.
   - Carefully dissolve in water bath or microwave.
   - Cool it to about 40 °C.
2. Pour the agarose into a tray, insert comb and remove any air bubbles.
3. Allow the gel to solidify at room temperature; this will take approximately 20 min.
4. Place the gel into the electrophoresis chamber and cover it with 1× TAE buffer.
5. Add loading dye to the digested PCR product and load it into a well.
6. Add a DNA ladder (e.g., 50 bp ladder) in the first or last well of the gel.
7. Connect electrodes to the power source (5 V/cm) and run the gel, ensuring the dye does not run off the gel.
8. Stain the gel by soaking it in ethidium bromide working solution (0.5 μg/mL) (*see* **Note 12**) for 20–25 min, then wash in distilled water for 20–25 min.
9. Observe the gel on a transilluminator by visualizing DNA under UV light.
10. Take a picture and print and/or store in a computer or in the laboratory book after proper labelling.

3.4.3 Visualization of RFLP Products on a Polyacrylamide Gel

1. Prepare 25 mL of a 7% polyacrylamide gel by combining the following reagents in a clean 50 mL beaker in the indicated order: 17.83 mL ddH2O, 2.5 mL 10× TBE, 4.37 mL acrylamide/bis-acrylamide (29:1) solution, 310 μL ammonium persulfate, and 15.6 μL TEMED (*see* **Note 19**).

2. Gently mix and pour the gel mix between the glass plates of the gel casting chamber.
3. Add the comb and allow to polymerize for ~20 min (avoid direct sunlight).
4. Remove the comb carefully.
5. Clamp the gel-containing glasses in the vertical electrophoresis apparatus.
6. Cover with 1× TBE.
7. Gently clean the wells with an appropriate plastic pipette or with a pipette tip.
8. Remove air bubbles at the bottom between the glass plates.
9. Add loading dye to the sample containing the digested PCR product and load it into a well.
10. Add a DNA ladder (e.g., 50 bp ladder) in one well of the gel.
11. Connect the electrodes to the power source (7 V/cm) and run the gel ensuring the dye does not run off the gel.
12. Stain the gel by soaking it in ethidium bromide working solution (0.5 μg/mL) for 10–15 min, and wash in water for 1–2 min.
13. Observe the gel on a transilluminator by visualizing DNA under UV light.
14. Take a picture and print and/or store in a computer or in the laboratory book after proper labeling.
15. Compare RFLP patterns with reference profiles to assign the phytoplasma to a ribosomal group and subgroup [12] (*see* **Note 20**).

3.5 Sequencing and Virtual RFLP (Optional)

Virtual RFLP analysis can be performed using iPhyClassifier for automated pattern comparison after sequencing (*see* **Note 21**) of the PCR amplicon [14, 20].

1. Sanger sequence the 16S rRNA gene PCR amplicons for sequence homology analysis or virtual RFLP (*see* **Note 22**).
2. Analyze sequences using software (e.g., pDRAW32 (www.acaclone.com)) or online tool iPhyClassifier to simulate RFLP patterns and confirm group/subgroup assignment [14, 41].
3. Compare sequence similarity to delineate '*Candidatus* Phytoplasma' species [17–19].

4 Notes

1. Alternatively, a commercial DNA extraction kit can be used instead of the CTAB extraction following the manufacturer's instructions (e.g., DNeasy Plant Mini Kit, Qiagen).
2. Close reagent and sample tubes immediately after removing an aliquot. The following steps of the analysis must be performed in separate places with dedicated laboratory equipment (pipettes, tips, tubes, racks for tubes): reaction mix preparation (without DNA), adding DNA (UV chamber), and PCR product analysis.
3. Alternatively, 2× PCR master mix (e.g., PCR Master Mix (2×)) can be used.
4. A safer alternative can be used (e.g., SYBR Green, GelRed).
5. The 17 restriction enzymes that are used for differentiation of ribosomal groups and subgroups are: *Mse*I (*Tru*1I), *Alu*I, *Rsa*I, *Hha*I, *Hae*III, *Hpa*II, *Taq*I, *Hin*fI, *Sau*3AI, *Kpn*I, *Tha*I, *Bam*HI, *Dra*I, *Eco*RI, *Hpa*I, *Ssp*I, and *Bfa*I. Restriction enzymes should be selected based on the phytoplasma group being analyzed. The most informative restriction enzymes (e.g., *Alu*I, *Hha*I, *Rsa*I, *Taq*I, *Tru*1I) are recommended for initial group/subgroup identification [12].
6. For RFLP product separation, PAGE is preferred for resolving small fragments, while 2% agarose gel is suitable when few large, well-separated fragments are expected.
7. Collect samples from symptomatic tissues (e.g., leaf midribs, shoot phloem scraps) during periods in which high phytoplasma titer is expected (e.g., late summer for woody plants). For insect vectors, use adults or nymphs from known phytoplasma host plants.
8. Freezing plant samples in liquid nitrogen preserves DNA integrity. If immediate freezing is not possible, store at 4 °C for a shorter period (days, depending on plant material).
9. Ensure DNA is free of PCR inhibitors (e.g., polysaccharides, phenolics). If using CTAB, perform additional chloroform extractions if the aqueous phase is cloudy.
10. Optimize annealing temperatures for group-specific primers (Table 1). Perform a gradient PCR to identify the optimal annealing temperature if necessary.
11. Nested PCR increases sensitivity but also the risk of contamination. Include negative controls (no template DNA) and positive controls (known phytoplasma DNA) in each run.

12. Expected amplicon sizes: P1/P7 (1.8 kb), R16F2n/R16R2 (1.2 kb). If no bands are observed, check DNA quality, primer specificity, or perform nested PCR.
13. Measure the amount of agarose gel needed in relationship with the capacity of the tray (for exact amount see gel electrophoresis system manufacturer instruction), for small tray 80 mL should be enough, for larger about 120 mL or more will be needed.
14. In this protocol, Taq DNA polymerase employed for PCR amplification has loading dye in the 10× PCR buffer. If the PCR buffer used does not contain loading dye, it must be added to the PCR product before gel electrophoresis. Typically, mix 6 μL of PCR product with 1.5 μL of 5× loading dye before loading the sample into the gel wells.
15. In this protocol, post-staining was employed. Alternatively, if the pre-stained gel contains ethidium bromide or an alternative dye (e.g., SYBR Green, GelRed), the DNA will bind to the dye during electrophoresis, and you can directly proceed to visualization.
16. Amount of PCR product added to the RFLP reaction varies based on the PCR product concentration estimated by PCR amplicon band intensity in agarose gel visualization of PCR product.
17. Some restriction enzymes require different incubation temperatures. Follow manufacturer recommendations for incubation time and temperature.
18. The choice between agarose and polyacrylamide gel for RFLP analyses depends upon the length of expected fragments. Polyacrylamide gels (7%) provide better resolution for small fragments (<300 bp) compared to agarose gels (3%). Use a 50 bp ladder for accurate sizing.
19. Polyacrylamide gel polymerization begins when TEMED is added to the acrylamide/bisacrylamide solution.
20. RFLP profiles may vary due to inter-operon heterogeneity or mixed phytoplasma infections. Cloning of PCR products followed by sequencing, or applying NGS, enables detection of mixed infections and addresses challenges such as inter-operon heterogeneity by identifying and distinguishing multiple sequences within a single sample.
21. Sequencing and virtual RFLP are recommended for novel phytoplasmas or ambiguous RFLP profiles.
22. Alternatively, some NGS methods can be used. Particularly in the case of inter-operon heterogeneity or mixed phytoplasma infection.

References

1. Bertaccini A, Duduk B (2009) Phytoplasma and phytoplasma diseases: a review of recent research. Phytopathol Mediterr 48:355–378
2. Berges R, Rott M, Seemüller E (2000) Range of phytoplasma concentration in various plant hosts as determined by competitive polymerase chain reaction. Phytopathology 90:1145–1152. https://doi.org/10.1094/PHYTO.2000.90.10.1145
3. Bertaccini A, Marani F (1982) Electron microscopy of two viruses and mycoplasma-like organisms in lilies with deformed flowers. Phytopathol Mediterr 21:8–14
4. Cousin MT, Sharma AK, Isra S (1986) Correlation between light and electron microscopic observations and identification of mycoplasma-like organisms using consecutive 350 nm thin sections. J Phytopathol 115:368–374. https://doi.org/10.1111/j.1439-0434.1986.tb04350.x
5. Haggis GH, Sinha RC (1978) Scanning electron microscopy of mycoplasma-like organisms after freeze fracture of plant tissues affected with clover phyllody and aster yellows. Phytopathology 68:677–680
6. Zwolińska A, Krawczyk K, Borodynko-Filas N et al (2019) Non-crop sources of rapeseed Phyllody phytoplasma ('*Candidatus* Phytoplasma asteris': 16SrI-B and 16SrI-(B/L)L), and closely related strains. Crop Prot 119:59–68. https://doi.org/10.1016/j.cropro.2018.11.015
7. Lim PO, Sears BB (1989) 16S rRNA sequence indicates that plant-pathogenic mycoplasma-like organisms are evolutionarily distinct from animal mycoplasmas. J Bacteriol 171:5901–5906. https://doi.org/10.1128/jb.171.11.5901-5906.1989
8. Deng SJ, Hiruki C (1991) Amplification of 16S ribosomal-RNA genes from culturable and nonculturable mollicutes. J Microbiol Methods 14:53–61. https://doi.org/10.1016/0167-7012(91)90007-D
9. Davis RE, Lee I-M (1993) Cluster-specific polymerase chain reaction amplification of 16S rDNA sequences for detection and identification of mycoplasma-like organisms. Phytopathology 83:1008–1011. https://doi.org/10.1094/Phyto-83-1008
10. Lee I-M, Hammond RW, Davis RE et al (1993) Universal amplification and analysis of pathogen 16S rDNA for classification and identification of mycoplasma-like organisms. Phytopathology 83:834–842. https://doi.org/10.1094/Phyto-83-834
11. Seemüller E, Marcone C, Lauer U et al (1998) Current status of molecular classification of the phytoplasmas. J Plant Pathol 80:3–26
12. Lee I-M, Gundersen-Rindal DE, Davis RE et al (1998) Revised classification scheme of phytoplasmas based on RFLP analyses of 16S rRNA and ribosomal protein gene sequences. Int J Syst Evol Micr 48:759–774. https://doi.org/10.1099/00207713-48-4-1153
13. Lee I-M, Davis RE, Gundersen-Rindal DE (2000) Phytoplasma: phytopathogenic mollicutes. Ann Rev Microbiol 54:221–255. https://doi.org/10.1146/annurev.micro.54.1.221
14. Zhao Y, Wei W, Lee IM et al (2013) The iPhyClassifier, an interactive online tool for phytoplasma classification and taxonomic assignment. In: Phytoplasma: methods and protocols, pp 329–338. https://doi.org/10.1007/978-1-62703-089-2_28
15. Wei W, Davis RE, Lee I-M et al (2007) Computer-simulated RFLP analysis of 16S rRNA genes: identification of ten new phytoplasma groups. Int J Syst Evol Microbiol 57:1855–1867. https://doi.org/10.1099/ijs.0.65000-0
16. Wei W, Lee I-M, Davis RE et al (2008) Automated RFLP pattern comparison and similarity coefficient calculation for rapid delineation of new and distinct phytoplasma 16Sr subgroup lineages. Int J Syst Evol Microbiol 58:2368–2377. https://doi.org/10.1099/ijs.0.65868-0
17. Wei W, Zhao Y (2022) Phytoplasma taxonomy: nomenclature, classification, and identification. Biology. https://doi.org/10.3390/biology11081119
18. IRPCM (2004) Candidatus Phytoplasma', a taxon for the wall-less, non-helical prokaryotes that colonize plant phloem and insects. Int J Syst Evol Microbiol 54:1243–1255. https://doi.org/10.1099/ijs.0.02854-0
19. Bertaccini A, Arocha-Rosete Y, Contaldo N et al (2022) Revision of the '*Candidatus* Phytoplasma' species description guidelines. Int J Syst Evol Microbiol 72:005353. https://doi.org/10.1099/ijsem.0.005353
20. Schneider B, Seemüller E, Smart CD et al (1995) Phylogenetic classification of plant pathogenic mycoplasma-like organisms or phytoplasmas. In: Razin S, Tully J (eds)

Molecular and diagnostic procedures in mycoplasmology. Academic Press, New York

21. Lee I-M, Bertaccini A, Vibio M et al (1995) Detection of multiple phytoplasmas in perennial fruit trees with decline symptoms in Italy. Phytopathology 85:728–735. https://doi.org/10.1094/Phyto-85-728
22. Padovan AC, Gibb KS, Bertaccini A et al (1995) Molecular detection of the Australian grapevine yellows phytoplasma and comparison with grapevine yellows phytoplasmas from Italy. Aust J Grape Wine Res 1:25–31. https://doi.org/10.1111/j.1755-0238.1995.tb00074.x
23. Gundersen DE, Lee I-M (1996) Ultrasensitive detection of phytoplasmas by nested-PCR assays using two universal primer pairs. Phytopathol Mediterr 35:144–151
24. Lee I-M, Martini M, Bottner KD et al (2003) Ecological implications from a molecular analysis of phytoplasmas involved in an aster yellows epidemic in various crops in Texas. Phytopathology 93:1368–1377. https://doi.org/10.1094/PHYTO.2003.93.11.1368
25. Lee I-M, Gundersen-Rindal D, Davis RE et al (2004) Candidatus Phytoplasma asteris', a novel taxon associated with aster yellows and related diseases. Int J Syst Bacteriol 54:1037–1048. https://doi.org/10.1099/ijs.0.02843-0
26. Skrzeczkowski LJ, Howell WE, Eastwell KC (2001) Bacterial sequences interfering in detection of phytoplasma by PCR using primers derived from the ribosomal RNA operon. Acta Hortic 550:417–424. https://doi.org/10.17660/ActaHortic.2001.550.62
27. Lorenz KH, Schneider B, Ahrens U et al (1995) Detection of the apple proliferation and pear decline phytoplasmas by PCR amplification of ribosomal and nonribosomal DNA. Phytopathology 85:771–776. https://doi.org/10.1094/Phyto-85-771
28. Heinrich M, Botti S, Caprara L et al (2001) Improved detection methods for fruit tree phytoplasmas. Plant Mol Biol Rep 19:169–179. https://doi.org/10.1007/BF02772160
29. Gibb KS, Padovan AC, Mogen BD (1995) Studies on sweet potato little-leaf phytoplasma detected in sweet potato and other plant species growing in Northern Australia. Phytopathology 85:169–174. https://doi.org/10.1094/Phyto-85-169
30. Namba S, Kato S, Iwanami S et al (1993) Detection and differentiation of plant-pathogenic mycoplasma-like organisms using polymerase chain reaction. Phytopathology 83:786–791. https://doi.org/10.1094/Phyto-83-786
31. Lee I-M, Gundersen DE, Hammond RW et al (1994) Use of mycoplasmalike organism (MLO) group-specific oligonucleotide primers for nested-PCR assays to detect mixed-MLO infections in a single host plant. Phytopathology 84:559–566. https://doi.org/10.1094/Phyto-84-559
32. Christensen NM, Nicolaisen M, Hansen M (2007) Distribution of phytoplasmas in infected plants as revealed by real-time PCR and bioimaging. Mol Plant Microbe Interact 17:1175–1184. https://doi.org/10.1094/MPMI.2004.17.11.1175
33. Hodgetts J, Boonham N, Mumford R et al (2009) Panel of 23S rRNA gene-based real-time PCR assays for improved universal and group-specific detection of phytoplasmas. Appl Environ Microbiol 75:2945–2950. https://doi.org/10.1128/AEM.02610-08
34. Dickinson M (2015) Loop-mediated isothermal amplification (LAMP) for detection of phytoplasmas in the field. In: Lacomme C (ed) Plant pathology. Methods in molecular biology, vol 1302. Humana Press, New York
35. Davis RE, Jomantiene R, Kalvelyte A et al (2003) Differential amplification of sequence heterogeneous ribosomal RNA genes and classification of the 'Fragaria multicipita' phytoplasma. Microbiol Res 158:229–236. https://doi.org/10.1078/0944-5013-00201
36. Duduk B, Calari A, Paltrinieri S et al (2009) Multigene analysis for differentiation of aster yellows phytoplasmas infecting carrots in Serbia. Ann Appl Biol 154:219–229. https://doi.org/10.1111/j.1744-7348.2008.00285.x
37. Schneider B, Seemüller E (1994) Presence of two sets of ribosomal genes in phytopathogenic mollicutes. Appl Environ Microbiol 60:3409–3412. https://doi.org/10.1128/aem.60.9.3409-3412.1994
38. Marcone C, Lee I-M, Davis RE et al (2000) Classification of aster yellows-group phytoplasmas based on combined analyses of rRNA and tuf gene sequences. Int J Syst Evol Microbiol 50:1703–1713. https://doi.org/10.1099/00207713-50-5-1703
39. Martini M, Lee I-M, Bottner KD et al (2007) Ribosomal protein gene-based phylogeny for finer differentiation and classification of phytoplasmas. Int J Syst Evol Microbiol 57:2037–2051. https://doi.org/10.1099/ijs.0.65013-0
40. Mitrović J, Kakizawa S, Duduk B et al (2011) The groEL gene as an additional marker for finer differentiation of '*Candidatus*

Phytoplasma asteris'-related strains. Ann Appl Biol 159:41–48. https://doi.org/10.1111/j.1744-7348.2011.00472.x

41. Zhao Y, Wei W, Lee I-M et al (2009) Construction of an interactive online phytoplasma classification tool, iPhyClassifier, and its application in analysis of the peach X-disease phytoplasma group (16SrIII). Int J Syst Evol Microbiol 59:2582–2593. https://doi.org/10.1099/ijs.0.010249-0

Chapter 4

Imp Gene Analysis for Diagnosis and Genotyping of Flavescence Dorée Related Phytoplasmas

Luisa Filippin, Graziana Da Rold, and Elisa Angelini

Abstract

Imp is a gene coding for the immunodominant membrane protein Imp that is supposed to be involved in host–pathogen interactions. It was identified in some '*Candidatus* Phytoplasma' species and showed a considerable sequence variability. For this characteristic it can be exploited to complement conventional phytoplasma classification based on the conserved 16S rRNA gene, enhancing the differentiation of closely related phytoplasma strains. This protocol describes a method for amplification and sequence analysis of the *imp* gene of Flavescence dorée (FD) related phytoplasmas, that can be useful for detecting and genotyping the strains involved, for example, in the complex FD epidemiology.

Key words Diagnostics, Genotyping, Epidemiology, Membrane protein, Nested PCR, Sanger sequencing

1 Introduction

The main methods for revealing phytoplasma presence in plants or insects are based on specific DNA amplification through polymerase chain reaction (PCR). A lot of phytoplasma genes are used as targets for PCR. Taxonomy of phytoplasmas relies on the highly conserved 16S rRNA gene, and both universal and group-specific primers for this region are available [1]. The combination of PCR and subsequent restriction length polymorphism analyses (RFLP) allows the classification of phytoplasmas at 16Sr group and subgroup levels [2, 3]. Over time, other genomic regions have been considered for investigating the phytoplasma genomes. Less conserved, non-ribosomal genes permitted to outline differences among 16Sr subgroups and concurred to the definition of the new '*Candidatus* Phytoplasma' species, together with distinctive biological, phytopathological and genetic properties, as indicated in taxonomic guidelines [4]. Their relevance applies not only for phylogenetic studies but also for disease management, as well as for

Katrin Janik and Mattia Tabarelli (eds.), *Phytoplasma: Methods and Protocols*, Methods in Molecular Biology, vol. 3008, https://doi.org/10.1007/978-1-0716-5104-9_4,

quarantine and epidemiological purposes. Some of those molecular markers are housekeeping genes, such as *tuf*, *rp*, *groEL*, *map*, *secA*, *secY*; more recently, many studies addressed more variable genes encoding surface proteins such as *hflB* [5], *amp* [6], *imp* [7], *stamp* [8], *vmp1* [9, 10], *vmpA*, and *vmpB* [11]; the collected genetic variability data were used, for example, for achieving a finer differentiation of closely related phytoplasma strains, mapping their geographical distribution, and defining their ecological niches [12].

The *imp* gene of phytoplasmas encodes for a single membrane protein, which is thought to be involved in the host-phytoplasma interaction. It belongs to the so-called immunodominant membrane proteins (IDPs), a group of proteins that comprises a major portion of the total cellular membrane proteins [13]. They are classified in three distinct types: immunodominant membrane proteins (IMP), immunodominant proteins A (IdpA), and antigenic membrane protein (AMP) [6]. These proteins are not homologous and differ in their aminoacidic sequences, genomic localization, and predicted structure. Imp protein has a transmembrane domain at its N-terminus, while the rest of the hydrophilic residues are exposed to the cytoplasm of their insect or plant hosts [14].

Thus far, the *imp* gene has been detected in strains of phytoplasmas belonging to '*Candidatus* Phytoplasma asteris' [15], '*Ca.* P aurantifoliae' [16], '*Ca.* P. australiense' [17], '*Ca.* P. mali' [18], '*Ca.* P. oryzae' [15], '*Ca.* P. pruni' [15], '*Ca.* P prunorum' and '*Ca.* P. pyri' [19], and 16SrV groups [20, 21]. *Imp* gene shows a marked sequence variability, significantly higher than that observed in housekeeping genes [7, 22, 23]. Moreover, a high number of mutations are non-synonymous, although they do not appear to affect the predicted protein structure [14]. This suggests that the *imp* gene is subject to strong selective pressure, supporting a possible role in phytoplasma adaptation to the environment; indeed, interaction studies demonstrated that the encoded protein can bind to protein extracts from insect vectors as well as to plant actin [24]. Thanks to its high variability, the *imp* gene is particularly suitable as a genetic marker for a deep differentiation among closely related phytoplasma strains, and was successfully used for investigating intraspecific sequence polymorphisms in strains of '*Ca.* P. aurantifoliae', '*Ca.* P. mali', '*Ca.* P. pruni', '*Ca.* P. prunorum' '*Ca.* P. pyri' species, and 16SrV group [21, 23, 25–29].

Here we describe the protocol for the detection and differentiation of phytoplasmas belonging to the 16SrV-C, -D, and -E subgroups, based on *imp* gene amplification and sequence analysis. The reported primers were originally designed for genotyping Flavescence dorée (FD) and closely related phytoplasmas but are also capable of amplifying strains of '*Ca.* P. rubi' species (16SrV-E subgroup) [20, 30]. FD is an epidemic and quarantine grapevine disease in Europe, associated with phytoplasmas of the 16SrV-C and –D ribosomal phylogenetic subgroups. The vector of FD phytoplasmas is the grapevine phloem-feeder leafhopper *Scaphoideus titanus*. During the past years, many studies highlighted the complex epidemiology of FD, including confirmed and potential secondary vectors, alternative host plants, and a considerable genetic diversity of FD-related phytoplasma strains [7, 29, 31–35]. The primers described here were able to detect *imp* genes of phytoplasmas present in both insect and plant samples of many species. As most of the host plants are woody species, in which the phytoplasma titer is often low [36], the protocol includes a second primer pair for nested or semi-nested PCR, enhancing its sensitivity and reliability even for problematic samples. The successive *imp* sequence analysis can contribute to discovering new pieces of the FD epidemiology puzzle.

2 Materials

Prepare all solutions using analytical grade reagents and deionized, distilled, sterile ultrapure water (ddH_2O). For PCR and sequencing, use ultrapure, nuclease-free sterile water. Aliquot stock PCR reagents and store them at −20 °C for long-term conservation; "in-use" aliquots of $MgCl_2$ and PCR 10× buffer can be kept at 4–8 °C. Stock aliquots should be prepared in a room separated from the one used for DNA amplification. All plastic microfuge tubes must be sterile and nuclease-free.

2.1 PCR and Preparation for Sanger Sequencing

1. Thin-wall, PCR-suitable, 0.5 mL or 0.2 mL tubes; 0.2 mL strip or single tubes for sequencing.
2. 1.5 mL microcentrifuge tubes.
3. Ultrapure, nuclease-free water.
4. 20 μM forward and reverse primers (Table 1 and Fig. 1).
5. *Taq* Polymerase together with a suitable buffer and $MgCl_2$ solution.

Table 1
Primer sequences used in this protocol for amplification of the *dnaD-imp-pyrG* genomic region

Primer name	Primer sequence	Length (nt)	Position
dnaDFDf1	5′-TAG AGA GAA TTT TAG GCC ACG-3′	21	*dnaD*
dnaDFDf2	5′-ATA GAA AAT AAC GAT AAA GCA G-3′	22	*dnaD*
pyrGFDr1	5′-AAT AAT GAA GAA CAA TTA CCT G-3′	22	*pyrG*
pyrGFDr2	5′-TCA AGA CCT TTT AAA CCA CAC CC-3′	23	*pyrG*

The primers were designed in the flanking regions of the *imp* gene: forward primers on *dnaD* gene, reverse primers on *pyrG* gene

Fig. 1 Primer positions on the *dnaD-imp-pyrG* genomic region

6. dNTPs mix solution, at a final concentration of 10 mM each.
7. Template DNA: total DNA obtained from plants or insects (*see* **Note 1**), positive controls (i.e., DNA from phytoplasma reference strains or obtained from confirmed infected plants or insect samples, belonging to 16SrV-C, D, or E subgroups).
8. Filter tips.
9. DNA-degrading solution (e.g., a 1:10 (v:v) bleach:water solution or a commercial DNA cleaner).
10. Gloves.

2.2 Agarose Gel Electrophoresis

1. Standard agarose for gel electrophoresis.
2. ddH_2O.
3. 5× TBE buffer (stock): mix 54.48 g Tris base, 27.8 g boric acid, and 10 mL of 0.5 M EDTA, pH 8.0, in a final volume of 1000 mL ddH_2O. Dilute to 0.5× TBE with ddH_2O (*see* **Note 2**).
4. 6× gel-loading buffer Bromophenol blue: mix ddH_2O with 30% (v/v) glycerol and 0.25% (w/v) bromophenol blue powder in a glass beaker; aliquot 1 mL into 1.5 mL tubes. Keep one tube at 4 °C and store the remaining ones at −20 °C.
5. DNA ladder of the appropriate size range, typically 100 bp or 1 kb.
6. Non-mutagenic nucleic acid stain (*see* **Note 3**).

2.3 Equipment and Instruments

1. Micropipettes P2, P10, P20, P200, P1000, with appropriate tips.
2. Repeater pipette.
3. Vortex mixer.
4. Microcentrifuge.
5. Thermal cycler.
6. Balance.
7. Glass bottle or flask (100 or 250 mL, depending on the volume of the agarose solution).
8. Horizontal gel electrophoresis apparatus with gel caster, tray, and combs.
9. Microwave oven.
10. Power supply.

11. A gel image system for UV illumination, imaging, and documentation of stained agarose gel.
12. An.abi file software reader.

3 Methods

3.1 PCR

PCR is highly sensitive and therefore prone to DNA contamination—a risk that is amplified in nested PCR. Clean the workbench with a DNA-degrading solution before setting up the reactions to reduce the risk of contamination. Always wear gloves and keep your hands clean. Use filter tips in every step.

1. Thaw the PCR reagents at 4 °C, e.g., in the fridge (except for the Taq polymerase, that must be kept at −20 ° C until use). Vortex all the components except the polymerase, then spin briefly the tubes.
2. Prepare 0.2 or 0.5 mL PCR-tubes (*see* **Note 4**). The number of tubes must correspond to the number of samples to be analyzed, plus the positive and the negative controls. Label each tube with a marker, indicating the sample number and any other relevant information (primer pairs, number of the PCR, etc).
3. In a 1.5 mL tube, prepare the reagent mix for the direct PCR, following the order indicated in Table 2. To calculate the final volumes of each reagent, multiply the single quantities by the

Table 2
Volumes of PCR reagents for one reaction, both for direct and nested or semi-nested PCR

	Direct PCR		Nested/semi-nested PCR	
Reagents	Volume (μL)	Final concentration	Volume (μL)	Final concentration
Deionized sterile water, nuclease-free	14.45	–	13.85	–
10× PCR buffer without $MgCl_2$	2	1×	2	1×
$MgCl_2$ solution, 50 mM	0.6	1.5 mM	1.2	3 mM
dNTPs solution, 10 mM	0.6	0.3 mM	0.6	0.3 mM
Forward primer, 20 μM	0.6	0.6 μM	0.6	0.6 μM
Reverse primer, 20 μM	0.6	0.6 μM	0.6	0.6 μM
Taq polymerase, 5 U/μL	0.15	0.0375 U/μL	0.15	0.0375 U/μL
Template	1		1	
Total volume	20		20	

number of tubes plus one (*see* **Note 5**). The forward and reverse primers are dnaDFDf1 and pyrGFDr1, respectively.

4. Gently mix the mixture by pipetting or flicking the tube (avoid vortexing) and spin down all the volume in a microcentrifuge.
5. Open the tube caps and aliquot 19 μL of the solution into each tube using a single-channel P20 or a repeater pipette, then close the caps.
6. Pipette 1 μL of DNA template into each tube (*see* **Note 6**), opening the tube one by one; close the tube, change the tip, and continue with the next templates. Set up a negative control by substituting the DNA with 1 μL of water. Mix and spin.
7. Place the tubes in a thermal cycler (*see* **Note 7**). The PCR conditions are as follows: initial denaturation at 94 °C for 3 min; 35 amplification cycles of 94 °C for 30 s, 57 °C for 30 s, and 66 °C for 1 min; followed by a final elongation step at 66 °C for 7 min (*see* **Note 8**).
8. Keep the PCR products at 4 °C for short-term (up to a maximum of a few days), or at −20 °C for long-term storage.
9. If a high phytoplasma titer is expected, proceed with the agarose gel analysis (*see* Subheading 3.2 and **Note 9**). For samples with low phytoplasma concentration, perform a semi-nested or nested PCR as described in the following steps.
10. Prepare one 0.5 mL tube for each of the tubes used in the first round of PCR. Add 49 μL of nuclease-free water to each tube, and 1 μL of each PCR product to the corresponding water aliquots to obtain a 1:50 dilution. Vortex and centrifuge. Use 1 μL of those dilutions as the template for the following nested PCR.
11. Prepare a new set of tubes as described in **step 2** above (the composition of the master mix is indicated in Table 2). Use the primers dnaDFDf1/pyrGFDr2 for semi-nested and dnaDFDf2/pyrGFDr2 for nested PCR (*see* **Note 10**). Calculate the cumulative volumes as done for the first PCR.
12. Mix gently the mixture, briefly spin down, and place the tubes in the thermal cycler. Use the same PCR conditions as for the direct PCR.

3.2 Agarose Gel Electrophoresis

1. Prepare the 1% w/v agarose solution in 0.5× TBE in a glass bottle or flask, without filling the container more than 50% of its volume. The volume has to be adequate to obtain a 4–6 mm thick gel (*see* **Note 11**).
2. Dissolve agarose in a microwave oven (*see* **Note 12**).
3. Gently shake the mixture until the agarose is completely dissolved. Then, allow it to cool to approximately 60 °C.

4. Assemble the casting components (gel caster, tray, comb); add the nucleic acid stain to the agarose solution at the appropriate dilution, then pour it into the tray and remove any air bubbles (*see* **Note 13**); let it solidify for approximately 20–30 min.
5. Insert the gel with the tray in a suitable electrophoresis chamber filled with 0.5× TBE solution. Be sure that the gel is fully submerged. Carefully remove the comb.
6. Transfer 5 μM of each PCR product into a separate 0.2 mL tube and add 1 μM of 6× loading dye to each.
7. Mix the two ingredients by pipetting with a P20 or P10 and load the total 6 μM into the wells.
8. Load the DNA ladder in the first and/or the last well.
9. Assemble electrophoretic apparatus and connect electrodes to the power supply.
10. Run the gel at a voltage of 5–6 V/cm, where *cm* refers to the distance in centimeters between the two electrodes. Stop the run when the blue front of the loading dye has migrated approximately 75–90% of the total gel length.
11. Examine the gel by ultraviolet light on a transilluminator. For FD-D, the expected amplicon sizes are 940 bp for direct PCR, 865 bp for semi-nested PCR, and 830 bp for nested PCR, respectively (Fig. 2).
12. Take a picture of the gel and store it.

3.3 Preparation of PCR Products for External Sanger Sequencing and Sequence Analysis

1. Perform a PCR-cleanup step using a commercial kit. For Sanger sequencing, the DNA should be free from salts, proteins, dNTPs, primers, RNA.
2. After DNA purification, quantify its concentration. While we recommend a DNA quantification based on an agarose gel run, other methods are also suitable (*see* **Note 14**).
3. Prepare a 1% agarose gel as indicated in the previous paragraph. The use of thin combs (0.75 mm thick) is advised to obtain sharper bands.
4. Load the positive samples at least in two 1:2 dilutions (e.g., 1 and 2 μL) to minimize pipetting errors (*see* **Note 15**). Load a DNA ladder as a reference (*see* **Note 16**). Run the gel as described in the previous subsection.
5. Quantify sample DNAs by comparing the band intensities of your DNA with the intensity of the ladder (*see* **Note 17**).
6. A common and accessible way to obtain the sequence of a PCR product is to rely on an external company that offers a Sanger sequencing service. Prepare your samples for sequencing following the instructions provided by the sequencing service (*see* **Note 18**).

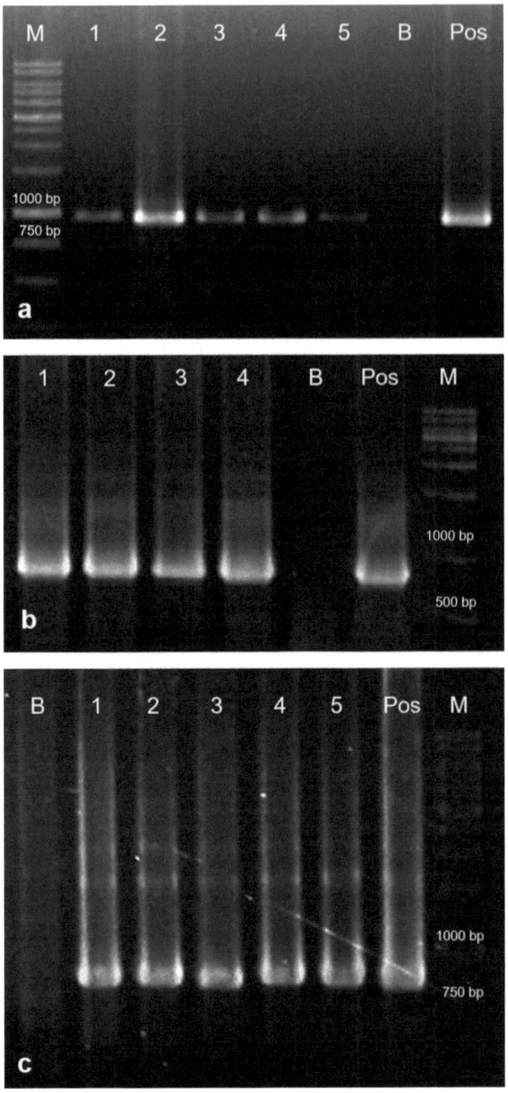

Fig. 2 PCR bands amplified by different primer combinations. (**a**) dnaDFDF1/pyrGFDr1 primers (direct PCR); (**b**) dnaDFDF1/pyrGFDr2 primers (semi-nested PCR); (**c**) dnaDFDF2/pyrGFDr2 primers (nested PCR). Lane 1–4/5: FD-infected grapevine samples; M: DNA marker (1 kb DNA ladder); B: blank; Pos: positive control

7. Visualize the Sanger sequencing chromatogram with a software capable of reading.abi files (*see* **Note 19**).
8. From the amplified *dnaD-imp-pyrG* genomic region, only the *imp* gene sequence will be employed for phylogenetic analysis. Use an online tool such as ORF Finder (NCBI) to identify the region coding the Imp protein and trim the consensus nucleotide sequence accordingly.

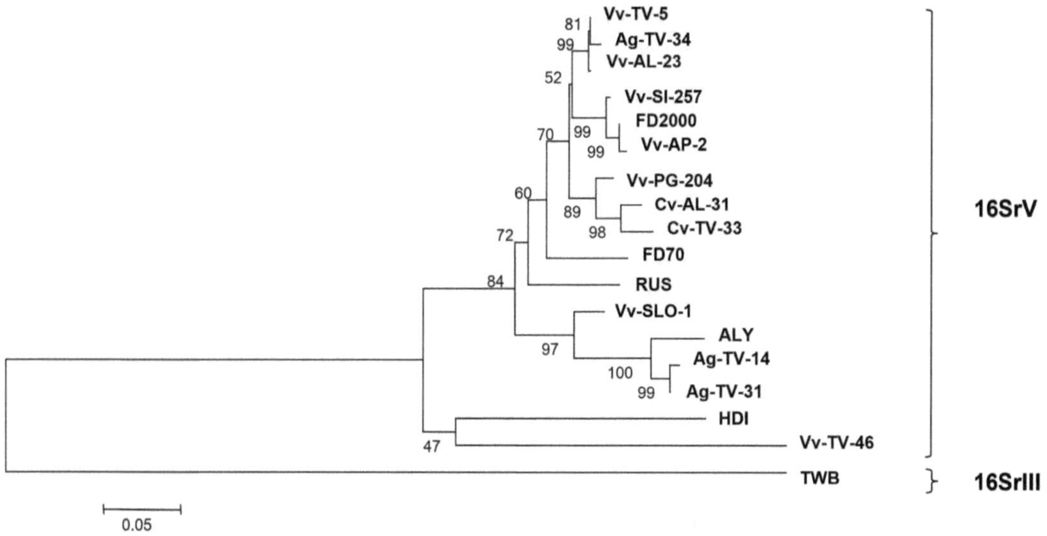

Fig. 3 Neighbor-joining tree of *imp* sequences built with MEGA software. Bar length is proportional to the number of base substitutions per site. Numbers on the branches are the confidence/bootstrap values obtained for 1000 replicates. Host: Ag, ALY: *Alnus glutinosa*; Cv: *Clematis vitalba*; Vv, FD2000, FD70: *Vitis vinifera*; RUS (Rubus Stunt): *Rubus fruticosus*; HDI (Hemp dogbane): *Apocynum cannabinum*; TWB (Tsuwabuki witches'-- broom): *Solanum tuberosum*. The *imp* gene sequence of TWB was obtained from GenBank database (accession number AB469014)

9. The obtained consensus sequences can be used for genetic diversity or phytoplasma phylogeny analyses (Fig. 3) (*see* **Note 20**).

4 Notes

1. The DNA can be extracted through a CTAB-based protocol [37, 38] or a commercial kit.
2. Calculate a volume of 0.5× TBE sufficient to prepare the gel and fill the electrophoresis cell.
3. The staining in a solution of 0.5 mg/mL Ethidium Bromide in water has historically been used as an alternative for DNA visualization. Ethidium Bromide is highly sensitive but is also a strong mutagen; therefore, it must be handled with extreme caution, and all contaminated material must be disposed of as hazardous waste. Safer, non-mutagenic alternatives are now available and strongly recommended.
4. Handle the tubes from the bottom, close the cap without touching the inner side, and arrange them in a fitting rack.
5. It is necessary to calculate a slightly higher volume of master mix because a small part of the mix is adsorbed by the plastic

tubes and tips. To reduce volume loss due to plastic adsorption, the use of low-retention plastics is recommended.

6. The extracted DNA of some samples may contain inhibitors, i.e., molecules that interfere with PCR amplification. This is quite common, for example, with leaf samples collected at the end of the season. Often, the presence of inhibiting compounds is associated with DNA extracts that appear colored. To verify the presence of PCR inhibition, a "test-PCR" can be performed by adding 1 μL of a positive DNA control to the reaction tube, along with the DNA sample to be tested. If no band is detected, it is confirmed the presence of inhibitory molecules in the sample, which prevented the amplification of the positive DNA control. To overcome the problem, it is usually sufficient to dilute 1:10 or 1:100 the DNA sample in water and use the dilutions as a PCR template. Alternatively, DNA can be precipitated in sodium acetate, and the pellet resuspended in a new buffer.
7. To reduce the risk of DNA contamination, it is recommended to perform the following steps in different rooms: PCR setup, amplification in a thermal cycler, and gel electrophoresis.
8. The expected amplicon size is 940 bp for FD-D. Slight differences in amplicon size are expected, due to the considerable length variability of *imp* gene; for example, in a set of 40 *imp* sequences from 16SrV-C, D, E subgroup phytoplasmas, the gene length varied from 432 to 471 nt [39].
9. If running the direct PCR products in the agarose gel, it is recommended not to use the same PCR samples as templates for nested or semi-nested PCR. This is because of the high risk of contamination from DNA aerosols in the electrophoresis room, where tubes containing amplicons are frequently handled and opened. Instead, repeat the first PCR and use that product for the second one.
10. Both nested and semi-nested PCRs are methods where the DNA template is the product of a first (direct) PCR. In nested PCR, both primers anneal within the first PCR product; in semi-nested PCR, only one primer is internal to the first amplicon, while the second one is the same used in the direct run. In our experience, nested PCR gives good results when testing grapevine samples; the risk of non-specific bands is higher with spontaneous plants and insect DNA as a template. In these cases, the use of semi-nested PCR significantly reduced the presence of undesired amplicons.
11. The volume of the agarose solution depends on the casting apparatus; it may vary from 30 mL for a gel tray 7 × 10 cm to 120 mL for a tray 15 × 15 cm.

12. The use of a microwave is easy and fast, but may be hazardous: the agarose solution may become superheated and boil over when whirled; for this reason, use a low to medium setting and do not tightly close bottles or flasks. As an alternative, agarose can be melted on a magnetic hot plate with a stir bar.
13. The stain can be added directly to the melted agarose solution before casting the gel; in this case, post-run staining is not required, and the gel can be examined during the electrophoresis. The gel tray may be placed on the UV transilluminator for a check of the DNA migration, then submerged again in the cell for continuing the run. Alternatively, after the run is finished, soak the gel in a solution containing the nucleic acid stain, prepared according to the manufacturer's instructions, for about 30 min.
14. Other methods for DNA quantification include absorbance-based spectrophotometry and fluorometry. Each method has its own limitations: quantification on agarose gel allows visualization of DNA integrity or residual RNA, but not proteins. Conversely, spectrophotometry can reveal protein contamination but not DNA or RNA contaminants. None of these methods can detect the presence of residual salts.
15. If the amplicon concentration is very high, load a diluted or smaller volume of the original PCR sample to avoid signal saturation.
16. These DNA ladders contain DNA fragments of known DNA sizes and masses and allow mass estimation of DNA samples by comparison.
17. It is best to compare the band intensity of the fragment of interest with the ladder band closest in size.
18. Usually, a dried PCR product is requested in a certain amount based on the DNA size. Each sample will be sequenced in a single strand with a primer, to be sent together with the amplicon. For good quality sequence data, have your DNA fragments sequenced double strand with both primers used for PCR amplification.
19. A free reader of.abi files is Chromas (www.technelysium.com.au). For the assembly of multiple sequences, more advanced software is needed. We use the set of DNA sequence assembly tools preGap5/Gap5 by Staden-Package (http://staden.sourceforge.net), which allows the generation of a consensus sequence for each sample.
20. Among the available tools, MEGA [40] is widely used: it contains a tool for performing multiple alignments based on CLUSTALW and for building phylogenetic trees with different methods. The reliability of the output is assessed by a bootstrap

test with 500 or 1000 replicates. On the other hand, for more rigorous phylogenetic analyses or when working with large datasets, the use of more specialized software, such as BEAST [41], MrBayes [42], RAMxML [43], or IQ-TREE [44], is recommended. These are highly reliable and precise, although more complex to use.

References

1. Bertaccini A, Paltrinieri S, Contaldo N (2019) Standard detection protocol: PCR and RFLP analyses based on 16S rRNA gene. In: Musetti R, Pagliari L (eds) Phytoplasmas, Methods in molecular biology, vol 1875. Humana Press, New York. https://doi.org/10.1007/978-1-4939-8837-2_7
2. Lee I-M, Gundersen-Rindal DE, Davis RE et al (1998) Revised classification scheme of phytoplasmas based on RFLP analyses of 16S rRNA and ribosomal protein gene sequences. Int J Syst Bacteriol 48:1153–1169. https://doi.org/10.1099/00207713-48-4-1153
3. Zhao Y, Wei W, Davis RE et al (2010) Recent advances in 16S rRNA gene-based phytoplasma differentiation, classification and taxonomy. In: Weintraub PG, Jones P (eds) Phytoplasmas: genomes, plant hosts and vectors. CAB International, Wallingford. https://doi.org/10.1079/9781845935306.0064
4. The IRPCM Phytoplasma/Spiroplasma Working Team—Phytoplasma Taxonomy Group (2004) "Candidatus Phytoplasma", a taxon for the wall-less, non-helical prokaryotes that colonize plant phloem and insects. Int J Syst Evol Microbiol 54:1243–1255. https://doi.org/10.1099/ijs.0.02854-0
5. Schneider B, Seemüller E (2009) Strain differentiation of "*Candidatus* Phytoplasma mali" by SSCP- and sequence analyses of the *hfl*B gene. J Plant Pathol 91:103–112
6. Kakizawa S, Oshima K, Namba S (2006) Diversity and functional importance of phytoplasma membrane proteins. Trends Microbiol 14: 254–256. https://doi.org/10.1016/j.tim.2006.04.008
7. Danet JL, Balakishiyeva G, Cimerman A et al (2011) Multilocus sequence analysis reveals the genetic diversity of European fruit tree phytoplasmas and supports the existence of inter-species recombination. Microbiology 157: 438–450. https://doi.org/10.1099/mic.0.043547-0
8. Fabre A, Danet JL, Foissac X (2011) The "stolbur" phytoplasma antigenic membrane protein gene *stamp* is submitted to diversifying positive selection. Gene 472:37–41. https://doi.org/10.1016/j.gene.2010.10.012
9. Cimerman A, Pacifico D, Salar P et al (2009) Striking diversity of *vmp1*, a variable gene encoding a putative membrane protein of the "stolbur" phytoplasma. Appl Environ Microbiol 75:2951–2957. https://doi.org/10.1128/AEM.02613-08
10. Fialová R, Válová P, Balakishiyeva G et al (2009) Genetic variability of Stolbur phytoplasma in annual crop and wild plant species in South Moravia. J Plant Pathol 91:411–416
11. Malembic-Maher S, Desqué D, Khalil D et al (2020) When a Palearctic bacterium meets a Nearctic insect vector: genetic and ecological insights into the emergence of the grapevine Flavescence dorée epidemics in Europe. PLoS Pathog 16:e1007967. https://doi.org/10.1371/journal.ppat.1007967
12. Martini M, Quaglino F, Bertaccini A (2019) Multilocus genetic characterization of phytoplasmas. In: Bertaccini A, Oshima K, Kube M, Rao G (eds) Phytoplasmas: plant pathogenic bacteria—III. Springer, Singapore. https://doi.org/10.1007/978-981-13-9632-8_9
13. Kakizawa S, Oshima K, Nishigawa H et al (2004) Secretion of immunodominant membrane protein from onion yellows phytoplasma through the Sec protein-translocation system in *Escherichia coli*. Microbiology 150:135–142. https://doi.org/10.1099/mic.0.26521-0
14. Konnerth A, Krczal G, Boonrod K (2016) Immunodominant membrane proteins of phytoplasmas. Microbiology 162:1267–1273. https://doi.org/10.1099/mic.0.000331
15. Kakizawa S, Oshima K, Ishii Y et al (2009) Cloning of immunodominant membrane protein genes of phytoplasmas and their *in planta* expression. FEMS Microbiol Lett 293:92–101. https://doi.org/10.1111/j.1574-6968.2009.01509.x
16. Yu Y, Yeh K, Lin C (1998) An antigenic protein gene of a phytoplasma associated with sweet potato witches' broom. Microbiology 144:

1257–1262. https://doi.org/10.1099/00221287-144-5-1257

17. Tran-Nguyen LT, Kube M, Schneider B et al (2008) Comparative genome analysis of "*Candidatus* Phytoplasma australiense" (subgroup tuf-Australia I; rp-A) and "*Ca.* Phytoplasma asteris" strains OY-M and AY-WB. J Bacteriol 190:3979–3991. https://doi.org/10.1128/jb.01301-07
18. Berg M, Davies DL, Clark MF et al (1999) Isolation of the gene encoding an immunodominant membrane protein of the apple proliferation phytoplasma, and expression and characterization of the gene product. Microbiology 145:1937–1943. https://doi.org/10.1099/13500872-145-8-1937
19. Morton A, Davies DL, Blomquist CL et al (2003) Characterization of homologues of the apple proliferation immunodominant membrane protein gene from three related phytoplasmas. Mol Plant Pathol 4:109–114. https://doi.org/10.1046/j.1364-3703.2003.00155.x
20. Da Rold G, Filippin L, Borgo M et al (2010) Characterization of the *imp* gene in "flavescence dorée" and related phytoplasmas. In: Bertaccini A, Laviña A, Torres E (eds) Current status and perspectives of phytoplasma disease research and management, Abstract book of the combined meeting of WG 1-4, Sitges, 2010
21. Schneider B, Hüttel B, Zübert C et al (2020) Genetic variation, phylogenetic relationship and spatial distribution of "*Candidatus* Phytoplasma ulmi" strains in Germany. Sci Rep 10:21864. https://doi.org/10.1038/s41598-020-78745-w
22. Neriya Y, Sugawara K, Maejima K et al (2011) Cloning, expression analysis, and sequence diversity of genes encoding two different immunodominant membrane proteins in poinsettia branch-inducing phytoplasma (PoiBI). FEMS Microbiol Lett 324:38–47. https://doi.org/10.1111/j.1574-6968.2011.02384.x
23. Marcone C, Pierro R, Palmieri C (2024) Occurrence, impact, and multilocus sequence analysis of alder yellows phytoplasma infecting common alder and Italian alder in southern Italy. Microorganisms 12:1140. https://doi.org/10.3390/microorganisms12061140
24. Rossi M, Samaržija I, Šeruga Musić M et al (2019) Diversity and functional importance of phytoplasma membrane proteins. In: Bertaccini A, Oshima K, Kube M, Rao G (eds) Phytoplasmas: plant pathogenic bacteria—III. Springer, Singapore. https://doi.org/10.1007/978-981-13-9632-8_5
25. Siampour M, Izadpanah K, Galetto L et al (2013) Molecular characterization, phylogenetic comparison and serological relationship of the Imp protein of several "*Candidatus* Phytoplasma aurantifolia" strains. Plant Pathol 62:452–459. https://doi.org/10.1111/j.1365-3059.2012.02662.x
26. Al-Ghaithi AG, Al-Subhi AM, Al-Mahmooli IH et al (2018) Genetic analysis of "*Candidatus* Phytoplasma aurantifolia" associated with witches' broom on acid lime trees. PeerJ 6:e4480. https://doi.org/10.7717/peerj
27. Bohunická M, Valentová L, Suchá A et al (2017) Identification of 17 "*Candidatus* Phytoplasma pyri" genotypes based on the diversity of the *imp* gene sequence. Plant Pathol 67:971–977. https://doi.org/10.1111/ppa.12805
28. Da Rold G, Filippin L, Malembic-Maher S et al (2011) The *imp* gene in "flavescence dorée" and related phytoplasmas from grapevine, *Clematis* sp and alder. In: Brown BR, Bertaccini A (eds) 18th International congress of the International Organization for Mycoplasmology IOM, Chianciano Terme, 2011
29. Filippin L, De Prà V, Zottini M et al (2011) Nucleotide sequencing of *imp* gene in phytoplasmas associated to "flavescence dorée" from *Ailanthus altissima*. Bull Insectol 64(Supplement):S49–S50
30. Trivellone V, Ripamonti M, Angelini E et al (2019) Evidence suggesting interactions between immunodominant membrane protein Imp of Flavescence dorée phytoplasma and protein extracts from distantly related insect species. J Appl Microbiol 127:1801–1813. https://doi.org/10.1111/jam.14445
31. Filippin L, Jović J, Cvrković T et al (2009) Molecular characteristics of phytoplasmas associated with *Flavescence dorée* in clematis and grapevine and preliminary results on the role of *Dictyophara europaea* as a vector. Plant Pathol 58:826–837. https://doi.org/10.1111/j.1365-3059.2009.02092.x
32. Lessio F, Bocca F, Alma A (2019) Development, spatial distribution, and presence on grapevine of nymphs of *Orientus ishidae* (Hemiptera: Cicadellidae), a new vector of Flavescence dorée phytoplasmas. J Econ Entomol 112:2558–2564. https://doi.org/10.1093/jee/toz177
33. Angelini E, Squizzato F, Lucchetta G et al (2004) Detection of a phytoplasma associated with grapevine *Flavescence dorée* in *Clematis vitalba*. Eur J Plant Pathol 110:193–201. https://doi.org/10.1023/B:EJPP.0000015361.95661.37

34. Arnaud G, Malembic-Maher S, Salar P et al (2007) Multilocus sequence typing confirms the close genetic interrelatedness of three distinct "flavescence dorée" phytoplasma strain clusters and group 16SrV phytoplasmas infecting grapevine and alder in Europe. Appl Environ Microbiol 73:4001–4010. https://doi.org/10.1128/AEM.02323-06
35. Mehle N, Jakoš N, Mešl M et al (2019) Phytoplasmas associated with declining of hazelnut (*Corylus avellana*) in Slovenia. Eur J Plant Pathol 155:1117–1132. https://doi.org/10.1007/s10658-019-01839-3
36. Berges R, Rott M, Seemüller E (2000) Range of phytoplasma concentration in various plant hosts as determined by competitive polymerase chain reaction. Phytopathology 90:1145–1152. https://doi.org/10.1094/PHYTO.2000.90.10.1145
37. Angelini E, Clair D, Borgo M et al (2001) Flavescence dorée in France and Italy—occurrence of closely related phytoplasma isolates and their near relationships to Palatinate grapevine yellows and an alder yellows phytoplasma. Vitis 40:79–86. https://doi.org/10.5073/vitis.2001.40.79-86
38. Rizzoli A, Belgeri E, Jermini M et al (2021) *Alnus glutinosa* and *Orientus ishidae* (Matsumura, 1902) share phytoplasma genotypes linked to the "Flavescence dorée" epidemics. J Appl Entomol 145:1015–1028. https://doi.org/10.1111/jen.12933
39. Filippin L, Angelini E. European Patent no. EP2918685
40. Kumar S, Tamura K, Nei M (1994) MEGA—molecular evolutionary genetics analysis software for microcomputers. Comput Appl Biosci 10:189–191. https://doi.org/10.1093/bioinformatics/10.2.189
41. Suchard MA, Lemey P, Baele G et al (2018) Bayesian phylogenetic and phylodynamic data integration using BEAST 1.10. Virus Evol 4: vey016. https://doi.org/10.1093/ve/vey016
42. Uche-Ikonne O, Dondelinger F, Palmer T (2021) Software Application Profile: Bayesian estimation of inverse variance weighted and MR-Egger models for two-sample Mendelian randomization studies—mrbayes. Int J Epidemiol 50:43–49. https://doi.org/10.1093/ije/dyaa191
43. Stamatakis A, Ludwig T, Meier H (2005) RAxML-III: a fast program for maximum likelihood-based inference of large phylogenetic trees. Bioinformatics 21:456–463. https://doi.org/10.1093/bioinformatics/bti191
44. Nguyen LT, Schmidt HA, von Haeseler A et al (2015) IQ-TREE: a fast and effective stochastic algorithm for estimating maximum-likelihood phylogenies. Mol Biol Evol 32: 268–274. https://doi.org/10.1093/molbev/msu300

Chapter 5

Triplex qPCR for Grapevine Yellows Phytoplasmas Combined with a Universal Endogenous Control

Luca Fontanesi, Anna Pedroncelli, Mattia Tabarelli, Andreas Gallmetzer, and Katrin Janik

Abstract

Flavescence dorée and *Bois noir* are two major grapevine yellows diseases affecting European viticulture. Accurate, early detection of the associated bacterial agents—*Flavescence dorée* phytoplasma (FDp) and '*Candidatus* Phytoplasma solani'—in both grapevines and insect vectors is critical for effective disease monitoring and control. Quantitative PCR (qPCR) is a highly sensitive molecular tool widely used for phytoplasma detection, often implemented in multiplex formats for simultaneous identification of multiple targets. However, the integration of internal quality controls is essential to validate DNA extraction efficiency and rule out PCR inhibition. While species-specific endogenous controls are commonly used, their limited host range restricts assay flexibility. This protocol describes a novel triplex qPCR assay enabling the simultaneous detection of FDp, '*Ca. P. solani*', and a universal endogenous control based on a conserved region of the eukaryotic 28S rRNA gene. This universal control allows broad applicability across diverse eukaryotic hosts, including both grapevine and key insect vectors, facilitating robust and flexible diagnostics in a single reaction setup.

Key words Grapevine yellows, Phytoplasma, Flavescence dorée, Bois noir, '*Candidatus* Phytoplasma solani', Real-time PCR, Multiplex qPCR, Endogenous control, 28S rRNA, Diagnostic

1 Introduction

Flavescence dorée and *Bois noir* are two major grapevine yellows diseases in European viticulture. Both diseases are associated with phytoplasma infections and cause significant economic losses in affected regions [1]. Reliable detection of these pathogens in both grapevines and their insect vectors is essential for effective disease management and containment. For these reasons, accurate molecular diagnostic techniques, such as quantitative PCR (qPCR), are crucial for the effective management of phytoplasma diseases. Such molecular methods are powerful and sensitive tools that allow the detection of phytoplasma even at low concentrations. Several protocols describing the simultaneous detection of more

Katrin Janik and Mattia Tabarelli (eds.), *Phytoplasma: Methods and Protocols*, Methods in Molecular Biology, vol. 3008, https://doi.org/10.1007/978-1-0716-5104-9_5,

than one phytoplasma species or other phytopathogens by qPCR-multiplexing have been developed so far [2–4]. A crucial hallmark of PCR analyses is quality control, in which the presence and integrity of the extracted DNA is analyzed and the absence of PCR inhibition can be determined. Quality controls are often performed as a PCR of a known target, present in the DNA sample. These known targets can be exogenous (i.e., added) template nucleic acids ("spike controls") or internal, endogenous targets on the co-extracted host DNA. In contrast to exogenous controls, endogenous quality controls allow to confirm successful extraction of intact DNA, and a PCR-inhibition check in one step.

For the detection of Grapevine flavescence dorée phytoplasma (FDp) and '*Candidatus* Phytoplasma solani' ('*Ca.* P solani'), endogenous controls have been developed for the host plant *Vitis vinifera* and the main insect vectors, *Scaphoideus titanus* and *Hyalesthes obsoletus*, respectively [4–6].

However, the use of species-specific endogenous controls is a limitation to a broader application of multiplex qPCR, because the controls restrict the assay's applicability to single organisms. On the contrary, a universal endogenous control that amplifies a host DNA target of a broad range of organisms abolishes this issue. Here, we present a new triplex real-time qPCR set-up to detect FDp and '*Ca.* P solani' in combination with a universal endogenous control that amplifies a stretch of the eukaryotic 28S rRNA DNA sequence [7, 8].

2 Materials

Prepare and store all reagents at room temperature (unless indicated otherwise). Diligently follow all waste disposal regulations when disposing of waste materials.

2.1 DNA Extraction from Vitis Leaves

1. Plant material, 50–100 mg of *Vitis* leaf stalks or leaf midribs (*see* **Note 1**).
2. Tungsten carbide beads, 3 mm.
3. DNA extraction kit, suitable for plant material (*see* **Note 2**).
4. CTAB Extraction Buffer: mix in sterile water: 2.5% (w/v) Cetyltrimethylammoniumbromid (CTAB), 100 mM Tris, pH 8.0, 1.4 M NaCl, 50 mM EDTA, pH 8.0, 1% (w/v) PVP-40. Add 0.5% (w/v) ascorbic acid (add directly before use).
5. RNase A, 100 mg/mL.
6. Ethanol, 100% (v/v) or 96% (v/v).
7. Reaction tubes (2 mL).
8. Water bath.

Table 1
Sequences of primers and probes used in this protocol

Name	Sequence (5′-3′)	Specificity
UNI28S-fwd UNI28S-rev UNI28S-P	CTACTATCTAGCGAAACC AYTAGAGTCAAGCTCAAC **HEX**-AAA[+G][+A][+A]G[+A][+C][+C] [+C][+T]-**BHQ1**	28S rRNA [8] Endogenous control
FD-fwd FD-rev FDgen_probe	TTATGCCTTATGTTACGTCTTCTATTGTTA TCTCCTTGTTCTTGCCATTCTTT **FAM**-ACCTTTTGACTCAATTGA-**BHQ1**	sec Y gene [7] Detection of FDp
BN-fwd BN-rev BNgen_probe	AAGCAGGTTTAGCGATGGTTGT TGGTACCGTTGCTTCATCATTT **Cy5**-TTAATACCACCTTCAGGAAA-**BHQ3**	Stol11 genomic fragment [7] Detection of '*Ca*. P. solani'

Fluorophores and quenchers are indicated in bold in the probe sequences. A "+" prior to a nucleotide in brackets within a sequence indicates that the nucleotide is a locked nucleotide (LNA)

9. Wet ice, for sample cooling.
10. Bead mill homogenizer (*see* **Note 3**).
11. Microcentrifuge tubes (2 mL).
12. Benchtop centrifuge with a rotor compatible with 2 mL tubes, for centrifugation at max. 20,000× *g*.

2.2 Real-Time PCR Detection

1. DNA samples extracted from the leaf tissue mentioned in Subheading 2.1.
2. Thermal cycler for quantitative real-time PCR (*see* **Note 4**).
3. Plasmids for standard curve (*see* **Note 5**).
4. Primers and Probes (Table 1 and **Note 6**).
5. Multiplex qPCR master mix (*see* **Note 7**).

3 Methods

Carry out all procedures at room temperature unless otherwise specified. The first part of the protocol focuses on DNA extraction from leaves. The second part describes a hydrolysis probe-based qPCR multiplex technique (*see* **Note 8**) for the rapid and accurate detection of phytoplasma DNA.

3.1 DNA Extraction from Grapevine Leaves

The following protocol is an adaptation using the columns and reagents of the DNeasy Mini Plant Kit from Qiagen, Hilden, Germany, in combination with self-made buffers (*see* **Note 2**).

1. Cut the leaf stalks or the midrib into very small pieces and weigh 50–100 mg of plant material (*see* **Note 1**). Transfer it into a 2 mL reaction tube. Add two tungsten carbide beads to each tube containing the plant material.
2. Add 800 μL of CTAB-Buffer Lysis Buffer (supplemented with ascorbic acid) to each tube. Disrupt the samples using a bead mill homogenizer at 30.0 Hz for 5–10 min until all samples are homogenized.
3. Add 4 μL of 100 mg/mL RNase A to each tube.
4. Incubate the tubes at 65 °C for 30 min in a water bath.
5. Add 260 μL of Buffer P3 to each tube, mix, and incubate the tubes on ice for 10 min.
6. Centrifuge the lysate at 20,000× *g* for 10 min to form a pellet.
7. Transfer the lysate into a QIAshredder spin column (part of the above-mentioned DNA extraction kit) placed in a 2 mL collection tube and centrifuge for 2 min at 20,000× *g*.
8. Transfer 500 μL of the flow-through to a new 2 mL tube without disturbing the pellet.
9. Add 750 μL (1.5 volumes) of Buffer AW1 (part of the above-mentioned DNA extraction kit) to the flow-through and mix by pipetting.
10. Transfer 650 μL of the mixture to a DNeasy Mini spin column (part of the above-mentioned DNA extraction kit) placed in a 2 mL collection tube. Centrifuge at 6000× *g* for 1 min.
11. Discard the flow-through and repeat **steps 8–10** with the remaining sample and use the same spin column in **step 10** (*see* **Note 9**).
12. Add 500 μL of Buffer AW2 (part of the above-mentioned DNA extraction kit) to the column and centrifuge for 1 min at 6000× *g*. Discard the flow-through.
13. Add 500 μL of 100% (or 96%) ethanol and centrifuge for 3 min at 20,000× *g*.
14. Transfer the spin column to a new 1.5 or 2 mL microcentrifuge tube.
15. Aliquot and preheat 200 μL/sample AE Buffer (part of the above-mentioned DNA extraction kit) in a water bath at 65 °C to increase DNA yield (*see* **Note 10**).
16. Add 100 μL of preheated AE Buffer directly to the column matrix, incubate at room temperature for 5 min, and centrifuge at 6000× *g* for 1 min.
17. Repeat the elution **step 16** to get a total volume of 200 μL of eluted DNA per column.

Table 2
Pipetting scheme for the triplex qPCR

Reagent	Stock concentration	Working concentration	Volume per reaction (μL)
iQ Multiplex Powermix (*see* **Note 7**)	2X	1X	10
UNI28-fwd	20 μM	0.4 μM	0.4
UNI28-rev	20 μM	0.4 μM	0.4
UNI28S-P	5 μM	0.2 μM	0.8
FD-fwd	20 μM	0.9 μM	0.9
FD-rev	20 μM	0.9 μM	0.9
FDgen_probe	5 μM	0.2 μM	0.8
BN-fwd	20 μM	0.9 μM	0.9
BN-rev	20 μM	0.9 μM	0.9
BNgen_probe	5 μM	0.2 μM	0.8
Nuclease-free water			1.2
Subtotal			18
Template DNA			2.0
Total			20

3.2 Real-Time qPCR Detection

1. Analyze each sample in duplicates using the primers and probes listed in Table 1 (*see* **Notes 11 and 12**).
2. Prepare the reaction mixture by adding the PCR ingredients listed in Table 2.
3. Cycler conditions: lid temperature 105 °C, initial denaturation at 95 °C for 3 min, 45 cycles as follows: 95 °C for 15 s, 60 °C for 1 min.

3.3 Interpretation of qPCR Results

1. Confirm that each target (FDp-FAM, CaPsol-Cy5, and 28S-HEX) shows clear, exponential amplification curves that plateau, indicating successful amplification. In Fig. 1, typical standard curves for the multiplexed detection of FDp (Fig. 1a), '*Ca.* P. solanis' (Fig. 1b) and 28S (Fig. 1c) in different concentrations are depicted.
2. Ensure that the baseline signal remains flat (i.e., no amplification) in wells containing non-template or negative controls (*see* **Note 13**).
3. If a standard curve is analyzed in parallel: Use the qPCR software to extract Cq values for each standard dilution (*see* **Note 14**). Plot the Cq values (*y*-axis) versus the log of the

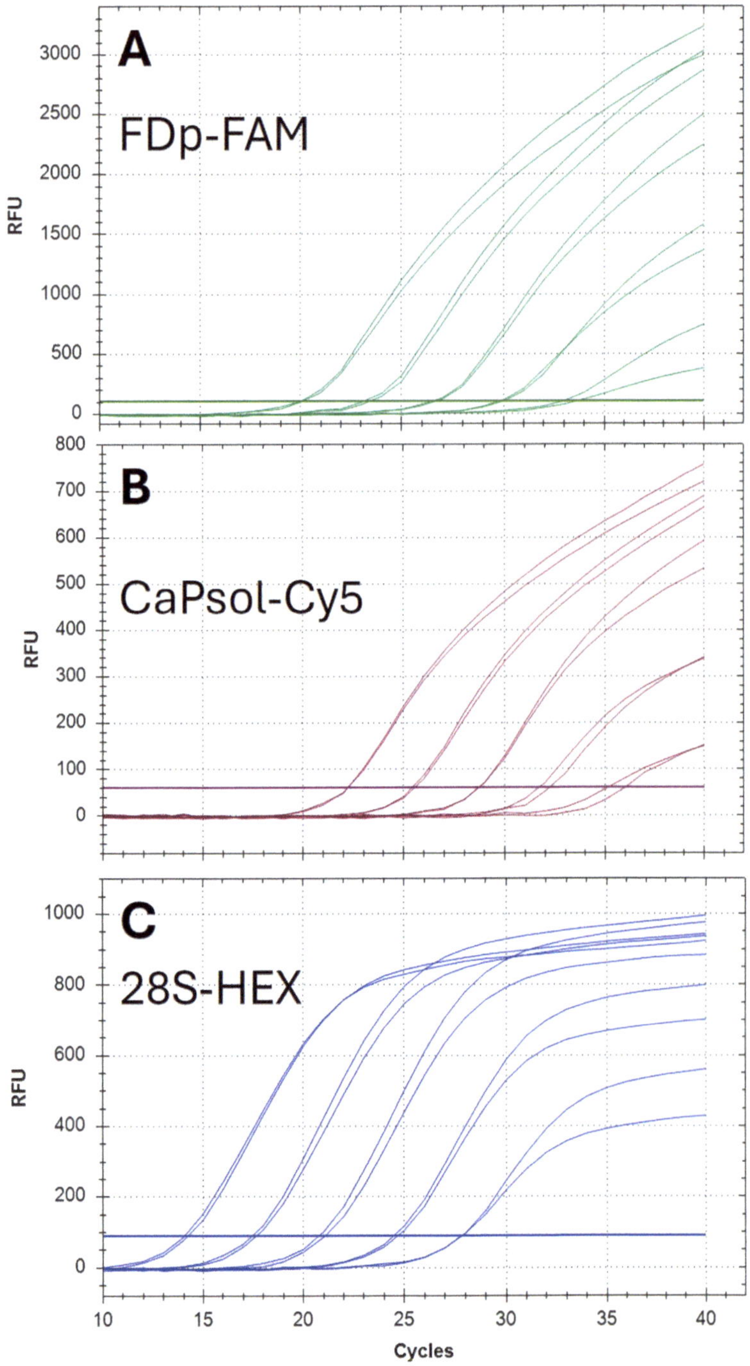

Fig. 1 Amplification curves from a triplex qPCR assay targeting Flavescence dorée phytoplasma (FDp), '*Candidatus* Phytoplasma solani' (CaPsol), and the 28S rRNA gene as the internal control. Fluorescence signals are displayed separately for each target: (**a**) FDp, (**b**) CaPsol, and (**c**) 28S. Ten-fold serial dilutions of plasmid templates containing the respective subcloned amplicons were analyzed in duplicate to assess assay sensitivity and dynamic range (*see* **Note 5**)

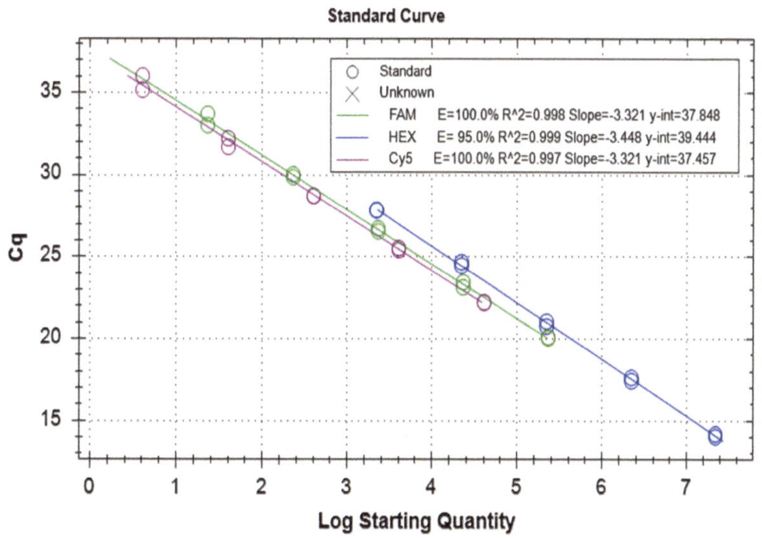

Fig. 2 Relationship between Cq and target quantity in serial dilutions of plasmids used as positive controls for PCR amplification. The standard curve graphs plot the quantification cycle (Cq) against the logarithm of the starting quantity (Log Starting Quantity) of the template in the PCR sample. It includes data points for three fluorescent dyes: FDp-FAM (green), 28S-HEX (blue), and CaPsol-Cy5 (purple). Each dye has its own linear regression line, with associated statistical parameters such as efficiency (*E*), R^2 value, slope, and y-intercept

starting quantity (*x*-axis) for each assay (Fig. 2) and assess linearity and calculate PCR efficiency using the slope: Efficiency (%) $= (10^{(-1/\text{slope})} - 1) \times 100$ (*see* **Note 15**).

4. Quantify the amount of template DNA in samples by comparing their Cq values to the standard curve or by using the $2^{-\Delta\Delta Ct}$ method (*see* **Notes 5, 16, 17, and 18**).
5. An example including a detailed evaluation and interpretation of the results of a multiplex qPCR of four different samples is given in Fig. 3.

4 Notes

1. Phytoplasma colonize the phloem in the midrib and the stalk of the leaf. Thus, the extraction from the leaf's midrib or stalk is recommended rather than DNA extraction from the leaf blade.
2. DNA extraction kits from different suppliers can be used. We use the DNeasy Plant Mini Kit from Qiagen, Hilden, Germany, and get reliable results. Keep in mind that DNA extraction efficiency, the DNA yield, and the presence of impurities that might affect PCR performance might vary when using kits from different suppliers.

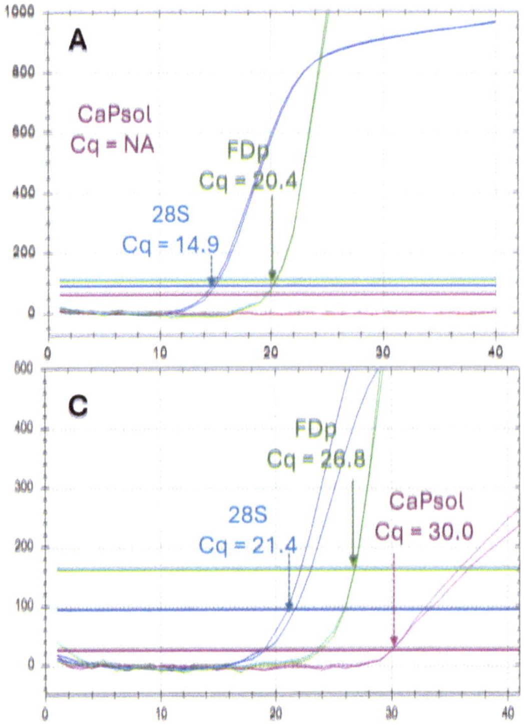

Fig. 3 Multiplex qPCR detection of FDp, '*Candidatus* Phytoplasma solani' (CaPsol), and 28S rRNA gene (28S) targets in four different, representative samples. Amplification curves are shown for FDp (green), CaPsol (purple), and the internal control 28S rDNA (blue). (**a**) Sample is positive for FDp only. (**b**) Sample is positive for CaPsol only. (**c**) Sample is positive for both FDp and CaPsol. (**d**) Sample is negative for both FDp and CaPsol. All four samples are positive for 28S, confirming the presence and integrity of amplifiable DNA. The different Cq values for 28S reflect varying input DNA amounts among samples. Exact target quantification can be performed by comparing the Cq values of the different samples to the respective standard curves in Fig. 2

3. We use the TissueLyser II from Qiagen, Hilden, Germany; however, any bead mill homogenizer suitable for the sample size is suitable.
4. The thermal cycler must be equipped with the appropriate excitation light sources and filters to detect the used fluorescent probes. We use the CFX Connect Real-Time PCR Detection System and the CFX Opus Real-Time PCR System from Bio-Rad, Hercules, CA, United States, for the detection of the fluorescent probes described in this protocol. If other fluorophore-quencher combinations are used make sure that a thermal cycler with the appropriate detection system is used.
5. For the absolute quantification of phytoplasma DNA, a standard curve with defined numbers of target copies must be analyzed in parallel. The plasmids contain the subcloned fragments of the above-described PCR-amplicons for FDp (pJET1.2-FD), '*Ca.* P. solanis' (pJET1.2-BN) and 28S rRNA gene (pJET1.2-28S). We used the plasmids pJET1.2-FD

(range 2.31 × 10^5–10^1 copies/reaction, pJET1.2-BN (range 4.03 × 10^4–4.03 × 10^0 copies/reaction), and pJET1.2-28S (range 2.21 × 10^7–2.21 × 10^3 copies/reaction) for the analyses shown in Figs. 1 and 2. The ranges are chosen to cover the Cq values expected in the sample analysis.

The Cq signal of the sample can then be compared to the standard curve, and the target quantity in the sample can be determined. We subcloned the PCR amplicon to generate pJET1.2 plasmids for standard quantification [8]. For relative quantification without the use of standard curves, the $2^{-\Delta\Delta CT}$ method can be used [9]. If the $2^{-\Delta\Delta CT}$ method is used, normalize the target gene abundance (i.e., FD and BN) to the DNA amount (i.e., 28S). That allows the relative quantification of target DNA in reference to total DNA in the sample and allows to compare results between different PCR results without the use of standard curves. Keep in mind that the abundance of 28S rRNA gene copies in a genome might vary between organisms. Thus, comparison of 28S-normalized target gene abundance between different host organisms is not recommended [8].

6. The fluorophores and quenchers can be adapted to be suitable for other real-time qPCR cyclers. As fluorophores, we use Hexachlorofluorescein (HEX), Carboxyfluorescein (FAM), and Cyanine (Cy5). In our study, we utilized the quenchers BHQ1 and BHQ3 for our probes. However, it is important to note that the choice of fluorophores and quenchers can vary depending on the suppliers from whom the probes are purchased. When using fluorophores with different fluorophore/quencher combinations, it is crucial to ensure that the emission wavelengths of the fluorophores do not overlap. This precaution is necessary to guarantee clear discrimination between the different targets being analyzed. Proper selection and validation of these components in combination with the used qPCR multiplex master mix are essential to maintain the accuracy and reliability of the multiplex assays.
7. Use a reagent master mix designed for the simultaneous amplification and detection of multiple targets in a single qPCR. We use the 2X iQ Multiplex Powermix from Bio-Rad, Hercules, CA, United States, but any other reagent mix suitable for your detection system can be used.
8. We use TaqMan probes with fluorophore and quencher combinations that are suitable for our thermal cycler detection system (*see* **Note 6**). Other probe systems or fluorophore-quencher combinations can be used. However, they must be adapted according to the respective reaction chemistry and the detection system.

9. The same spin column can be used to increase the amount of DNA bound to the column's silica-matrix and thus enhance the DNA-concentration in the eluate.
10. Instead of the buffer, DNA can also be diluted in pre-warmed, nuclease-free water. The choice of the eluate depends on the downstream application and how long the DNA sample is supposed to be stored. DNA is more stable in buffers containing EDTA, but EDTA can negatively affect PCR efficiency.
11. Add a non-template control (nuclease-free water instead of the template DNA) to detect contamination and primer–dimer formation, a negative control (DNA from a sample known to be not infected) to check for non-specific amplification or background signal, and a positive control (DNA from a sample known to be infected). If you are running standard curves for absolute target quantification in parallel, a separate positive control is not strictly necessary.
12. The primers and probes listed in the protocol are not strictly specific to FDp and '*Ca*. P. solani'. Beside FDp, the *secY* (FDp) primers also amplify Elm Yellows phytoplasma (EY1 phytoplasma isolate) [7]. Stol11 primers amplify Stolbur SE phytoplasma from the 16SrXII group [7]. Since only FDp and '*Ca*. P. solani' is known to infect grapevine, it can be assumed that a positive PCR result indicates the presence of these respective phytoplasma. However, it is recommended to confirm FDp-positive samples by amplifying and sequencing the phytoplasmal *16S rRNA* gene and the 16S–23S intergenic spacer region using the P1/P7 primer pair, followed by sequence analysis with iPhyClassifier [10, 11].
13. This confirms the absence of contamination or primer–dimer artifacts.
14. The Cq is the cycle number where the fluorescence signal curve (relative fluorescence units, RFU) exceeds the determined threshold. Setting the threshold correctly in qPCR is critical for accurate and reproducible Cq (Ct) determination. The threshold line should intersect the exponential phase of the respective amplification curves. Thus, the threshold must be individually placed for each PCR (i.e., each fluorescent curve) in the multiplex. The qPCR evaluation software provides an automatic calculation of the threshold, but it is recommended to carefully check the threshold position and—if necessary—adapt it manually. Standard curves provide a critical reference for setting the threshold and determining the range of optimal PCR efficiency.
15. PCR efficiency can be determined using the standard curve or serial dilutions of a positive sample. Figure 1 shows the raw amplification data that were used to assess the PCR

performance for each of the runs in the multiplex PCR shown in Fig. 2. An efficiency of 100% indicates that the amount of template DNA is doubled in every cycle. We recommend an efficiency of 95–105% and an R^2 value (an indicator of the reliability of the standard curve fit) of ≥0.98.

16. Only samples with a Cq within the range of the standard curve can be quantified using the standard curve. Quantification using Cq values outside the standard curve range is not recommended, as PCR efficiency cannot be reliably assumed beyond the validated standard range.
17. Cq (quantification cycle) values are influenced by multiple run-specific variables, such as reagent batches, thermal cycler calibration, fluorescence detection sensitivity, and threshold settings. Even slight differences in these conditions between qPCR runs can shift Cq values, making direct comparisons unreliable.
18. Absolute quantification using a standard curve or relative quantification via the $2^{-\Delta\Delta Cq}$ method (*see* **Note 5**) is recommended because both approaches account for variability in amplification efficiency and experimental conditions, enabling accurate and reliable comparison of target DNA levels across samples and runs. Comparing Cq values (across different runs) can lead to inaccurate conclusions about target abundance. For valid comparisons, all samples should be run in the same qPCR experiment under identical conditions or normalized using standard curves or included in every run.

Acknowledgements

We thank Christine Kerschbamer for technical support and Urban Spitaler and Sabine Oberhofer for scientific discussions about grapevine yellowing diseases.

References

1. Bertaccini A, Duduk B, Paltrinieri S et al (2014) Phytoplasmas and phytoplasma diseases: a severe threat to agriculture. Am J Plant Sci 5:1763–1788. https://doi.org/10.4236/ajps.2014.512191
2. Abou-Jawdah Y, Aknadibossian V, Jawhari M et al (2019) Real-time PCR protocol for phytoplasma detection and quantification. Methods Mol Biol 1875:117–130. https://doi.org/10.1007/978-1-4939-8837-2_9
3. Linck H, Krüger E, Reineke A (2017) A multiplex TaqMan qPCR assay for sensitive and rapid detection of phytoplasmas infecting Rubus species. PLoS One 12:e0177808. https://doi.org/10.1371/journal.pone.0177808
4. Pelletier C, Salar P, Gillet J et al (2009) Triplex real-time PCR assay for sensitive and simultaneous detection of grapevine phytoplasmas of the 16SrV and 16SrXII-A groups with an endogenous analytical control. Vitis 48:87–89. https://doi.org/10.5073/vitis.2009.48.87-95
5. Gallmetzer A, Innerebner G, Roschatt C et al (2020) Dynamics of *Scaphoideus titanus* population in southern South Tyrol (Italy) and

detection of grapevine Flavescence dorée phytoplasma in the insect with a multiplex real-time PCR. Vitis 59:85–89. https://doi.org/10.5073/VITIS.2020.59.85-89

6. Fahrentrapp J, Michl G, Breuer M (2013) Quantitative PCR assay for detection of Bois noir phytoplasmas in grape and insect tissue. Vitis 52:85–89. https://doi.org/10.5073/vitis.2013.52.85-89
7. Hren M, Boben J, Rotter A et al (2007) Real-time PCR detection systems for Flavescence dorée and Bois noir phytoplasmas in grapevine: comparison with conventional PCR detection and application in diagnostics. Plant Pathol 56: 785–796. https://doi.org/10.1111/j.1365-3059.2007.01688.x
8. Mittelberger C, Obkircher L, Oberkofler V et al (2020) Development of a universal endogenous qPCR control for eukaryotic DNA samples. Plant Methods 16:341. https://doi.org/10.1186/s13007-020-00597-2
9. Livak KJ, Schmittgen TD (2001) Analysis of relative gene expression data using real-time quantitative PCR and the 2(−ΔΔC(T)) method. Methods 25:402–408. https://doi.org/10.1006/meth.2001.1262
10. Smart CD, Schneider B, Blomquist CL et al (1996) Phytoplasma-specific PCR primers based on sequences of the 16S-23S rRNA spacer region. Appl Environ Microbiol 62: 2988–2993. https://doi.org/10.1128/aem.62.8.2988-2993.1996
11. Zhao Y, Wei W, Lee I-M et al (2013) The iPhyClassifier, an interactive online tool for phytoplasma classification and taxonomic assignment. Methods Mol Biol 938:329–338. https://doi.org/10.1007/978-1-62703-089-2_28

Chapter 6

Rapid and Sensitive Detection of Phytoplasma Diseases Using a CRISPR/Cas12a DETECTR Assay Combined with Isothermal Recombinase Polymerase Amplification

Wei Wei, Yinong Yang, and Justin Shih

Abstract

Our protocol outlines a DNA endonuclease-targeted CRISPR trans reporter (DETECTR) assay, which combines CRISPR/Cas12a technology with isothermal Recombinase Polymerase Amplification (RPA) for the rapid and specific detection of phytoplasma diseases in plants. This isothermal method utilizes RPA to amplify the target DNA fragment from the genomic DNA of phytoplasmas, followed by incubation with Cas12a nuclease and CRISPR RNAs (crRNAs) specifically designed to target unique phytoplasma DNA sequences. Upon initial cleavage of the amplified target DNA, Cas12a gains enzymatic activity to indiscriminately cleave single-stranded fluorescent oligonucleotide reporters, generating a fluorescent signal for highly sensitive detection of the pathogen. The protocol provides detailed instructions on: (i) sample collection and preparation; (ii) assay reaction setup, including RPA and Cas12a detection steps; (iii) reaction and detection conditions; and (iv) guidelines for accurately interpreting fluorescence data to detect phytoplasma DNA. This protocol is designed for researchers and agricultural professionals to effectively adopt and implement this advanced diagnostic technique.

Key words CRISPR RNAs (crRNAs), Protospacer adjacent motif (PAM), Lateral flow immunostrips, Field-deployable

1 Introduction

Phytoplasmas are wall-less bacteria with small, AT-rich genomes that have emerged as a growing concern in global agriculture. These pathogens invade plant phloem tissues, leading to various debilitating diseases and substantial economic losses [1]. The spread of phytoplasmas is primarily facilitated by sap-feeding insect vectors, such as leafhoppers and psyllids [2]. However, they can also be transmitted through grafting or vegetative propagation of infected plants [3]. Early detection of phytoplasmas is critical for effective disease management, as it allows for timely intervention to prevent the spread of these diseases.

Katrin Janik and Mattia Tabarelli (eds.), *Phytoplasma: Methods and Protocols*, Methods in Molecular Biology, vol. 3008, https://doi.org/10.1007/978-1-0716-5104-9_6,

Since phytoplasmas cannot be cultured in vitro, traditional methods for identifying phytoplasmas relied on observable plant symptoms, host plant range, and insect vector specificity [4]. However, the development of molecular techniques, such as polymerase chain reaction (PCR), restriction fragment length polymorphism (RFLP) profiling, and DNA sequencing, has significantly improved the accuracy and speed of phytoplasma detection [5, 6].

The CRISPR-/Cas12a-based diagnostic assay is a recent technical advancement in phytoplasma detection. It offers a highly sensitive and rapid method for the early detection of phytoplasmas [7]. Cas12a nuclease is particularly well-suited for targeting the AT-rich genomes of phytoplasmas due to its requirement for an AT-rich protospacer adjacent motif (PAM) sequence (5′-TTTN-3′) [8–10]. This specificity makes it highly effective for detecting key genetic markers like the 16S rRNA or *SecY* gene, commonly used to identify different phytoplasma strains. Upon binding to its target DNA, Cas12a not only cleaves the double-stranded DNA but also becomes activated to cleave the single-stranded oligonucleotide reporter and release amplified detectable signals, such as fluorescence, in diagnostic assays [8–10].

DETECTR assay [11] further enhances the Cas12a detection system by integrating it with isothermal amplification techniques like RPA [8]. This combination allows specific and supersensitive detection of pathogen DNA under isothermal conditions (37 °C), making it more portable and adaptable to both laboratory and field settings. Therefore, the RPA/Cas12a DETECTR assay is expected to facilitate the early detection of phytoplasmas and the effective management of phytoplasma-associated diseases. In this protocol, we demonstrate the application of the Cas12a-based DETECTR assay using potato purple top (PPT) phytoplasma ('*Candidatus* Phytoplasma trifolii'-related strain [12, 13]) as an example, providing a step-by-step guide to its implementation.

2 Materials

2.1 Biological Material

Potatoes infected with potato purple top phytoplasma (originally collected from the Columbia Basin, Washington State [12]).

2.2 Reagents

The products and kits specified in this protocol have been validated for optimal performance under the conditions described. Substitution with equivalent products from other suppliers may be feasible; however, users should ensure functional equivalence and make any necessary adjustments to the protocol to accommodate variations in reagent composition or performance characteristics.

1. Commercial, silica-matrix-based, plant DNA extraction kit (e.g., DNeasy Plant Mini Kit).

2. Commercial cloning kit (e.g., pCR2.1-TOPO™), to generate a targeted DNA fragment.
3. High-yield T7 RNA synthesis kit, for in vitro transcription (e.g., HiScribe T7 High Yield RNA Synthesis Kit).
4. Spin column-based RNA clean-up kit (e.g., RNA Clean & Concentrator Kit).
5. Lba Cas12a enzyme (e.g., EnGen® Lba Cas12a (Cpf1)) and appropriate compatible reaction buffer.
6. Isothermal amplification kit compatible with recombinase polymerase amplification (RPA) (e.g., TwistAmp® Liquid Basic).
7. Fluorescently labeled oligonucleotide reporters.
8. Lateral Flow Assay Kit (e.g., HybriDetect—Universal Lateral Flow Assay Kit).
9. dNTPs (1.2 mM).
10. Magnesium acetate: 14 mM in water.
11. Nuclease-free water.
12. Liquid nitrogen.
13. 95–100% Ethanol.
14. 1.5% (w/v) agarose gel: dissolve 1.5 g of molecular biology grade agarose in 100 mL of TAE buffer and heat it in a microwave. Allow it to cool, add SYBR Safe dye, and cast the gel in a tray.

2.3 Equipment

1. Mortar and pestle, or a pestle for 1.5 mL microcentrifuge tubes.
2. PCR thermocycler.
3. Microcentrifuge, capable of reaching 16,110× *g* (maximum speed).
4. Fluorescence microplate reader, suitable for the measurement of the selected fluorophores (e.g., BioTek Synergy H1 multimode microplate reader).
5. UV transilluminator or gel documentation system.
6. Water bath.

2.4 Consumables

1. 384-well microplates, suitable for fluorescence detection.
2. Pipettes (P10, P20, P100, P200, and P1000) and filter tips (0.1–10 μL, 10–100 μL, and 100–1000 μL).
3. 1.5 mL nuclease-free microcentrifuge tubes.
4. UV transilluminator or gel documentation system.
5. Ice for sample cooling.

3 Methods

The detailed DETECTR assay protocol is described below, along with a corresponding diagram that visually outlines the diagnostic method, as shown in Fig. 1.

3.1 Sample Preparation and DNA Extraction

1. Collect plant tissue from phytoplasma-infected plants. Ensure that samples are fresh and avoid contamination (*see* **Note 1**). PPT phytoplasma-infected potato leaf samples were used in this protocol, while phytoplasma-free potatoes served as negative controls.
2. Grind 100 mg of potato leaf tissue infected with PPT-phytoplasma in liquid nitrogen using a mortar and pestle until it becomes a fine powder. Then, transfer it to a 1.5 mL

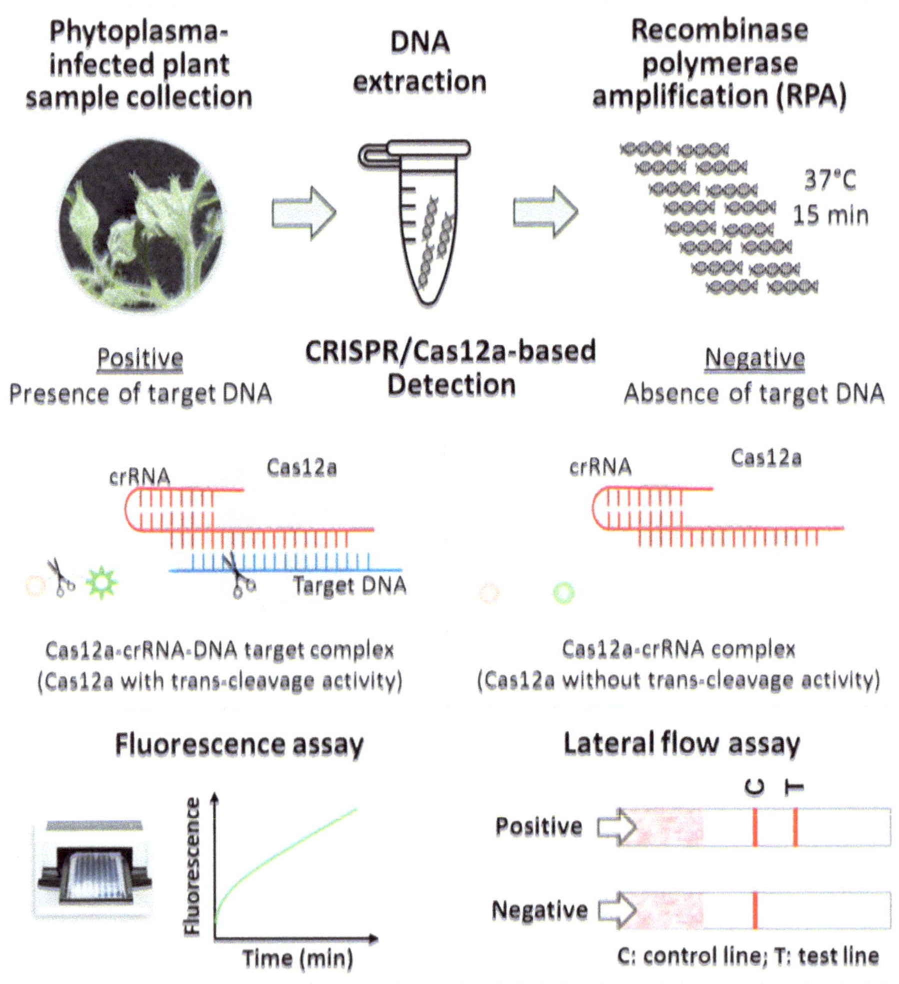

Fig. 1 Schematic overview of the isothermal RPA/Cas12a DETECTR assay for phytoplasma detection

microcentrifuge tube. Alternatively, grind the tissue in a 1.5 mL microcentrifuge tube using an appropriate pestle.

3. Extract DNA from the potato samples using an appropriate DNA extraction Kit. The following steps are based on the use of DNeasy Plant Mini kit (*see* **Note 2**).
4. Add 400 μL of lysis buffer and 20 μL of Proteinase K. Mix thoroughly by vortexing.
5. Incubate the mixture in a water bath at 65 °C for 10 min.
6. Add 10 μL of RNase A, and incubate at room temperature for 2 min.
7. Add 130 μL of P3 buffer, mix by vortexing, and incubate on ice for 5 min.
8. Centrifuge at 16,110× *g* for 5 min. Transfer the supernatant to a new tube.
9. Add 1.5 volumes of buffer AP3/E and mix by pipetting. Then, apply the mixture to the spin column.
10. Wash the column two times with 500 μL of AW buffer.
11. Elute the DNA with 100 μL of AE buffer, following the manufacturer's instructions.

3.2 Target Sequence Selection, PCR Amplification, and Cloning

1. Amplify the 16S–23S rDNA fragment from the samples infected with PPT phytoplasma using the primer pair P1A/16S-SR (*see* Table 1, [14]).
2. Clone the amplified fragment into an appropriate plasmid vector (e.g., pCR2.1-TOPO™, *see* **Note 2**) to produce a target DNA plasmid (Pppt–ITS) following the manufacturer's instructions (*see* **Note 3**).

Table 1

Sequences of PCR primers, RPA primers, T7 promoter, and crRNA template that are described in this protocol for Cas12a DETECTR assay combined with RPA for PPT phytoplasma detection

Primer and other sequences		Sequence	Reference
PCR primers	P1A	5′-AACGCTGGCGGCGCGCCTAATAC-3′	[1]
	16S-SR	5′-GGTCTGTCAAAACTGAAGATG-3′	[1]
RPA primers	Forward	5′-TGATAAGCGTGAGGTCGGTGGTTCAAGTCC-3′	[7]
	Reverse	5′-TCTTAGTGCCAAGGCATCCACTATATGCCC-3′	[7]
T7 promoter		5′-TAATACGACTCACTATAG-3′	[7]
crRNA template[a]		5′-GGTCCTGCTTAAGAAGTTCTTTATCTACACT TAGTAGAAATTACT **ATAGTGAGTCGTATTA**-3′	[7]

[a]The synthesized single stranded DNA oligonucleotide was used as a template for in vitro transcription of crRNA. The T7 promoter sequence is bold. For more details, *see* **Note 5**

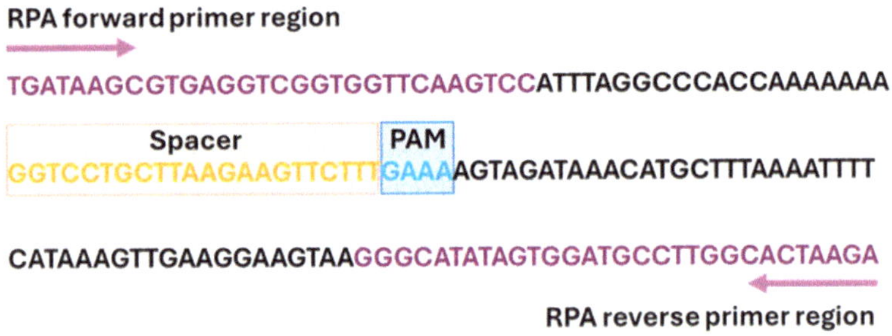

Fig. 2 The 16S–23S rDNA gene fragment of the potato purple top phytoplasma. The sequence includes the following features: RPA primer regions, a protospacer adjacent motif (PAM) sequence, and a CRISPR RNA (crRNA) spacer region

3. Select the detection target (*see* **Note 4**).
4. Design the specific RPA primers and crRNA spacer sequences for the RPA and DETECTR assays (Table 1; Fig. 2; *see* **Note 4**).

3.3 crRNA Preparation and In Vitro Transcription (IVT)

Prepare crRNA, either by having it synthesized by a specialized company or by using an RNA Synthesis Kit. The following steps are based on the HiScribe™ T7 High Yield RNA Synthesis Kit protocol, with minor adjustments (*see* **Note 2**).

1. Synthesize a single-stranded DNA (ssDNA) oligonucleotide containing a T7 promoter, followed by the reverse complement of the desired crRNA (Table 1).
2. Anneal the ssDNA template with a complementary T7 forward primer to generate a double-stranded promoter region (*see* **Note 5**).
3. Add 2 μL of 10X Reaction Buffer, included in the kit, to a nuclease-free reaction tube.
4. Add 8 μL of 100 mM NTP Mix (25 mM of ATP, GTP, CTP, and UTP each).
5. Add 1 μL each of T7 forward primer and the single-stranded oligonucleotide DNA template (Table 1).
6. Add 2 μL of T7 RNA Polymerase Mix (included in the kit).
7. Add 6 μL nuclease-free water to bring the total volume to 20 μL.
8. Mix gently by flicking the tube or pipetting up and down, and incubate the reaction at 37 °C for 2 h or longer in a PCR thermocycler (*see* **Note 6**).
9. Add 68 μL nuclease-free water to the reaction.
10. Add 10 μL of 10X DNase I buffer (included in the kit) to the reaction.

11. Add 2 μL of DNase I to the reaction and incubate at 37 °C for 30 min.
12. Purify the resulting crRNA using an appropriate RNA Clean-up kit. The following steps are based on the protocol of the RNA Clean & Concentrator Kit (*see* **Note 2**), with minor adjustments.
13. Add two volumes of RNA Binding Buffer (included in the kit) to the RNA sample.
14. Add an equal volume of 95–100% ethanol to the RNA sample and mix well.
15. Transfer the mixture to the spin column placed in a collection tube. Centrifuge at 12,000× *g* for 30 s, and discard the flow-through.
16. Add 400 μL RNA Prep Buffer (included in the kit) to the column, centrifuge at 12,000× *g* for 30 s, and discard the flow-through.
17. Add 700 μL RNA Wash Buffer (included in the kit) to the column, centrifuge at 12,000× *g* for 30 s, and discard the flow-through.
18. Add 400 μL RNA Wash Buffer to the column and centrifuge at 12,000× *g* for 1 min to remove any residual wash buffer.
19. Transfer the column to a clean, nuclease-free tube.
20. Add 30 μL of RNase-free water (included in the kit) directly to the column matrix and let it sit for 1 min.
21. Centrifuge at 12,000× *g* for 30 s to elute the crRNA.
22. Collect and store the eluted crRNA at −80 °C (*see* **Note 7**).

3.4 In Vitro Cleavage Assay to Assess the Activity and Specificity of crRNA

1. Mix 50 nM target pPPT–ITS plasmid DNA (from Subheading 3.2), 50 nM Lba Cas12a, and 50 nM IVT crRNA (from Subheading 3.3) in 20 μL Cas12a enzyme reaction buffer (from Subheading 2.2, **item 10**).
2. Incubate at 37 °C for 1 h in a PCR thermocycler.
3. Mix the reaction product and incubate at room temperature for 10 min.
4. Analyze the cleavage product on a 1.5% agarose gel and interpret the results (*see* **Note 8**).

3.5 Integrated Cas12a Assay with RPA for Detection of Phytoplasma Target DNA

1. Prepare a 20-μL RPA reaction mixture by combining 10 ng target pPPT–ITS plasmid DNA (from Subheading 3.2, **step 2**, as a positive control) or infected plant genomic DNA (from Subheading 3.1), 1.2 mM dNTPs (from Subheading 2.2, **item 10**), 0.4 μM of each RPA primer (specific to the PPT phytoplasma target, *see* **Note 9**; Table 2; Fig. 2), and the proprietary enzyme and reaction components provided in an appropriate

Table 2
Cas12a DETECTR assay reagents and their volumes for diagnostic assay setup

Cas12a assay reagents	Volume (μL)	Final concentration
Nuclease-free water	12.75	NA
NEBuffer 2.1 (10X)	2	1X
LbCas12a (1 μM)	1	50 nM
crRNA (1 μM)	1.25	62.5 nM
Fluorescent oligonucleotide reporter (1 μM)	1	50 nM
RPA product	2	NA
Total	*20*	

isothermal amplification kit, according to the manufacturer's instructions (*see* **Note 2**).

2. Add 14 mM of magnesium acetate to initiate the reaction.
3. Incubate the RPA reaction mixture at 37 °C in a PCR thermocycler for 15 min.
4. After incubation, take 2 μL of the RPA reaction product for the Cas12a DETECTR assay.
5. Prepare a 20-μL Cas12a DETECTR reaction mixture (Table 2) by adding 50 nM of LbCas12a (e.g., EnGen® Lba Cas12a), 62.5 nM of crRNA (from Subheading 3.3), 50 nM of custom fluorescent oligonucleotide reporter (*see* **Note 10**) and adjusting the volume with a compatible reaction buffer for LbCas12a (such as NEB 2.1 buffer, supplied with EnGen® Lba Cas12a).
6. Incubate the Cas12a DETECTR reaction mixture at 37 °C in a PCR thermocycler or the microplate reader required for the subsequent step (*see* **Note 11**).
7. Place the above reaction mixture in a 384-well fluorescent microplate reader and record the fluorescence intensity (FAM: excitation 495 nm, emission 525 nm) every 2 min for 2 h.
8. Compare fluorescence over the course of the reaction with the negative control (disease-free potato DNA).
9. Determine the presence of PPT phytoplasma target DNA.

3.6 Lateral Flow Assay

Synthesize the fluorescent oligonucleotide reporter containing a fluorophore at one end and often biotin on the other with modifications (*see* **Note 10**), compatible with a lateral flow assay. The following steps are based on the HybriDetect lateral flow assay (*see* **Notes 2** and **12**).

1. Prepare the DETECTR assay as previously described (please refer to Subheading 3.5) using the RPA reaction products (from Subheading 3.5, **step 4**) to amplify the target DNA (Fig. 2) and the Cas12a detection components to cleave the oligonucleotide reporter (*see* Subheading 3.5; Table 2).
2. Incubate the reaction mixture at 37 °C for 30 min in a PCR thermocycler.
3. After incubation, transfer 5 μL of the DETECTR reaction product into a 1.5 mL microcentrifuge tube.
4. Add 100 μL hybridization buffer (included in the kit) to the tube.
5. Incubate the mixture at room temperature for 5 min.
6. Dip the bottom end (*see* **Note 13**) of the single lateral flow immunostrip (included in the kit) into the reaction mixture.
7. Allow the strip to develop by leaving it in the mixture for 5–10 min (*see* **Note 14**).
8. Observe and interpret the results based on the visual detection on the immunostrip.
9. Visualize the results: two bands indicate positive detection (presence of the target DNA), while a single control band indicates a negative result.

4 Notes

1. To avoid contamination when collecting plant tissue from phytoplasma-infected plants, use sterilized tools and wear disposable gloves, changing them between samples. Collect fresh samples and place each in a sterile container immediately after collection. Ensure tools are re-sterilized between samples, and label each container with relevant details. Store and transport samples in clean, insulated coolers to maintain their integrity.
2. The products mentioned in these steps worked well in the described experimental setting. However, similar products from other companies or suppliers might be equally suitable for the method described.
3. To prevent cross-contamination during PCR amplification and cloning, use separate work areas for sample preparation and post-PCR handling and filtered pipette tips. Regularly clean work surfaces and include a no-template control to monitor for cross-contamination in each PCR run.
4. This protocol involves selecting the target DNA by aligning the 16S–23S rDNA fragment sequences from reference strains of known '*Ca.* Phytoplasma trifolii', and identifying a region of

around 150 bp unique to the PPT phytoplasma strain for specific target amplification and Cas12a targeting.

5. crRNA guides the Cas12 enzyme to a specific double-stranded DNA (dsDNA) sequence of PPT phytoplasma. The synthetic single-stranded DNA (ssDNA) oligonucleotide contains:
   (a) A T7 promoter sequence (5′-TAATACGACTCACTATA-3′, Table 1) is recognized by the T7 RNA polymerase to initiate transcription.
   (b) The reverse complement of the crRNA sequence is placed immediately downstream of the promoter.

   The full ssDNA oligo is as follows:

   5′-GGTCCTGCTTAAGAAGTTCTTTATCTACACTTAGTAGAAATTACT **ATAGTGAGTCGTATTA**-3′.

   (T7 promoter in bold, reverse complement of crRNA is upstream, Table 1).

   Although this oligo is single-stranded, transcription by T7 RNA polymerase typically requires a double-stranded T7 promoter. To address this, the oligo is annealed with a short complementary primer: 5′--TAATACGACTCACTATAG-3′.

6. For maximum yield, incubation can be extended up to 4 h. The reaction can be stopped by adding 1 μL of 0.5 M EDTA, pH 8.0.
7. For long-term storage, crRNA should be preserved at −80 °C, which stabilizes its structure and maintains its functional integrity. Proper aliquoting into multiple tubes minimizes freeze-thaw cycles, thereby reducing degradation. Under these conditions, crRNA can remain biologically active for at least 1 year, and potentially longer with optimal handling.
8. Interpret the results of the in vitro cleavage assay as follows:
   (a) Successful cleavage: after electrophoresis, distinct bands on the gel indicate successful cleavage of the target DNA by the crRNA/Cas12a system. The fragment sizes match the expected cut sites, confirming the accuracy of the gRNA/crRNA design.
   (b) No cleavage: if no cleavage is observed, it suggests that the crRNA/Cas12a did not cut the DNA.
   (c) Partial cleavage: both the full-length target DNA and smaller fragments are observed on the gel, indicating the crRNA/Cas12a is active but may not be fully efficient.
   (d) Off-target cleavage: unexpected fragments are observed, suggesting that crRNA/Cas12a cuts at unintended DNA sites.

9. Multiple RPA primers were designed to identify a primer pair with optimal performance, as different primers vary in efficiency and specificity. The testing and evaluation of RPA primers enables the selection of the primer pair that provides the highest sensitivity and accuracy for reliable phytoplasma detection.
10. Fluorescent oligonucleotide reporter:
    (a) Fluorescence readout for microplate reader: 5′--Fluorophore- Nucleotide Sequence -Quencher-3′, e.g., 5′-56-FAM/TTA TT/3IABkFQ/-3′.
    (b) Lateral flow assay: 5′-Fluorophore-Nucleotide Sequence-Biotin-3′, e.g., 5′-56-FAM/TTA TT/3Bio/-3′.

    Ensure the oligonucleotide reporter is of high purity (with HPLC purification) to avoid non-specific signals during the assay.
11. Given the high cost of a plate reader, a real-time PCR thermocycler can be used as an alternative for fluorescence detection. In a plate reader with the possibility for sample incubation and fluorescence detection, such as the BioTek Synergy H1 microplate reader, incubation and fluorescent measurement can be performed in the same instrument.
12. HybriDetect lateral flow immunostrips: Streptavidin is commonly used at the test line to capture biotin-labeled oligonucleotide reporter. Anti-fluorophore antibodies on the control line validate the presence of fluorophore-labeled reporters, ensuring assay accuracy. For example, anti-FAM antibody is used for FAM-labeled oligonucleotides.
13. Ensure that the strip is only submerged to the level indicated in the manufacturer's instructions (usually just below the line or arrow on the strip) to avoid false results caused by oversaturation and non-specific binding.
14. Allow the liquid to be absorbed into the strip through capillary action. During this time, the fluorescent oligonucleotide reporter, if cleaved by Cas12a, will bind to the specific antibodies on the strip, leading to a visible band at the test line.

Acknowledgments

We acknowledge the funding and support from the USDA Agricultural Research Service's National Plant Disease Recovery System (NPDRS) and Project Number 8042-22000-320-00D. This work was also supported by the USDA National Institute of Food and Agriculture and Hatch Appropriations under Project# PEN04842 and Accession #7005064.

References

1. Wei W, Shao J, Zhao Y et al (2024) *i*PhyDSDB: phytoplasma disease and symptom database. Biology 13:657. https://doi.org/10.3390/biology13090657
2. Weintraub PG, Beanland L (2006) Insect vectors of phytoplasmas. Annu Rev Entomol 51: 91–111. https://doi.org/10.1146/annurev.ento.51.110104.151039
3. Ivanauskas A, Inaba J, Zhao Y et al (2024) Differential symptomology, susceptibility, and titer dynamics manifested by phytoplasma-infected periwinkle and tomato plants. Plants 13:787. https://doi.org/10.3390/plants13060787
4. Wei W, Zhao Y (2022) Phytoplasma taxonomy: nomenclature, classification, and identification. Biology 11:1119. https://doi.org/10.3390/biology11081119
5. Lee IM, Gundersen-Rindal DE, Davis RE et al (1998) Revised classification scheme of phytoplasmas based on RFLP analyses of 16S rRNA and ribosomal protein gene sequences. Int J Syst Bacteriol 48:1153–1169. https://doi.org/10.1099/00207713-48-4-1153
6. Zhao Y, Wei W, Davis RE et al (2021) The agent associated with blue dwarf disease in wheat represents a new phytoplasma taxon, "*Candidatus* Phytoplasma tritici". Int J Syst Evol Microbiol 71:004604. https://doi.org/10.1099/ijsem.0.004604
7. Wheatley MS, Wang Q, Wei W et al (2022) Cas12a-based diagnostics for potato purple top disease complex associated with infection by "*Candidatus* Phytoplasma trifolii"-related strains. Plant Dis 106:2039–2045. https://doi.org/10.1094/PDIS-09-21-2119-RE
8. Chen JS, Ma E, Harrington LB et al (2018) CRISPR-Cas12a target binding unleashes indiscriminate single-stranded DNase activity. Science 360:436–439. https://doi.org/10.1126/science.aar6245
9. Zetsche B, Gootenberg JS, Abudayyeh OO et al (2015) Cpf1 is a single RNA-guided endonuclease of a class 2 CRISPR-Cas system. Cell 163:759–771. https://doi.org/10.1016/j.cell.2015.09.038
10. Zhang Y, Zhang Y, Xie K (2020) Evaluation of CRISPR/Cas12a-based DNA detection for fast pathogen diagnosis and GMO test in rice. Mol Breed 40:11. https://doi.org/10.1007/s11032-019-1092-2
11. Broughton JP, Deng X, Yu G et al (2020) CRISPR–Cas12-based detection of SARS-CoV-2. Nat Biotechnol 38:870–874. https://doi.org/10.1038/s41587-020-0513-4
12. Munyaneza JE, Crosslin JM, Jensen AS et al (2005) Update on the potato purple top disease in the Columbia Basin. In: Proceedings of the 44th annual Washington State potato conference, Moses Lake, WA, pp 57–70
13. Wei W, Davis RE, Bauchan GR et al (2019) New symptoms identified in phytoplasma-infected plants reveal extra stages of pathogen-induced meristem fate-derailment. Mol Plant Microbe Interact 32:1314–1323. https://doi.org/10.1094/MPMI-01-19-0035-R
14. Lee IM, Martini M, Marcone C et al (2004) Classification of phytoplasma strains in the elm yellows group (16SrV) and proposal of "*Candidatus* Phytoplasma ulmi" for the phytoplasma associated with elm yellows. Int J Syst Evol Microbiol 54:337–347. https://doi.org/10.1099/ijs.0.02697-0

Chapter 7

On-Site Universal Phytoplasma Detection by 23S rRNA-Gene-Targeted Loop-Mediated Isothermal Amplification (LAMP)

Mako Akahori, Akio Miyazaki, Hiroaki Koinuma, Nozomu Iwabuchi, Kensaku Maejima, Shigetou Namba, and Yasuyuki Yamaji

Abstract

This chapter describes a genus-universal, specific, sensitive, rapid, and simple detection system for phytoplasmas using universal loop-mediated isothermal amplification (LAMP) primers targeting the 23S rRNA gene. Diagnosis with this system consists of a simple boiling DNA extraction method (95 °C, 10 min) and DNA amplification by the LAMP assay (64 °C, 30–60 min), with a total time of about 60 min. This system enables on-site diagnosis, as it proceeds at a constant temperature, is easy to operate, requires no experimental devices or electricity, and generates little experimental waste. Moreover, the simple boiling DNA extraction method can be applied to a variety of sample types, including herbaceous plants, woody plants, and insect vectors.

Key words CTAB, Insect vector, Loop-mediated isothermal amplification (LAMP), On-site detection, Phytoplasma, Simple DNA extraction, Universal detection

1 Introduction

Current control measures for phytoplasma diseases in agricultural crops still rely on insect vector control and the removal of infected plants, as practical bactericides and resistant cultivars are still lacking [1–3]. To limit the damage induced by phytoplasmas, their infections must be detected as early as possible. Since wild plants around agricultural land can serve as phytoplasma reservoirs [4], their monitoring is a key step in understanding and breaking the infection cycle. Reliable diagnostic techniques that can be applied to various sample types in the field are therefore critical for the effective control of phytoplasma diseases.

Genetic diagnostic techniques for the detection of phytoplasmas consist mostly of polymerase chain reaction (PCR), quantitative PCR (qPCR), and loop-mediated isothermal amplification

Katrin Janik and Mattia Tabarelli (eds.), *Phytoplasma: Methods and Protocols*, Methods in Molecular Biology, vol. 3008,
https://doi.org/10.1007/978-1-0716-5104-9_7,

(LAMP) [5]. Unlike many plant-pathogenic bacteria, all known species in the genus '*Candidatus* Phytoplasma' are plant pathogens [6–8] and their plant host ranges often overlap [9, 10]. Moreover, the control measures for phytoplasma diseases are the same, regardless of the species classification. Therefore, genus-universal detection techniques are useful in the diagnosis of phytoplasma diseases and have been reported for PCR and qPCR targeting the 16S rRNA gene (rDNA) or 23S rDNA [11–15]. There are also many phytoplasma detection techniques based on loop-mediated isothermal amplification (LAMP), a method [16, 17] that allows the ultrasensitive, rapid, and simple detection of plant and animal pathogens. In most cases, however, LAMP-based techniques were designed for the 16S rDNA [5] of a specific or a limited number of species [10, 18, 19]. LAMP assays that were validated to universally detect the genus '*Ca.* Phytoplasma' have yet to be developed. Recently, we designed a phytoplasma universal LAMP primer set for 23S rDNA, which had never been used as a LAMP amplification target in phytoplasmas [20]. 23S rDNA has high potential as a target for designing universal LAMP primers due to its sequence length, which is approximately twice as long as that of 16S rDNA, and the accumulation of sequence information through recent high-throughput genome sequencing studies.

Our novel primer set, CaPU23S-4, was able to amplify the 23S rDNA of 31 tested '*Ca.* Phytoplasma' species within 30 min, thus covering almost all of the diverse genus '*Ca.* Phytoplasma'. The primers were highly specific and did not generate false-positive results for healthy plants and other bacteria, including those closely related to phytoplasmas [20]. The detection sensitivity of the developed LAMP assay was far superior to that of conventional PCR methods [12, 15]. Furthermore, the assay has great potential for the discovery of new phytoplasma diseases and unidentified vectors. A simple boiling DNA extraction method applicable to the LAMP assay was developed in parallel and was successfully applied to a variety of sample types, including herbaceous plants, woody plants, and insects. The combination of this DNA extraction method (95 °C, 10 min) and the CaPU23S-4-mediated LAMP assay (64 °C, 30–60 min) enabled the establishment of a universal detection system for phytoplasmas from plants and insects, allowing a diagnosis in about 60 min even in the field. This chapter describes how to perform the simple boiling DNA extraction method and the LAMP assay, as well as the precautions that need to be taken.

2 Materials

2.1 Reagents and Solutions

1. STE buffer: 100 mM NaCl; 10 mM Tris–HCl, pH 8.0; 1 mM EDTA. Store at ambient temperature.

Table 1
CaPU23S-4 LAMP primer sequences for the universal detection of phytoplasmas

Primer name	Sequence (5′-3′)
CaPU23S-4-F3	GATGTCGGCTCATCGCAT
CaPU23S-4-B3	CATAGCTACTCAGCTGTGGC
CaPU23S-4-FIP	CTGAACCCAGCTCGCGTTCCCCTGGAGCTGGAGAAGGT
CaPU23S-4-BIP	CCGTCGTGGGCGTTGGAAATTAACTGAAACACCAGCGGTATA
CaPU23S-4-LF	AACAGCCCAACCCTTGGA
CaPU23S-4-LB	GAACTGTCCCTAGTATGAGAAGACC

2. Optional: Polyvinylpolypyrrolidone (PVPP), if working with some woody plants (*see* **Note 1**).
3. Phytoplasma universal LAMP primer set (CaPU23S-4) (Table 1, Fig. 1) [20] (*see* **Note 2**). LAMP primers were synthesized, desalted, and purified (*see* **Note 3**). Store at −30 °C.
4. LAMP reagents:
   (a) 1 M Tris-HCl, pH 8.8. Store at −30 °C.
   (b) 1 M KCl, 1 M $MgSO_4$. Store at −30 °C.
   (c) 1 M $(NH_4)_2SO_4$. Store at −30 °C.
   (d) Tween 20. Store at 4 °C.
   (e) 5 M betaine. Store at 4 °C.
   (f) 100 mM of each dNTPs. Store at −30 °C.
   (g) 8 U/μL *Bst* DNA polymerase. Store at −30 °C.
   (h) Fluorescent detection reagent (FD, Eiken Chemical, Tokyo, Japan). Store at −30 °C (*see* **Note 4**).
   (i) Mineral oil (molecular biology grade). Store at −30 °C.
   (j) Distilled, nuclease-free water. Store at 4 °C.
5. Positive control DNA sample (total DNA from phytoplasma-infected plants or insects).
6. 70% (v/v) ethanol or a lighter for sterilization (*see* **Note 5**).

2.2 Consumables

1. Diagnostic samples: usually the base of the leaf midrib (5 × 2 mm) for plants, but 50 mg of trunk sawdust of coconut palm (*see* **Note 6**), and the whole body or one hind leg of insects (*see* **Note 7**).
2. Razors or tweezers to cut out diagnostic samples.
3. Optional: micro-spatulas for adding PVPP, if working with some woody plants (*see* **Note 1**).

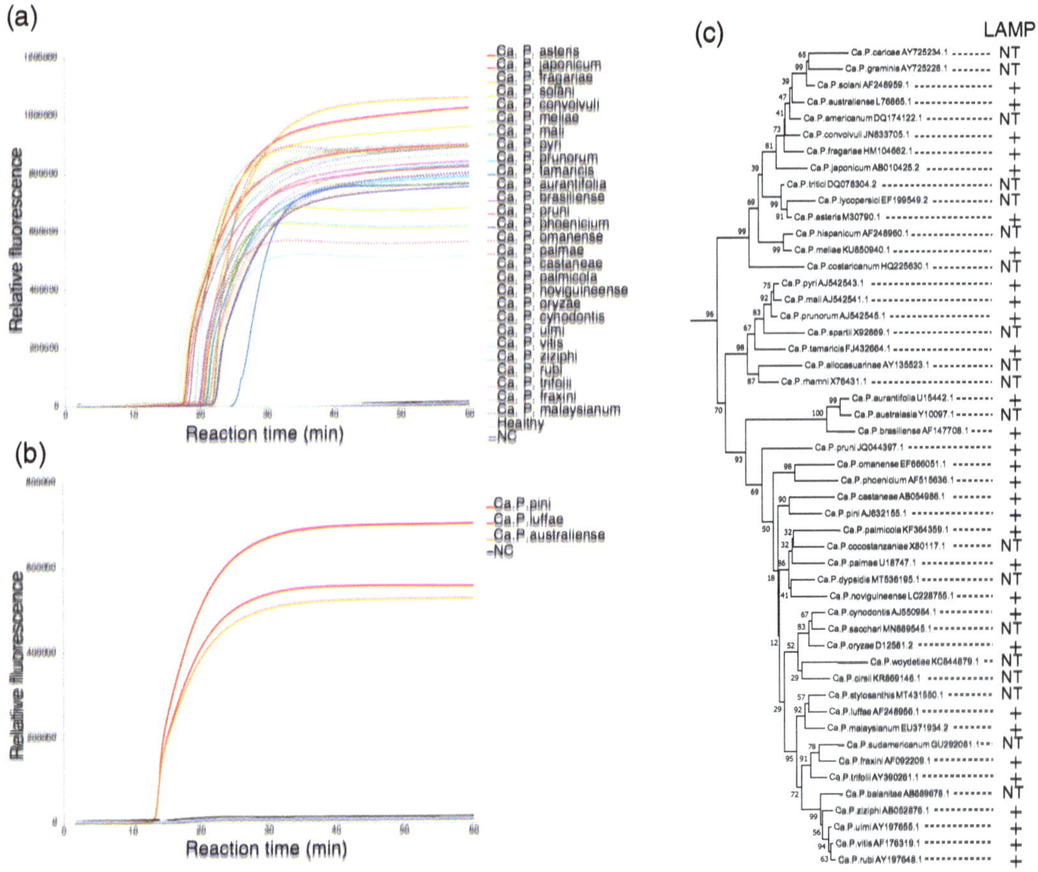

Fig. 1 Detection coverage of different phytoplasma species using the universal LAMP assay. (**a**, **b**) LAMP amplification curves using the CaPU23S-4 primer set. LAMP assays were performed for 31 phytoplasma species. The results were obtained with DNA (10 ng) from plants infected with each of 28 phytoplasma species (**a**) and with the artificially synthesized DNA (10 ng) based on three species ('*Ca*. P. australiense', '*Ca*. P. pini', '*Ca*. P. luffae') (**b**). LAMP reactions were performed at 64 °C for 60 min using Genie I (OptiGene Ltd. Horsham, UK). Total DNA (10 ng) from healthy *Glebionis coronaria* (Healthy) and distilled water (NC) served as negative controls for the LAMP reaction template. (**c**) Detection coverage of the universal LAMP assay as reflected in a phylogenetic tree based on the 16S rDNA of phytoplasmas. The phylogenetic tree was constructed with *Acholeplasma laidlawii* BN1-JA1 (JN935888) and *Bacillus subtilis* (NR112116) as outgroups. The numbers on the branches are the percentages of bootstrap values obtained for 500 replicates. Species positive for LAMP detection are indicated by +; untested species are indicated by NT. Note that the absence of a minus sign (−) indicating LAMP-negative is because all species tested were LAMP-positive. No amplification was detected in the negative control for this test (see black line "NC" very close to the *x*-axis)

4. 1.5 and 0.2 mL tubes (or similar).
5. Pipettors (with volume range of 1.0–100 μL) and appropriate filter tips.

2.3 Equipment

1. Drills (drill bit about 10 mm in diameter), to drill holes in the trunks of coconut palms to obtain sawdust.
2. Equipment capable of maintaining a constant temperature, such as a heating block or thermos. When using a thermos, a thermometer, and a floating tube rack for 1.5 and 0.2 mL tubes are also needed.
3. Ice pack and cold storage box (or similar).
4. Optional: UV light source.

3 Methods

3.1 Simple Boiling DNA Extraction for On-Site Diagnosis

1. Prepare 1.5 mL tubes for each sample. In addition to the number of diagnostic samples, prepare an additional 1.5 mL tube for the negative control of the DNA extraction.
2. Add 100 μL of STE buffer in a 1.5 mL tube for every leaf midrib or whole insect sample, 1000 μL for each coconut palm sawdust sample (*see* **Note 6**), or 10 μL for every single hind leg of an insect to be analyzed (*see* **Note 7**). For some woody plants, adding one micro-spatula scoop of PVPP to 100 μL of STE buffer will improve DNA extraction from the leaf midrib (*see* **Note 1**). To avoid contamination, use filter tips to aliquot the STE buffer.
3. Add the sample to the respective 1.5 mL tube containing the STE buffer and tap gently to soak the entire sample. To increase extraction efficiency for whole insects, gently crush the insect body in the STE buffer with a tip.
4. Incubate the tubes at 95 °C for 10 min using a heating block or thermos (*see* **Note 8**).
5. Collect the supernatant as simple extracted DNA in a new 1.5 mL tube. The extracted DNA can be stored at −30 °C until use.

3.2 Universal LAMP Assay

1. Prepare 0.2 mL tubes for the LAMP reaction mixture and place them on an ice pack. In addition, prepare three 0.2 mL tubes for one positive and two negative controls of the LAMP reaction.
2. Prepare the LAMP reaction solution as described in Table 2. To take the pipetting error into account, calculate one additional sample (*see* **Note 9**). To prevent contamination, use filter tips and close the tube cap immediately after aliquoting the reaction mixture. Aliquot 24 μL of the LAMP mastermix for each sample and control.
3. Add 1 μL of distilled water (as negative control 1), 1 μL of STE buffer incubated at 95 °C for 10 min without a diagnostic

Table 2
Composition of LAMP reaction solution per sample

LAMP reagent stock solution	Volume/sample (μL)	Final concentration
1 M Tris-HCl, pH 8.8	0.25	20 mM
1 M KCl	0.25	10 mM
1 M $MgSO_4$	0.2	8 mM
1 M $(NH_4)_2SO_4$	0.25	10 mM
Tween 20	0.025	0.1% (v/v)
5 M betaine	0.004	0.8 M
100 mM of each dNTPs	1.4	1.4 mM of each dNTP
8 U/μL *Bst* DNA polymerase	1.0	0.32 U/μL
Fluorescent detection reagent	1.0	1×
CaPU23S-4-F3 (10 μM)	0.5	0.2 μM
CaPU23S-4-B3 (10 μM)	0.5	0.2 μM
CaPU23S-4-FIP (10 μM)	4.0	1.6 μM
CaPU23S-4-BIP (10 μM)	4.0	1.6 μM
CaPU23S-4-LF (10 μM)	2.0	0.8 μM
CaPU23S-4-LB (10 μM)	2.0	0.8 μM
Distilled nuclease-free water	6.62	–
Total	*24.0*	

sample (as negative control 2), or 1 μL of each extracted sample DNA. Mix briefly by pipetting.

4. Add 20 μL mineral oil to each tube and close the caps immediately.
5. Finally, add 1 μL of positive control DNA sample and mix the reaction solution briefly by pipetting. Add 20 μL of mineral oil and close the caps immediately (*see* **Note 10**).
6. To amplify the partial 23S rDNA of phytoplasmas, incubate at 64 °C for 30–60 min using a heating block or a thermos.
7. DNA amplification is detected by colorimetric changes in the reaction mixture or changes in the fluorescence intensity under UV light. The colorimetric change/fluorescence should be observed immediately after completion of the LAMP reaction. If a colorimetric change/fluorescence is observed in the positive control and not in negative controls 1 and 2, the LAMP diagnosis is considered successful (Fig. 2) (*see* **Note 11**).

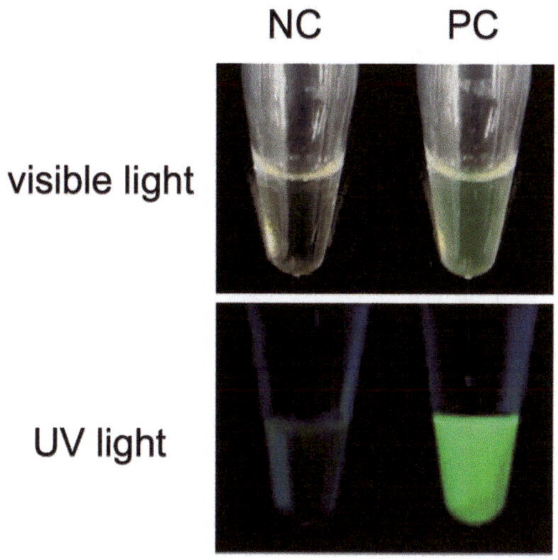

Fig. 2 The expected colorimetric/fluorescence change by the LAMP assay. Colorimetric changes under visible light and fluorescence changes under UV light are shown. PC: Positive control, NC: Negative control

4 Notes

1. The addition of PVPP to the STE buffer should be considered when LAMP assays for woody plants are not successful using the normal boiling extraction protocol. Since woody plants contain high levels of polyphenols and polysaccharides that inhibit nucleic acid amplification, removal of these components prior to DNA extraction is recommended [21]. PVPP facilitates DNA extraction from woody plants by removing plant-derived polyphenols that form complexes with nucleic acids and interfere with their amplification [21, 22]. In a previous study, the normal boiling extraction protocol failed to produce LAMP amplification from infected *E. zollingeri* samples [20]. After the addition of PVPP—in amounts consistent with the method of Hanaoka and Fukuda [23]—to 100 μL of STE buffer using a micro-spatula, phytoplasma was successfully detected in infected *E. zollingeri* samples (Table 3). No false-positives due to the addition of PVPP were identified in the noninfected samples. Note that centrifugation is not necessary because PVPP sinks to the bottom of the tube, and the supernatant is collected after boiling.
2. The 31 phytoplasmas detected by CaPU23S-4 corresponded to all eight 16S-groups (I–VIII) (Fig. 1) and 22 16Sr-RFLP-groups [20]. No false positives were observed for 14 bacterial species, including those closely related to phytoplasmas, such as *Bacillus* (phylum Firmicutes), *Mycoplasma*, *Mesoplasma*,

Table 3
Results of the universal LAMP assay using DNA extracted by the simple boiling method and other detection methods

Sample species and name		Phytoplasma infection (phytoplasma species)		DNA extraction methods			
				Boiling		Purified DNA[a]	FTA
Herbaceous plant	Periwinkle	Healthy		–		–	–
	(*Catharanthus roseus*)	Infected	('*Ca*. P. aurantifolia')	+		+	+
	Chrysanthemum	Healthy		–		–	–
	(*Glebionis coronaria*)	Infected	('*Ca*. P. asteris')	+		+	+
	Banana	Healthy		–		–	–
	(*Musa* sp.)	Infected	('*Ca*. P. noviguineense')	+		+	+
Woody plant	Hydrangea	Healthy		–		–	–
	(*Hydrangea macrophylla*)	Infected	('*Ca*. P. japonicum')	+		+	+
	Cassava	Healthy		–		–	–
	(*Manihot esculenta*)	Infected	('*Ca*. P. pruni')	+		+	+
	Elaeocarpus	Healthy		–	(–[b])	–	–
	(*Elaeocarpus zollingeri*)	Infected	('*Ca*. P. malaysianum')	–	(+[b])	+	+
	Coconut palm	Healthy		–[c]		–[c]	NT
	(*Cocos nucifera*)	Infected	('*Ca*. P. noviguineense')	+[c]		+[c]	NT
Insect vector	*Macrosteles striifrons*	Healthy		–	(–[d])	–	–
		Infected	('*Ca*. P. asteris')	+	(+[d])	+	+

NT not tested

[a]Purified DNA was obtained using a CTAB-based methods from plants and a column-based method from insects [20]

[b]Results obtained with the addition of PVPP

[c]DNA was extracted from the wood sawdust of *C. nucifera* trunk

[d]DNA was extracted from a single hind leg of *M. striifrons*

Ureaplasma, *Spiroplasma*, and *Acholeplasma* (class Mollicutes). Moreover, the LAMP assay using CaPU23S-4 has a detection sensitivity approximately 1000–10,000 times higher than that of conventional PCR using P1/P7 primers [12, 15]. As the LAMP assay using CaPU23S-4 targets 23S rDNA, it can also be employed to confirm the detection results obtained with conventional 16S rDNA-targeted methods.

3. Standard primer synthesis can be applied for the LAMP assay regardless of the manufacturer, as long as the oligonucleotides are desalted and purified.
4. The product mentioned in this step worked well in the experimental setting described. However, similar products from other companies or suppliers may be equally suitable for the method described.
5. Sterilize (by spraying with 70% (v/v) ethanol or lightly searing with a lighter) the parts of the razor or drill that touched the sample before taking the next diagnostic sample, to prevent cross-contamination.
6. In coconut palm (*C. nucifera*), trunk sawdust rather than leaves is commonly used for phytoplasma detection [24], because trunk samples are easier to collect than the leaves at the tops of these tall trees; also, the phytoplasma concentration is significantly higher in the trunk than in the leaves [25]. To ensure an accurate diagnosis, we optimized the amount of sawdust and STE buffer by testing trunk sawdust to STE buffer ratios of 1, 5, 10, and 50 mg/mL. A concentration of 50 mg of sawdust per 1 mL of STE buffer was found to be the optimal ratio for LAMP amplification and to avoid false negatives [20]. The CTAB-based extraction method for coconut palms used in previous studies required soaking the sawdust in CTAB extraction buffer for 48 h prior to DNA extraction by phenol-chloroform and alcohol precipitation [24]. With our boiling method, DNA extraction takes only 10 min, allowing for a significant time saving in the diagnosis.
7. Phytoplasmas accumulate in various organs of vector insects, including the alimentary canal, hemolymph, and salivary glands [26, 27]. While the detection of phytoplasmas from insects has conventionally required DNA extraction from their whole bodies [4, 28], recent studies have shown that phytoplasmas can be detected using column-purified DNA from legs containing hemolymph [29] and with simple boiling-extracted DNA from a single hind leg [20]. For the latter, collect the insect in a tube and place it on an ice pack to stop its movement. Then, remove a single hind leg using razors or tweezers, either visually or under a stereomicroscope, ensuring that the cut side is thoroughly immersed in 10 μL of STE buffer prior to boiling extraction. For simple boiling DNA extraction using the whole

body, collect the insect in a tube, place it on an ice pack to stop its movement, soak the insect in 100 μL of STE buffer, and gently crush it with a tip prior to boiling extraction.

8. DNA extraction takes about 3–4 h in the conventional CTAB-based method and about 1.5 h in the conventional FTA card-based method, which involves blotting crushed plant tissues onto a chemically treated card that captures and preserves nucleic acids. However, the simple boiling method takes about 10 min [20], making it suitable for rapid diagnosis in the field. The FTA card-based method is also superior if the DNA samples are to be analyzed at remote sites. Since both the simple boiling DNA extraction method and conventional methods can be combined with the LAMP assay (Table 2), the optimal method should be selected according to the diagnostic purpose by referring to this study.
9. To facilitate preparation of the LAMP reaction solution, prepare 100 μL each of 5× working primer mix and 2× LAMP reagent premix, as described in Tables 4 and 5, respectively. Both working stocks can be stored at −30 °C until needed. From these working stocks, prepare the LAMP reaction solution as described in Table 6. A 25-μL reaction volume is recommended to minimize the effect of nucleic acid amplification inhibitors present in DNA extracted using the simple boiling method.
10. The positive control should always be prepared last, to avoid cross-contamination of the samples.
11. To prevent contamination, used tubes and tips must be hermetically sealed and discarded. In particular, after their use in the LAMP reaction, the tubes should never be opened, electrophoresed, or autoclaved.

Table 4
Pipetting scheme for the 5× working primer mix

5× working primer mix	Volume/sample (μL)
CaPU23S-4-F3 (100 mM)	1.0
CaPU23S-4-B3 (100 mM)	1.0
CaPU23S-4-FIP (100 mM)	8.0
CaPU23S-4-BIP (100 mM)	8.0
CaPU23S-4-LF (100 mM)	4.0
CaPU23S-4-LB (100 mM)	4.0
Distilled nuclease-free water	74.0
Total	*100.0*

Table 5
Pipetting scheme for the 2× LAMP reagent premix

2× LAMP reagent premix	Volume/sample (μL)
1 M Tris-HCl (pH 8.8)	4.0
1 M KCl	2.0
1 M $MgSO_4$	1.6
1 M $(NH_4)_2SO_4$	2.0
1% (v/v) Tween 20	20.0
5 M betaine	32.0
100 mM of each dNTPs	2.8
Distilled nuclease-free water	27.2
Total	*100.0*

Table 6
Pipetting scheme for the LAMP reaction solution

LAMP reaction solution	Volume/sample (μL)
2× LAMP reagent premix	12.5
5× working primer mix	5.0
8 U/μL *Bst* DNA polymerase	1.0
Fluorescent detection reagent	1.0
DNA template	1.0
Distilled nuclease-free water	4.5
Total	*25.0*

Acknowledgements

We are grateful to Drs. A. Bertaccini (DISTAL, Alma Mater Studiorum, University of Bologna, Bologna, Italy), M. Satoh (Institute of Socio-Arts and Sciences, University of Tokushima, Tokushima, Japan), W. Wei (Beltsville Agricultural Research Center, United States Department of Agriculture, Beltsville, United States), and the past and present members of the Chiba Prefectural Agriculture and Forestry Research Center for providing phytoplasma DNA samples. This work was supported by grants from

JSPS KAKENHI (nos. 20KK0132, 23KJ0530, 23 K13864, and 25H02311), JST SPRING, the University of Tokyo (no. JPMJSP2108), JST FOREST Program (no. PMJFR243Q) and JST/JICA SATREPS (no. JPMJSA2508).

References

1. Namba S (2019) Molecular and biological properties of phytoplasmas. Proc Jpn Acad Ser B Phys Biol Sci 95:401–418. https://doi.org/10.2183/pjab.95.028
2. Maejima K, Oshima K, Namba S (2014) Exploring the phytoplasmas, plant pathogenic bacteria. J Gen Plant Pathol 80:210–221. https://doi.org/10.1007/s10327-014-0512-8
3. Hogenhout SA, Oshima K, Ammar E-D et al (2008) Phytoplasmas: bacteria that manipulate plants and insects. Mol Plant Pathol 9:403–423. https://doi.org/10.1111/j.1364-3703.2008.00472
4. Malembic-Maher S, Desqué D, Khalil D et al (2020) When a Palearctic bacterium meets a Nearctic insect vector: genetic and ecological insights into the emergence of the grapevine Flavescence dorée epidemics in Europe. PLoS Pathog 16:e1007967. https://doi.org/10.1371/journal.ppat.1007967
5. Nair S, Manimekalai R (2021) Phytoplasma diseases of plants: molecular diagnostics and way forward. World J Microbiol Biotechnol 37:102. https://doi.org/10.1007/s11274-021-03061-y
6. IRPCM (2004) "Candidatus Phytoplasma", a taxon for the wall-less, non-helical prokaryotes that colonize plant phloem and insects. Int J Syst Evol Microbiol 54:1243–1255. https://doi.org/10.1099/ijs.0.02854-0
7. Bertaccini A, Duduk B (2009) Phytoplasmas and phytoplasma diseases: a review of recent research. Phytopathol Mediterr 48:355–378. https://doi.org/10.14601/Phytopathol_Mediterr-3300
8. Bertaccini A, Arocha-Rosete Y, Contaldo N et al (2022) Revision of the "*Candidatus* Phytoplasma" species description guidelines. Int J Syst Evol Microbiol 72:005353. https://doi.org/10.1099/ijsem.0.005353
9. Kumari S, Nagendran K, Rai AB et al (2019) Global status of phytoplasma diseases in vegetable crops. Front Microbiol 10:1349. https://doi.org/10.3389/fmicb.2019.01349
10. Yu S, Pan Y, Zhu H et al (2023) Universal, rapid, and visual detection methods for phytoplasmas associated with coconut lethal yellowing diseases targeting 16S rRNA gene sequences. Plant Dis 107:276–280. https://doi.org/10.1094/PDIS-05-22-0996-SC
11. Hodgetts J, Boonham N, Mumford R et al (2009) Panel of 23S rRNA gene-based real-time PCR assays for improved universal and group-specific detection of phytoplasmas. Appl Environ Microbiol 75:2945–2950. https://doi.org/10.1128/AEM.02610-08
12. Deng SJ, Hiruki C (1991) Genetic relatedness between two nonculturable mycoplasmalike organisms revealed by nucleic acid hybridization and polymerase chain reaction. Phytopathology 81:1475–1479. https://doi.org/10.1094/Phyto-81-1475
13. Ito T, Suzaki K (2017) Universal detection of phytoplasmas and *Xylella* spp. by TaqMan singleplex and multiplex real-time PCR with dual priming oligonucleotides. PLoS One 12:e0185427. https://doi.org/10.1371/journal.pone.0185427
14. Namba S, Kato S, Iwanami S et al (1993) Detection and differentiation of plant-pathogenic mycoplasmalike organisms using polymerase chain reaction. Phytopathology 83:786–791. https://doi.org/10.1094/Phyto-83-786
15. Schneider B, Seemüller E, Smart CD et al (1995) Phylogenetic classification of plant pathogenic mycoplasma-like organisms or phytoplasmas. In: Razin S, Tully JG (eds) Molecular and diagnostic procedures in mycoplasmology, vol I. Academic Press, San Diego, pp 369–380. https://doi.org/10.1016/B978-012583805-4/50040-6
16. Notomi T, Okayama H, Masubuchi H et al (2000) Loop-mediated isothermal amplification of DNA. Nucleic Acids Res 28:e63. https://doi.org/10.1093/nar/28.12.e63
17. Wong YP, Othman S, Lau YL et al (2018) Loop-mediated isothermal amplification (LAMP): a versatile technique for detection of micro-organisms. J Appl Microbiol 124:626–643. https://doi.org/10.1111/jam.13647
18. Obura E, Masiga D, Wachira F et al (2011) Detection of phytoplasma by loop-mediated isothermal amplification of DNA (LAMP). J Microbiol Methods 84:312–316. https://doi.org/10.1016/j.mimet.2010.12.011

19. Dickinson M (2015) Loop-mediated isothermal amplification (LAMP) for detection of phytoplasmas in the field. In: Lacomme C (ed) Plant pathology, Methods in molecular biology, vol 1302. Humana Press, New York, pp 99–111. https://doi.org/10.1007/978-1-4939-2620-6_8
20. Akahori M, Miyazaki A, Koinuma H et al (2024) Use of the 23S rRNA gene as a target template in the universal loop-mediated isothermal amplification (LAMP) of genomic DNA from phytoplasmas. Microbiol Spectr 27:e0010624. https://doi.org/10.1128/spectrum.00106-24
21. Kasajima I, Sasaki K, Tanaka Y et al (2013) Large scale extraction of pure DNA from mature leaves of *Cyclamen persicum* Mill. and other recalcitrant plants with alkaline polyvinylpolypyrrolidone (PVPP). Sci Hortic 164:65–72. https://doi.org/10.1016/j.scienta.2013.09.011
22. Koonjul PK, Brandt WF, Lindsey GG et al (1999) Inclusion of polyvinylpyrrolidone in the polymerase chain reaction reverses the inhibitory effects of polyphenolic contamination of RNA. Nucleic Acids Res 27:915–916. https://doi.org/10.1093/nar/27.3.915
23. Hanaoka S, Fukuda Y (2020) Easy and rapid method to extract DNA from the leaves of two larches, *Larix kaempferi* and *Larix gmelinii* var. *japonica*. Boreal For Res 68:39–41. (in Japanese). https://doi.org/10.24494/jfsh.68.0_39
24. Nipah JO, Jones P, Dickinson MJ (2007) Detection of lethal yellowing phytoplasma in embryos from coconut palms infected with Cape St Paul wilt disease in Ghana. Plant Pathol 56:777–784. https://doi.org/10.1111/j.1365-3059.2007.01623.x
25. Córdova I, Oropeza C, Puch-Hau C et al (2014) A real-time PCR assay for detection of coconut lethal yellowing phytoplasmas of group 16SrIV subgroups A, D and E found in the Americas. J Plant Pathol 96:343–352. https://doi.org/10.4454/JPP.V96I2.031
26. Lefol C, Lherminier J, Boudon-Padieu E et al (1994) Propagation of Flavescence dorée MLO (mycoplasma-like organism) in the leafhopper vector *Euscelidius variegatus* Kbm. J Invertebr Pathol 63:285–293. https://doi.org/10.1006/jipa.1994.1053
27. Sinha RC, Chiykowski LN (1967) Initial and subsequent sites of aster yellows virus infection in a leafhopper vector. Virology 33:702–708. https://doi.org/10.1016/0042-6822(67)90070-0
28. Matić S, Candian V, D'Errico C et al (2022) In-field LAMP detection of Flavescence dorée phytoplasma in crude extracts of the *Scaphoideus titanus* vector. Agronomy 12:1645. https://doi.org/10.3390/agronomy12071645
29. Koinuma H, Maejima K, Tokuda R et al (2020) Spatiotemporal dynamics and quantitative analysis of phytoplasmas in insect vectors. Sci Rep 10:4291. https://doi.org/10.1038/s41598-020-61042-x

Part II

Vector Biology and Interaction Studies: Insect Vectors and Their Interactions with Phytoplasmas

Chapter 8

Phytoplasma Transmission Trials via Insect Vectors Under Controlled Conditions

Elena Gonella and Alberto Alma

Abstract

The most relevant way for phytoplasma transmission, both in nature and in the laboratory, is via insect-mediated inoculation. Therefore, the setup of proper transmission trials is necessary to investigate vector–phytoplasma interactions under controlled conditions. Here we present protocols for laboratory transmission experiments with insect vectors, describing procedures for measuring the acquisition, latency, and inoculation phases. The use of plant sources from the field is presented along with the use of experimentally infected laboratory plants. Similarly, methods to measure the field infectivity of insects are discussed in conjunction with experiments involving laboratory-reared specimens for studies on all transmission phases. Inoculation is described involving the use of an artificial diet, the final host plant, or a model plant.

Key words Leafhoppers, Planthoppers, Psyllids, Acquisition, Latency, Inoculation, Artificial diet, Vector competence, Transmission efficiency

1 Introduction

Research on vector-mediated transmission of phytoplasmas is in constant progress, not only to increase knowledge on the pathogen–vector interaction, but also to investigate new associations to improve deciphering phytoplasma epidemiology. Indeed, the mobility of pathogen-transmitting insects exposes crops to new transmission cycles [1]. The role of known and putative vectors of phytoplasmas in disease spread is therefore crucial for improving disease management [2], especially considering that vector control is the main tool for limiting phytoplasma infections due to the serious complications of direct control of these pathogens in the plant using antimicrobials [3]. Phytoplasma transmission trials are the most effective method for investigating vector-mediated diffusion. Additionally, they provide conclusive results when studying the role of a certain insect as a vector for a specific phytoplasma. A widely recognized laboratory transmission method involves the use

Katrin Janik and Mattia Tabarelli (eds.), *Phytoplasma: Methods and Protocols*, Methods in Molecular Biology, vol. 3008, https://doi.org/10.1007/978-1-0716-5104-9_8,

of an artificial diet to study an insect's ability to transfer phytoplasma to a feeding medium and thus bypassing factors that are involved in pathogen establishment within the plant. Alternatively, trials can be conducted using the host plant or a model plant [4]. The latter option may be preferred for pathosystems with a woody host. The phenology of these hosts can be challenging to manage in a laboratory setting, and their symptom expression may occur slowly. If a model plant is used, it must be a suitable alternative host for both the vector and the phytoplasma, at least supporting the insect's survival for a few days. An additional parameter that can be modulated when designing a phytoplasma transmission trial is the number of transmission phases (i.e., acquisition, latency, and inoculation) to be studied. If the role of field-collected adult vector populations is investigated, only the inoculation of the pathogen to the plant can be examined. This is because acquisition and latency in the insect are expected to have been completed prior to insect collection. If the dynamics of the complete transmission process are studied, experiments should be designed starting from healthy nymphs (preferably coming from a lab rearing), where all transmission phases from acquisition to inoculation take place under controlled conditions.

2 Materials

Controlled transmission trials through insect vectors involve the use of materials for insect maintenance and for preparing the substrate for phytoplasma inoculation, which may vary depending on the experimental setup, the studied phytoplasma, and its insect and plant hosts. Here, materials for standard experiments with an artificial diet [5–8], as well as potted plants/seedlings [9–13], are listed.

1. Hacksaw.
2. 2 mL microcentrifuge tubes.
3. 1× TE Buffer, pH 8.0: 100 mM Tris-HCl, pH 8.0, 10 mM EDTA, pH 8.0. Autoclave before use.
4. 1× TE Buffer, pH 8.0, 5% sucrose (w/v). Sterilize by filtration.
5. Parafilm.
6. Rack suitable for centrifuge tubes.
7. Cotton plugs for modified centrifuge tubes.
8. Dark cardboard or tissue.
9. For the insect rearing cages: plexiglas cylinder ($h = 20$–25 cm; diameter = 9–12 cm), or commercial insect-proof net cage, or a sleeve cage of polythene net (length = 73 cm, diameter = 17 cm, at least 60-mesh grid) and a cylinder of a plastic net (length = 31 cm, diameter = 16.5 mm). *See* **Note 1**.

10. Filter paper.
11. Seedlings of healthy plants.
12. Climate-controlled chamber or greenhouse (optional).

3 Methods

3.1 Preparation of Artificial Feeding Systems

1. Use a hacksaw to cut the bottom of a 2 mL microcentrifuge tube, at a distance of 1 cm from the lowest extremity.
2. Place 300 μL of 1× TE Buffer, 5% sucrose solution on the cavity at the back of the modified tube lid.
3. Cut a parafilm square of 3 cm; carefully stretch the square in all directions and place it on top of the lid's back to close the cavity.
4. Carefully close the tube lid and trim away the parafilm in excess; at this stage, check if the parafilm properly seals the TE solution without any breaks, otherwise repeat this step. The system is now ready (Fig. 1).

3.2 Phytoplasma Inoculation to an Artificial Diet by Field-Collected Insects

1. Place a single putatively infectious insect (adult stage) in each artificial feeding system.
2. Close the open end of the tube with a wet small cotton plug to allow airflow and keep it humid.
3. Place the artificial rearing systems in a rack suitable for centrifuge tubes, covering the bottom with dark cardboard or tissue.

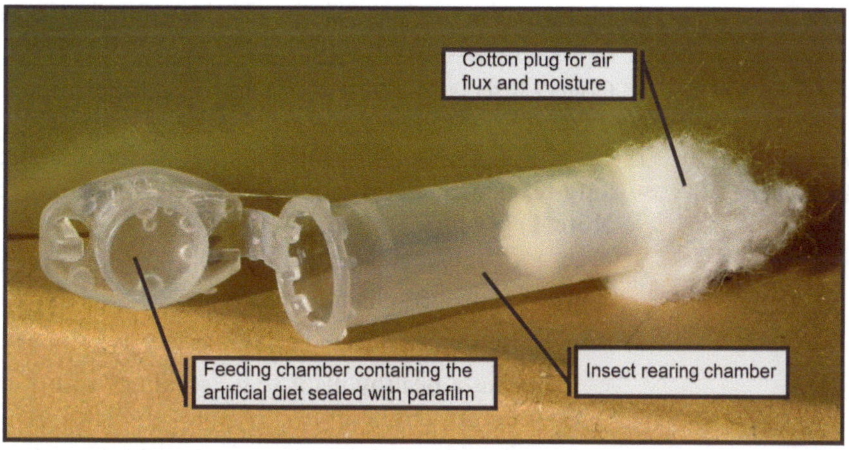

Fig. 1 Artificial feeding system used for phytoplasma transmission trials. Example of a modified microcentrifuge tubes containing a sterile sucrose solution in their lid (feeding chamber), sealed with parafilm. Insects are individually reared in the tube (rearing chamber)

4. Place the rack with tubes remaining horizontally; provide a light source directed towards the tubes' lids side containing the diet.
5. Rear insects for 24–48 h, checking cotton humidity daily.
6. Collect the insects and store them for diagnostic analyses (e.g., at −20 °C or −80 °C).
7. Collect the remaining diet with a micropipette and store it under the same conditions as the insects (*see* **Note 2**).

3.3 Maintenance of a Phytoplasma-Free Vector Population for Controlled Acquisition from Healthy Insects

An exemplificative protocol is presented to obtain a healthy rearing from monovoltine insects that feed on one or several woody hosts, with the overwintering phase (egg) spent under the plant bark (e.g., the vectors of Grapevine Yellows phytoplasmas *Scaphoideus titanus*, and *Orientus ishidae*) (*see* **Note 3**).

1. Prior to the beginning of the egg hatching period, prune plants from an infested area (or collect pruning residues, if available) (Fig. 2a). The most suitable plant part for oviposition of the selected species should be collected (for example, *S. titanus* mainly lays eggs on 2-year-old vine canes, Fig. 2b).
2. Place the pruned material into an insect-proof cage containing one or several potted plants suitable for sustaining newly emerging individuals for a few days (Fig. 2c, d). The amount of wood can be weighed before being caged, in order to normalize emergence data when estimating the population abundance of the vector in a certain area.
3. Maintain sufficient moisture for pruned branches, for example, by periodic watering, to support egg hatching.
4. Monitor the rearing daily by checking for the occurrence of newly emerged nymphs on the leaves of the supplied plants.
5. Collect insects to be used for controlled acquisition.

3.4 Phytoplasma Acquisition Source

1. Check for phytoplasma infection in the initial source (i.e., a long-term infected plant or field-collected infected vectors) (*see* **Note 4**) [14].
2. Place insects and plants in insect-proof cages, possibly in a climate-controlled chamber (*see* **Note 3**).
3. Rear continuously infected vectors either by:
   - Exposing a healthy plant (e.g., seedling) to 50–100 infectious adult vectors or
   - Exposing 50–100 healthy vectors (third or fourth instar nymphs) to an infected plant
4. Replace healthy plants with new seedlings every 2–3 weeks.
5. Add new healthy insects monthly to sustain the permanent infection.

Fig. 2 Set up of a phytoplasma-free rearing for univoltine insects with overwintering eggs laid under the bark. Rearing of healthy *Scaphoideus titanus* nymphs is provided as an example. (**a**, **b**) Pruning residuals are collected from *S. titanus*—infested vineyards in winter. (**c**, **d**) Canes are maintained in insect-proof cages, checking for the presence of neonates daily

3.5 Experimental Phytoplasma Acquisition and Latency

1. Collect healthy insects from a phytoplasma-free source, e.g., a laboratory rearing (*see* **Note 5**). Third or fourth instar nymphs (N3-N4) are the most suitable stage to initiate the experiment, except for trials specifically testing other life stages.
2. Place 50–100 specimens on a phytoplasma-infected plant for acquisition.
3. Rear the insects on the infected plant for 5–7 days to allow completing the minimum phytoplasma acquisition access period.
4. Transfer the insects to a healthy plant for 15–30 days to allow completing the latency access period. The duration of latency may vary, depending on specific traits of the vector–phytoplasma association (*see* **Note 6**).
5. Collect insects for use in subsequent steps (Fig. 3).

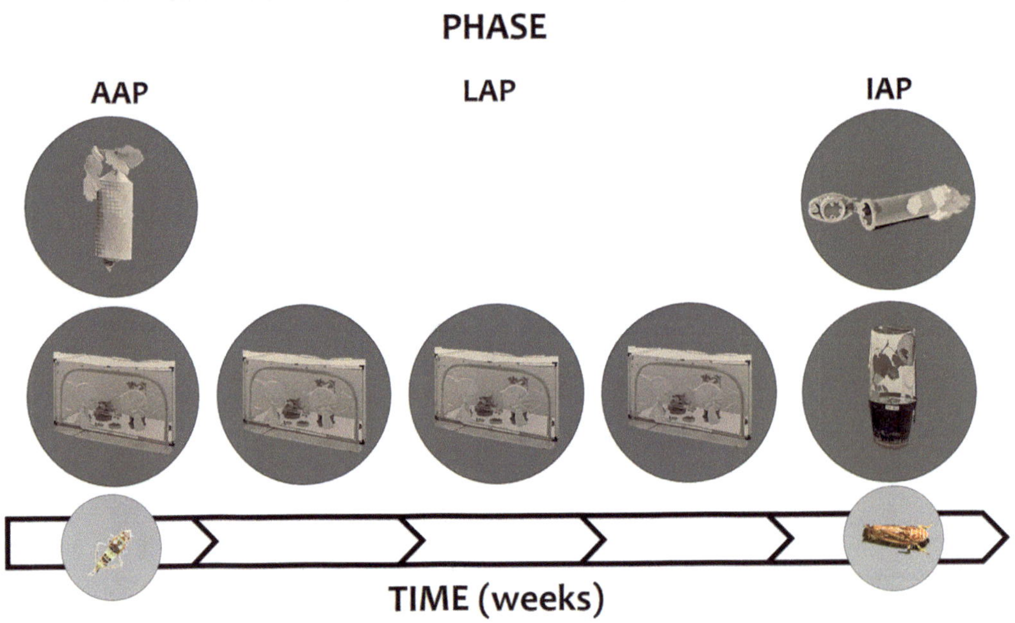

Fig. 3 Passages to perform a complete phytoplasma transmission trial. A transmission assay using *Scaphoideus titanus*, vector of Flavescence Dorée phytoplasma, is shown as an example. The experiment begins typically with the Acquisition Access Period (AAP), which can be performed either in the vineyard using a sleeve cage or in the laboratory using insect-proof cages. This is followed by a Latency Access Period (LAP), lasting a few weeks, usually performed under controlled conditions on a plant that supports vector survival. Finally, the Inoculation Access Period (IAP) is conducted either on an artificial diet or on the host plant. The total duration of the experiment includes the progression of insect post-embryonic development, with trials starting with nymphs and ending with the adult stage

3.6 Inoculation to a Host or Model Plant

1. Confirm the absence of phytoplasma in the plants to be used as a substrate for inoculation by molecular diagnosis (*see* **Note 7**).
2. Place putatively infectious insects in cages containing a healthy plant suitable for phytoplasma inoculation source (*see* **Note 8**).
3. Rear the insects on the healthy plant (*see* **Note 9**) for 1–7 days to allow completing the minimum phytoplasma inoculation access period, which may vary depending on the vector–phytoplasma association.
4. Collect insects and store them for diagnostic analyses (e.g., at −20 °C or −80 °C).
5. Grow the plant in a protected place (e.g., an insect-proof cage) for at least 20 days, to support phytoplasma colonization (*see* **Note 10**).
6. Monitor symptom appearance and collect appropriate plant samples for diagnostic analyses.

4 Notes

1. For trials with one to five insects on a single potted plant use a plexiglas cylinder (Fig. 4a): insert the cylinder into the topsoil, with filter paper covering the soil surface to facilitate counting of any dead specimens during the experiment.

   For trials with multiple potted plants and a larger number of insects (>20), use a commercial insect-proof net cage (available in several sizes) (Fig. 4b, c).

Fig. 4 Types of insect cages used for phytoplasma transmission trials. (**a**) Single specimens or small batches can be fed on seedlings or herbaceous plantlets to achieve faster plant colonization after inoculation. (**b**, **c**) Large insect groups can be fed on caged herbaceous (**b**) or woody (**c**) plants to allow contemporary phytoplasma acquisition by many specimens. (**d**) Phytoplasma acquisition from the field can be measured by isolating insects on sleeve cages

For trials with woody plants in the field, use a sleeve cage made of polythene net, supported by an inner cylinder of a plastic net. Secure the cage with a twist tie proximally around the branch, and distally over its tip (Fig. Close the cage on both sides around with a twist tie (Fig. 4d).

2. Continuous rearing of insect vectors for multiple generations is easily achieved for many leafhopper species, especially those living on herbaceous hosts and completing several generations per year. On the other hand, it may be more challenging for psyllids and cixiid planthoppers, due to the separate habitats of different life stages (e.g., different host or shelter plants, different plant parts colonized below and above ground). Moreover, even some paurometabolous Cicadellidae that do not require changing the host plant during their life cycle may be difficult to rear for several generations. This is especially true for monovoltine species that feed on one or several woody hosts and overwinter under the plant bark.
3. Laboratory transmission experiments may be conducted in a climatic chamber, where temperature, relative humidity, and light exposure are set according to the specific requirements of the insect and plant. Alternatively, trials can be conducted under greenhouse or outdoor conditions, if required.
4. Any phytoplasma-infected plant can be used as a source for phytoplasma acquisition, according to the phytoplasma selected for the experiment. The following essential conditions must be met:
   - The plant is a suitable host for the selected phytoplasma.
   - The plant sustains active feeding by the selected vector for at least 24–72 h. Therefore, even a plant species that is not included in the vector host range can be used [7, 15, 16]. In that case, preliminary trials should be performed to assess the maximum feeding period that supports the insect's survival. For the preliminary trials, groups of five to ten vector specimens are reared on a potted, healthy plant of the selected species, and the survival rate is checked daily.

   Preferable conditions for use of a plant as a phytoplasma acquisition source are the following:

   - The plant is fast-growing and easy to handle (e.g., herbaceous species).
   - The plant rapidly shows symptoms following phytoplasma infection.

   Alternatively, the phytoplasma source can be selected among field-infected plants, once the infection is confirmed by molecular diagnosis.

5. Controlled acquisition can be avoided if the experiment involves the use of field-collected adults to test field infectivity.
6. A plant species that does not support phytoplasma infection may be selected as a substrate during the latency period, to avoid superposition between acquisition and latency; alternatively (especially for monophagous or oligophagous vectors), the initial (or renewed) host plant can be used to maintain a continuous acquisition pressure.
7. If seedlings are used, molecular diagnosis to confirm the absence of phytoplasma can be avoided, except for phytoplasmas suspected or known to be seed-transmitted.
8. Ideally, the final host plant should be selected as a recipient for inoculation; however, in particular cases, alternative hosts can be selected. In this case, the same essential and preferential conditions described in **Note 3** apply.
9. According to the experimental design, the inoculation of phytoplasma into a plant can be performed using individual insects, as well as batches of five or more.
10. The time required to support sufficient phytoplasma colonization may vary depending on the plant; for a woody potted plant, up to several months may be necessary.

References

1. Alma A, Lessio F, Nickel H (2019) Insects as phytoplasma vectors: ecological and epidemiological aspects. In: Bertaccini A, Weintraub PJ et al (eds) Phytoplasmas: plant pathogenic bacteria – II transmission and management of phytoplasma - associated diseases. Springer Nature, Singapore, pp 1–25. https://doi.org/10.1007/978-981-13-2832-9_1
2. Weintraub PJ, Trivellone V, Krüger K (2019) The biology and ecology of leafhopper transmission of phytoplasmas. In: Bertaccini A, Weintraub PJ et al (eds) Phytoplasmas: plant pathogenic bacteria – II transmission and management of phytoplasma - associated diseases. Springer Nature, Singapore, pp 27–51. https://doi.org/10.1007/978-981-13-2832-9_2
3. Bertaccini A (2021) Containment of phytoplasma-associated plant diseases by antibiotics and other antimicrobial molecules. Antibiotics 10(11):1398. https://doi.org/10.3390/antibiotics10111398
4. Bosco D, Tedeschi R (2013) Insect vector transmission assays. In: Dickinson M, Hodgetts J (eds) Phytoplasma. Methods in molecular biology. Humana Press, Totowa, p 938. https://doi.org/10.1007/978-1-62703-089-2_7
5. Bressan A, Clair D, Sémétey O et al (2006) Insect injection and artificial feeding bioassays to test the vector specificity of Flavescence Dorée phytoplasma. Phytopathology 96(7): 790–796
6. Pinzauti F, Trivellone V, Bagnoli B (2008) Ability of *Reptalus quinquecostatus* (Hemiptera: Cixiidae) to inoculate stolbur phytoplasma to artificial feeding medium. Ann Appl Biol 153:299–305. https://doi.org/10.1111/j.1744-7348.2008.00261.x
7. Lessio F, Picciau L, Gonella E et al (2016) The mosaic leafhopper *Orientus ishidae*: host plants, spatial distribution, infectivity, and transmission of 16SrV phytoplasmas to vines. Bull Insectol 69(2):277–289
8. Mou DF, Di Lella B, Halbert SE et al (2022) Acquisition and transmission of the lethal bronzing phytoplasma by *Haplaxius crudus* using infected palm spear leaves and artificial feeding media. Phytopathology 112(10): 2052–2061. https://doi.org/10.1094/PHYTO-03-22-0079-R

9. Chuche J, Danet JL, Rivoal JB et al (2018) Minor cultures as hosts for vectors of extensive crop diseases: does *Salvia sclarea* act as a pathogen and vector reservoir for lavender decline? J Pest Sci 91:145–155. https://doi.org/10.1007/s10340-017-0885-5
10. Riedle-Bauer M, Paleskić C, Schönhuber C et al (2022) Vector transmission and epidemiology of '*Candidatus* Phytoplasma pyri' in Austria and identification of *Cacopsylla pyrisuga* as new pathogen vector. J Plant Dis Prot 129:375–386. https://doi.org/10.1007/s41348-021-00526-y
11. Rossi M, Galetto L, Bodino N et al (2023) Competition among Flavescence Dorée phytoplasma strains in the experimental insect vector *Euscelidius variegatus*. Insects 14(7):575. https://doi.org/10.3390/insects14070575
12. Cui W, Muñoz V, Navarrete M et al (2024) Insect vector and reservoir plant of 'Fragaria × ananassa' ohyllody phytoplasma (16SrXIII-F) in central region of Chile. Plant Dis 108(6): 1861–1868. https://doi.org/10.1094/PDIS-12-23-2552-RE
13. Rinklef A, Behrmann SC, Löffler D et al (2024) Prevalence in potato of '*Candidatus* Arsenophonus Phytopathogenicus' and '*Candidatus* Phytoplasma Solani' and their transmission via adult *Pentastiridius leporinus*. Insects 15(4):275. https://doi.org/10.3390/insects15040275
14. Pagliari L, Chuche J, Bosco D et al (2019) Phytoplasma transmission: insect rearing and infection protocols. In: Musetti R, Pagliari L (eds) Phytoplasmas. Methods in molecular biology, vol 1875. Humana Press, New York. https://doi.org/10.1007/978-1-4939-8837-2_3
15. Batlle A, Altabella N, Sabaté J et al (2008) Study of the transmission of stolbur phytoplasma to different crop species, by *macrosteles quadripunctulatus*. Ann Appl Biol 152:235–242. https://doi.org/10.1111/j.1744-7348.2007.00210.x
16. Cvrković T, Jović J, Krstić O et al (2022) Epidemiological role of *Dictyophara europaea* (Hemiptera: Dictyopharidae) in the transmission of '*Candidatus* Phytoplasma solani'. Horticulturae 8(7):654. https://doi.org/10.3390/horticulturae8070654

Chapter 9

A Workflow for Phytoplasma Discovery and Characterization in Potential Insect Vectors: From Non-destructive DNA Extraction and Vouchering to Next-Generation Sequencing

Valeria Trivellone, Yanghui Cao, and Christopher H. Dietrich

Abstract

Insect vectors remain unknown for most currently described phytoplasmas, and even known vectors are often understudied because voucher specimens essential to corroborating the species identification are rarely preserved and made available for subsequent analysis. However, because phytoplasmas often concentrate in the bodies of potential insect vectors that feed on infected plant phloem, recent studies have shown that screening DNA extracted from such insects is an efficient means for detecting and characterizing phytoplasmas in natural and anthropogenic ecosystems. Given uncertainties in morphology-based identification of insects and ongoing changes in insect taxonomy, an improved workflow is needed to integrate taxonomy, molecular biology, and epidemiology in host–pathogen research. Here we describe a newly established workflow to: (1) collect, handle, and preserve voucher specimens of potential vectors of phytoplasma; (2) non-destructively extract DNA from potentially infected insect specimens; (3) efficiently screen insect specimens for the presence of phytoplasmas; and (4) conduct downstream NGS analysis for multilocus characterization of phytoplasma strains and associated insect gut contents.

Key words Insect collection, NGS analysis, Non-destructive nucleic acid extraction, Pathogen, Databasing, Vouchers

1 Introduction

The large remaining gaps in knowledge of the identity of insect vectors of phytoplasmas worldwide accentuate the need for rigorous, repeatable protocols to preserve, screen, and analyze potential insect vector specimens more consistently [1]. Nearly, all previously known phytoplasmas were discovered by screening for their presence in plant tissue. Unfortunately, because most screenings have

---

Supplementary Information The online version contains supplementary material available at https://doi.org/10.1007/978-1-0716-5104-9_9.

Katrin Janik and Mattia Tabarelli (eds.), *Phytoplasma: Methods and Protocols*, Methods in Molecular Biology, vol. 3008, https://doi.org/10.1007/978-1-0716-5104-9_9,

focused on plants and transmission trials are difficult and time-consuming, vectors remain unknown for >85% of known phytoplasma strains. Moreover, these bacteria are difficult to detect consistently in plants due to their low concentrations and irregular distribution across an infected plant. Detection is also influenced by seasonal variation and environmental factors [2–4]. Thus, attempts to detect phytoplasmas almost always focus on plants showing "typical" symptoms of phytoplasma disease. Asymptomatic plants are rarely tested due to the expense and extremely low rates of detection. Thus, as a method for discovering new phytoplasmas, screening of plant tissues is highly inefficient due to the large number of plant samples needing to be processed, coupled with the often non-specific patterns of symptoms. The focus on plant screening also excludes the possibility of identifying the insect vectors that are crucial to phytoplasma transmission.

Because phytoplasmas are phloem-limited bacteria, and phloem-feeding hemipterans, including leafhoppers and psyllids, feed directly on phloem sap using piercing-sucking mouthparts, these insects regularly acquire phytoplasmas by feeding on infected plants. During feeding, phytoplasma concentration may increase in the insect's body and, in competent vectors, reaches higher than in insects that acquire phytoplasma cells only incidentally [5, 6], making detection easier as the phytoplasma concentrates in the body of a potential vector. Thus, screening specimens of phloem-feeding insects belonging to groups known to include phytoplasma hosts and vectors is an efficient method for discovering phytoplasmas and phytoplasma–insect associations [1, 7, 8].

According to the definition of the National Cancer Institute (www.cancer.gov) a biorepository is: "A facility that collects, catalogs, and stores samples of biological material, such as urine, blood, tissue, cells, DNA, RNA, and protein, from humans, animals, or plants for laboratory research." Museum biorepositories assure long-lasting preservation of material and data relevant to projects in several research fields, such as biodiversity, ecology, phylogenetics, evolutionary biology, population genetics, toxicology, environmental monitoring and pathology. As noted by Brooks [9] "parasitologists [should] deposit voucher specimens of all host species examined in the course of survey or inventory studies, including purely ecological field studies." Although a few studies pointed out the importance of catalog records and voucher specimens in the study of host–pathogen associations, this practice is still rarely applied routinely in the field of pathology [10, 11].

Previous efforts to screen for the presence of phytoplasmas in insects mostly focused on a few species in agroecosystems with ongoing phytoplasma disease outbreaks. Such studies have often overlooked the importance of preserving insect specimens in biorepositories and digitizing individualized information on insect-pathogen associations.

Until recently, classification and characterization of phytoplasmas (and many other bacteria) have relied almost exclusively on analysis of 16S rRNA gene sequence data. However, the use of this gene region alone has more recently been shown to provide inadequate phylogenetic and taxonomic discriminatory power [12–14]. This has spurred new efforts to incorporate complementary data from other parts of the genome into research on bacteria, including phytoplasmas [15–18], and to incorporate cost-effective next-generation sequencing technologies to overcome challenges in partial- or whole-genome sequencing for phytoplasmas and other groups of not yet culturable bacteria.

In a previous study, we used DNA samples extracted from leafhopper vouchers to compare standard Sanger sequencing and Illumina sequencing following anchored hybrid enrichment (AHE) [19]. For the 22 samples testing positive for phytoplasmas, AHE outperformed Sanger sequencing, yielding data for 20 samples, compared to only 16 for Sanger sequencing [7]. AHE also yielded phytoplasma sequence data for seven additional samples that did not meet the minimum qPCR threshold for positivity. Thus, for DNA samples obtained from single leafhopper specimens, AHE is more efficient and sensitive than traditional PCR and Sanger sequencing.

Moreover, in understanding phytoplasma epidemiology, a crucial step is documenting insect vector-plant associations and, as a consequence, inferring the potential risk of specific phytoplasma–plant associations. Recent studies have demonstrated that leafhoppers often retain plant DNA, presumably ingested during feeding, that can be detected and characterized by Sanger and next-generation sequencing approaches [20, 21]. For this reason, we adapted a widely used and cost-effective method for high-throughput next-generation DNA sequencing, AHE, to target molecular markers useful for characterizing phytoplasma strains and estimating their phylogenetic relationships while simultaneously detecting and characterizing plant DNA present in the insect samples, enabling us to identify potential host plants of both insects and phytoplasmas in DNA samples obtained from individual insect specimens. The AHE approach has been used successfully for phylogenomic analyses of a wide variety of organisms spanning the eukaryotes, including plants [22–28], animals [29–54], and fungi [55].

2 Materials

2.1 Insect Collection and Database

1. Specimens: single individuals preserved in 96% ethanol are stored in 2 mL screw–cap vials with "O" ring seal at −20 °C (short-term preservation) or −80 °C (long-term preservation).

Dry individuals are pinned and stored in insect drawers in collection cabinets.

2. Databasing: A relational database management system (RDBMS) and a tool to parse and clean the content of datasets such as OpenRefine [56] or R software [57].

2.2 Chelating Resin-Based DNA Extraction

The following amounts of working solutions are needed for non-destructive extraction of DNA from 100 insect specimens.

1. Lysis buffer: 20 mM Tris-HCl, 10 mM EDTA, 0.5% Sodium dodecyl sulfate (SDS), pH 8.0. For 10 mL lysis buffer, combine 0.2 mL 1 M Tris–HCl, 0.2 mL 0.5 M EDTA, 0.5 mL 10% SDS, and 9.1 mL ultrapure, nuclease-free water. The final pH is ~7.0 (*see* **Note 1**). Store at 4 °C and resuspend it before pipetting.
2. 100 μL 20 mg/mL Proteinase K.
3. 95% ethanol (EtOH). For temporary sample storage before and after DNA extraction.
4. 20% chelating resin-based working solution. Dissolve 2 g of iminodiacetic acid-based chelating resin (100–200 mesh, sodium form) in 10 mL of ultrapure, nuclease-free water. Store for up to 3 months at room temperature or 4 °C. Before use, shake the stock tube to resuspend the solution or mix gently using a magnetic stirrer at 200–500 rpm before pipetting.
5. 2.5% bleach working solution: dilute 91 mL of 8.25% concentrate bleach with deionized water to a final volume of 30 mL.

2.3 Real-Time PCR

1. A universal quantitative PCR (qPCR) master mix containing hot-start Taq DNA polymerase, dNTPs, optimized reaction buffer, and stabilizers for probe-based qPCR detection.
2. Universal primer pairs (p16S-fw 5′- CGTACGCAAGTATGAAACTTAAAGGA-3′, p16S-rv 5′-TCTTCGAATTAAACAACATGATCCA -3′) and a probe with 5′ labeled with the reporter dye FAM (6-carboxyfluorescein) and 3′ labeled with two quencher dyes, ZEN (a proprietary internal quencher) and IBFQ (Iowa Black Fluorescent Quencher) (5-′-FAM-TGACGGGAC-ZEN-TCCGCACAAGCG-IBFQ) [58].
3. ultrapure, nuclease-free water.
4. Phytoplasma positive control (*see* **Note 2**).

2.4 General Equipment

1. A benchtop fluorometer for quantifying nucleic acids, proteins, and other biomolecules using fluorescence-based detection with specific dyes for dsDNA, RNA, and protein assays (*see* **Note 3**).

2. Automated multi-channel liquid handling system for precise pipetting and sample preparation.
3. Automated multi-channel liquid handling system accessories, including pipette heads, on-bed tip bin, tip adapter blocks, tips blister refills, 24-column reservoir tray microplate.
4. Low-Profile 96-Well Skirted PCR Plates.
5. Optical sealing film for PCR plates.
6. Thermal cycler for quantitative PCR analysis.
7. Magnetic stirrer (*see* Subheadings 2.2, **item 4** and 3.2, **step 9**).
8. Thermostatic water bath (*see* Subheadings 3.2, **step 5** and 3.2, **step 11**).
9. Centrifuge (*see* Subheading 3.2, **step 6**).
10. Gel electrophoresis apparatus (*see* Subheading 3.2, **step 16**).
11. A microvolume absorbance-based spectrophotometer.

3 Methods

The overall workflow is summarized in Fig. 1.

1. Label each bulk sample with the collection event information (Fig. 2a).

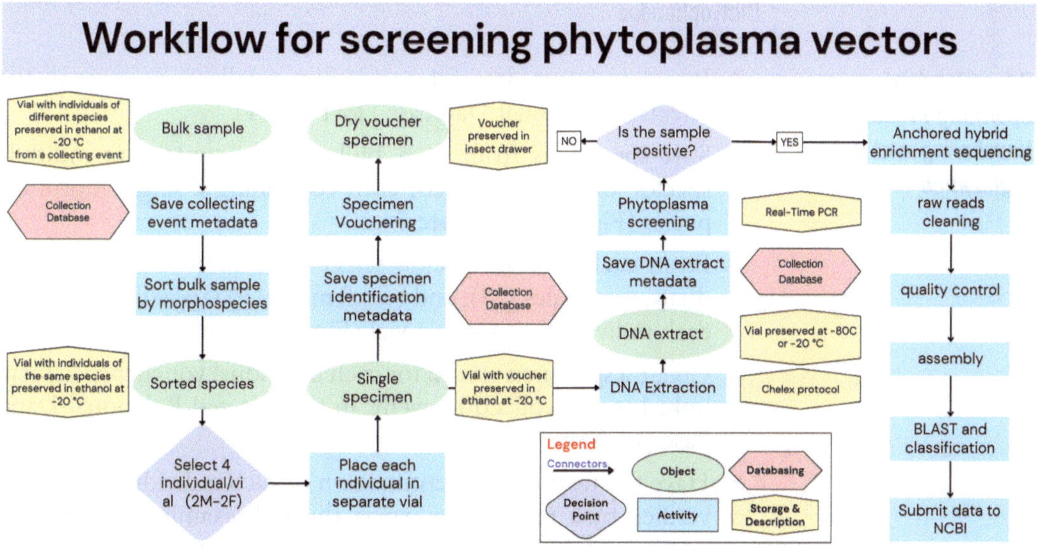

Fig. 1 Workflow for screening phytoplasma vectors and uncovering host–pathogen associations. Critical steps are archiving data (pink diamond shape), vouchering the specimens, storing objects (yellow shape, green oval), and analyzing genomic data

Fig. 2 Voucher specimens, from collection to long-term preservation at Illinois Natural History Survey Insect collection (University of Illinois at Urbana-Champaign). (**a**) Bulk sample labelled with the collection event information. (**b**) Sorted vial with a single morphospecies. (**c**) Individual specimen before DNA extraction. (**d**) Dry pinned voucher specimen belonging to the family Dictyopharidae

3.1 Standard Operating Procedures: Vouchering and Data Digitization

2. Insert and save data for each unique collection event in the selected Relational Database (*see* **Note 4**).
3. Sort specimens from each bulk sample containing multiple species into separate vials, each containing a single morphospecies (Fig. 2b) (*see* **Note 5**).
4. From each morphospecies, select a maximum of four individuals, two males (M), and two females (F) if possible and store each individual in a single vial (Fig. 2c) (*see* **Note 6**).
5. Create a specimen record for each individual voucher specimen, linked to the collection event, in the selected Relational Database (*see* **Note 7**).
6. Have the taxonomic identification of each individual specimen verified by a taxonomic expert through examination of diagnostic morphological structures (e.g., genitalia); submit specimens for long-term preservation (in ethanol at −20 °C or as dry pinned specimen, Fig. 2d) in the institution's insect collection.

3.2 DNA Extraction

1. If the extraction is from a dry insect voucher, re-hydrate the specimens in a humidifying chamber overnight. If the voucher is ethanol-preserved, transfer it to a paper towel for a few minutes to allow ethanol to evaporate.
2. Wash each specimen in 500 μL of a 2.5% bleach solution for 5 min at 7.5 °C (*see* **Note 8**).
3. Rinse the specimen 3 times with 250 μL of sterile water.
4. Place the specimen in a fresh 1.5 mL centrifuge tube and add 100 μL lysis buffer and 1 μL proteinase K (*see* **Note 9**).
5. Incubate the tubes at 55 °C overnight or for 48 h (*see* **Note 10**).
6. Centrifuge the tubes for 30 s at 11,000× *g*.
7. Pipet the liquid (DNA sample) into a fresh tube without disturbing the insect body.
8. Return the extracted specimen back into the original voucher vial. The individual is ready to be pinned or stored in ethanol for long-term preservation (*see* **Note 11**).
9. Keep the chelating resin-based working solution in a flask on a magnetic stirrer while pipetting.
10. Distribute 100 μL of chelating resin-based working solution directly from the agitator into the tube with the DNA sample (*see* **Note 12**).
11. Incubate the tube at 55 °C for 30 min. Do not shake the tube.
12. Invert the tube five times gently.
13. Let the solution cool down at room temperature for approximately 1 h.
14. Centrifuge for 1 min at 25,000× *g*.
15. Transfer the supernatant to a fresh centrifuge tube.
16. Quantify each DNA sample and evaluate quality through gel electrophoresis to assess the level of DNA integrity (*see* **Note 13**).
17. Transfer the DNA samples to a 96-deep well plate and proceed with the screening step (Subheading 3.3).

3.3 Real-Time PCR

To avoid contamination and human errors, it is recommended to load the Real-Time PCR plate in an automated multi-channel liquid handling system.

For the PCR solution, the amounts must be adapted as necessary according to the manufacturer's instructions. Reagents from a particular supplier are mentioned in **Note 14**, but this does not mean that these are the only reagents suitable for the methods described.

1. Prepare a real-time PCR solution sufficient for all DNA samples by adding all the reagents in the following order: ultrapure, nuclease-free water, forward and reverse primers (typically at a final concentration of 300 nM), probe (typically at a final concentration of 100 nM) and a premixed solution containing Taq DNA polymerase, dNTPs, buffer, and other necessary components for a PCR reaction. Mix thoroughly and transfer the solution to a single channel of the 24-column reservoir tray compatible with the pipetting machine (*see* **Note 14**).
2. Load the multi-channel liquid handling system according to the protocol bed layout by allocating one on-bed tip bin, two blister refills, one 24-column reservoir tray, 96-deep well plate with DNA samples, one fresh low-Profile 96-Well Skirted PCR plate.
3. Run the customized multi-channel liquid handling system protocol. The protocol loads the low-Profile 96-Well Skirted PCR plate with 6 μL real-time PCR reaction solution and 4 μL DNA template each well. The total loading time is 3 min and 15 s.
4. While the loading is running, set up the Real-Time PCR software with the following cycling conditions: 1 cycle at 60 °C for 30 s, 95 °C for 2 min, and 49 cycles at 95 °C for 15 s, 60 °C for 1 min [59]. Include appropriate controls for each run (*see* **Note 15**).
5. After the automated loading, seal the PCR plate and transfer it into the real-time PCR cycler and start the run.
6. Once the run is completed, remove the plate and throw it away. Save the resulting data.
7. By evaluating the curves, identify the positive samples according to an empirical cut-off Cq value (*see* **Note 16**).
8. Submit samples testing positive for the presence of phytoplasmas for DNA sequencing.

3.4 Anchored Hybrid Enrichment Sequencing

Due to the need for case-specific decisions based on the available genome resources and the existence of a pending patent application on the probe design, this section is presented in narrative format rather than in command mode or as a universally applicable protocol. While it may not be directly reproducible, this section provides a detailed example intended to serve as a guideline for developing similar strategies.

We use anchored hybrid enrichment (AHE) sequencing to obtain DNA sequence data from 129 phytoplasma genes and 5 plant genes. Anchored hybrid enrichment (AHE) uses DNA hybridization probes to capture and enrich specific regions of interest from the genome, incorporating multiple probes per locus to capture variable regions (including conserved exon regions and more variable flanking regions, including introns) useful for both coarse- and finer-scale phylogenetic resolution and characterization.

Locus Selection and Probe Design To improve the efficiency of data collection, specific probes were designed to target a total of 178 loci from 129 phytoplasma genes and 8 loci from or adjacent to 5 plant genes. The loci were identified using genomic resources publicly available on the Nucleotide Database and the Genome Database of National Center for Biotechnology Information (NCBI) up until August 21, 2021.

1. Forty-five (24 genes) out of 178 loci are widely used for classification and characterization of phytoplasmas (Supplementary Table 1, loci id from PhL1 to PhL45). Reference sequences of these genes were downloaded from the Nucleotide Database of NCBI as well as retrieved from the phytoplasma genomes listed in Supplementary Table 2. Sequences of each gene were aligned by MAFFT [60]. with an automatic selection of alignment algorithms, followed by manual checking. For the genes with highly variable sequences among different phytoplasma strains, we split the sequences into several well-aligned segments (here referred to as loci). The full alignments of these 24 genes were used for probe design.
2. The other 133 newly selected phytoplasma protein-coding loci (Supplementary Table 1, loci id from PhL46 to PhL178) were additionally selected from single-copy orthologs among 50 phytoplasma genomes available in the Genome database of NCBI (Supplementary Table 2). To find more of these target regions, putatively core orthologous genes across multiple proteomes were identified using OrthoFinder v2.5.1 [61]. Orthologs present in at least 40 of the analyzed genomes and with a minimum length of 120 base pairs (bp) were selected for further analysis. Sequences of each orthologous gene were aligned and checked manually as in **Step 1**. Alignment cleaning was performed by splitting into several short alignments to exclude ambiguous regions. Then the alignments containing <120 bp positions or >30 bp consecutive indels were discarded.
3. Probes were designed for 8 plant loci (Supplementary Table 1, loci id from PlL1 to PlL8) to retrieve five primary DNA barcode loci used in plants, three for the chloroplast ribulose-1,5-bisphosphate carboxylase-oxygenase (rbcL), maturase K (matK), the non-coding spacer region trnH-psbA, and the remaining five for two nuclear internal transcribed spacers (ITS1 and ITS2) [62]. Because the non-coding regions trnH-psbA, ITS1, and ITS2 are highly variable and not suitable for

probe design, we selected their adjacent conserved genes (trnH, psbA, 18S rDNA, 5.8S rDNA, 26S rDNA) as target regions.

4. A sliding window approach with a 120-bp window and 1 bp skip was performed, followed by BLASTN searches within each generated data metric, to estimate the pairwise sequence identity. Newly selected loci with an average pairwise sequence identity <0.7 were discarded.
5. Tiled probes were selected from the relatively conserved regions with a spacing of mostly 30–50 bp. GC content and homopolymers were checked for each probe using a custom script. Probes with % GC ranging from 0.2 to 0.65 were considered as qualified. Probes with a single homopolymer region >12 bp or three >6 bp homopolymer regions were excluded from the probe kit.
6. To avoid the probes targeting genomic regions of phytoplasma potential vectors or their primary bacterial endosymbionts, probes were compared with the genomic data for the leafhopper *Homalodisca vitripennis* (Germar) [63] and endosymbionts "*Candidatus* Karelsulcia muelleri" Moran and "*Candidatus* Nasuia deltocephalinicola" Noda [64] using BLASTN searches, and those with sequence similarity >0.6 were removed from the probe kit. The final probe kit contains 57,914 unique 120-bp probes. Biotinylated probes were synthesized by a commercial provider specialized in custom oligonucleotide synthesis. The quality of the probes was tested using screening procedures described by Lemmon et al. [19]. The AHE probe kit we currently use was designed based on phytoplasma genome resources available in 2021. Since then, some of the previously published genomes have been shown to contain serious errors (e.g., due to contamination), and several additional phytoplasma genomes have been published. Thus, refinements to our original probe design would benefit from incorporating newly available genome information. Alternatively, recent improvements in whole-genome sequencing protocols for phytoplasmas suggest that, in the near future, it may be possible to sequence whole phytoplasma genomes from DNA extracted from potential insect vectors. We routinely preserve the extracted genomic DNA that remains following our use of the screening and AHE sequencing procedures described here. These residual DNA samples will remain available for use in future sequencing projects.

Library Preparation Capture-based library preparations for Illumina sequencing were performed using a high-throughput workflow with proprietary chemistry by a specialized genomics service company. DNA is sheared to a mean fragment length of 500 bp, fragments are end-repaired and A-tailed, followed by incorporation of uniquely indexed Illumina adaptors and PCR enrichment. Samples are pooled equimolar and sequenced on an appropriate Illumina instrument, often NovaSeq (2 × 150, or 2 × 250 bp).

Sequence Data Processing Industry-standardized tools are utilized for many portions of data processing. Sequencing reads are demultiplexed using Illumina's BCL-Convert.

Bioinformatics Raw sequencing reads are further analyzed using the following pipeline.

1. Raw reads are processed using TrimmomaticPE [65] or bbduk.sh in BBMap to remove adaptors and poor-quality data.
2. Cleaned reads are assessed for quality using FASTQC [66] and then assembled using SPAdes [67] or similar software.
3. TBLASTN searches [68] are performed against these assemblies with the protein sequences of the target loci as queries and a cut-off E value of 10^{-5}.
4. Candidate sequences are used for BLASTX search against the database comprising the original query sequences. Sequences eligible for reciprocal best BLAST hits are further confirmed through TBLASTN against the standard nucleotide databases in GenBank, and only those with the best hit as a phytoplasma gene will be used for phylogenetic analysis. For plant genes, the classification will be confirmed through BLASTN, Barcode of Life Datasystem (BOLD) [69].
5. All the sequences of the *16S* gene of the detected phytoplasmas are further blasted against the i*Phy*Classifier database, and virtual RFLP analysis is carried out according to the protocol in Zhao et al. [70].

Raw read data (.fastq files) from AHE sequencing will be deposited in the NCBI SRA and made available to others wishing to conduct further comparative genomic analyses and screen for additional "by-catch" loci usually present in anchored hybrid short-read datasets that may include additional phytoplasma loci and/or allow assembly of partial phytoplasma genomes.

4 Notes

1. Using the stock solutions of each reagent at pH ~7.0 is suggested to avoid adjusting the pH.

2. As a positive control, we use DNA extracted from infected plants or insect vectors with a concentration between 10^8 and 10^2 Genome Units/μL.
3. Commercial dsDNA quantification assay kits are available and use a DNA-selective dye that binds specifically to dsDNA, emitting fluorescence proportional to the DNA amount. The fluorescence intensity is then measured using a fluorometer to determine the DNA concentration.
4. At a minimum, the collection event record should include the place name, geographic coordinates, date of collection, name of collector, collecting method, and a unique identifier for the bulk sample resulting from the collecting event. We use the open-source TaxonWorks platform (taxonworks.org), which provides web-accessible, user-friendly interfaces for storing biodiversity data, including taxonomic nomenclature and specimen occurrences. The data from the Illinois Natural History Survey Insect collection database is available here: https://www.gbif.org/dataset/68513375-3aa5-4f6f-9975-d97d56c21d61.
5. Assigning specimens in a bulk sample to morphospecies is accomplished by carefully examining the sample under a dissecting microscope and placing all specimens that appear to be identical in external appearance (i.e., size, shape and coloration) into a separate vial, taking into account that males and females of the same species may differ slightly in size and overall coloration (e.g., females may be slightly larger and lighter in color than males of the same species). Individual morphospecies initially sorted using these criteria may later be found to comprise multiple, closely related species separable only by characters of the genitalia. Therefore, it is very important to check and verify species identifications after the non-destructive DNA extraction step by examining the genitalia.
6. The number of individuals of each morphospecies selected for screening is somewhat arbitrary and depends on the particular goals and project budget.
7. The specimen record should include a tentative identification (e.g., family or genus name) and indicate the sex/developmental stage of the specimen. Insects are often difficult to identify without examining the genitalia, which may be visible only after clearing the abdomen. The non-destructive DNA extraction protocol clears the pigment from the exoskeleton and digests the internal soft tissues, leaving the sclerotized parts of the genitalia visible. Thus, verification of the species identity must usually be done following the DNA extraction step. A video

demonstration showing the procedure for creating a specimen record is available here: https://mediaspace.illinois.edu/media/t/1_j2gynrwp#.

8. Bleaching removes contaminant nucleic acids from external body surfaces, ensuring that any DNA extracted originated from within the insect body.
9. To improve penetration of the lysis buffer into the body cavity, the abdomen of the specimen is gently separated from the thorax prior to incubation. Also, the total amount of lysis buffer may vary with the size of the specimen, i.e., it is important that the entire specimen be immersed in the buffer.
10. Longer incubation times result in greater digestion and clearing of internal soft tissues, rendering the internal genitalia more clearly visible.
11. After extraction using this non-destructive method, the insect specimens consist of the intact exoskeleton, which may appear translucent without the internal soft tissues. The male genitalia, which usually need to be examined to identify species, will be at least partly visible through the exoskeleton and can be checked to confirm the species identification. Females are identified based on their overall similarity in appearance to males from the same bulk sample. Isolated females without associated males may not be identifiable to the species. The specimens may be dried and pinned or point-mounted for deposit in a pinned insect collection. The abdomen, if separated from the rest of the specimen, may be stored in a genitalia vial containing glycerin pinned below the specimen. Alternatively, for particularly small, delicate specimens, the entire specimen (body and abdomen) may be preserved in glycerin in a genitalia vial. The specimen should be labeled with the occurrence data (place and date of collection, name of collector) and a barcode or other label with the unique identifier linked to the specimen record in the collection database.
12. The amount of chelating resin solution varies with the initial amount of lysis buffer. For the step of sequencing using the AHE method, a TRIS-containing buffer must be used with a TRIS final concentration of at least 10 mM.
13. To complement or as an alternative to gel electrophoresis, both 260/280 and 260/230 absorbance ratios can be measured using a microvolume spectrophotometer. A 260/280 ratio below 1.8 indicates that the sample is likely contaminated with protein or other substances. In this case, a purification step using ethanol precipitation or another treatment to remove salts and other contaminants is suggested. A 260/230 ratio below 2.0 indicates that the sample is contaminated with chemical residues from the purification process, like

phenol, guanidine thiocyanate, etc. The quality and quantity of the DNA, as well as its suitability for downstream analysis, including AHE sequencing, were assessed in Brown et al. [71].

14. We prepare 606 μL reaction mix for 101 DNA samples using the following volumes: 65.65 μL ultrapure, nuclease free water, 15.15 μL of each primer (final concentration 300 nM), 5.55 μL of probe (final concentration 100 nM), and 505 μL TaqMan™ Universal PCR Master Mix. Reagents from a particular supplier are mentioned here solely to document the conditions that proved well in our setup. Comparable products from other suppliers may be equally suitable, provided the protocol is appropriately adjusted according to the manufacturer's guidelines.
15. PCR mix with water instead of DNA is used as a negative control, while DNA from a phytoplasma-positive sample is used as a positive control in all amplification reactions.
16. The quantification cycle (Cq) value is evaluated for each sample using the cut-off value reported by Trivellone et al. [1]; i.e., samples with a Cq value above 30.38 are considered positive for phytoplasma presence. All the samples that yielded a positive signal for the presence of bacterial 16S rRNA gene are then submitted for downstream analyses.

Acknowledgement

This work was supported by the U.S. National Science Foundation Grant DEB-2244871. The authors thank all the people that have been contributing to setting up and testing various steps of the workflow for the project, Morgan Brown, Chantelle Diener, Chistine Fink, Ava Gabrys, Lisa Kwan, Sara Ottati, Charlotte Szekeres. We also thank Tommy McElrath and Michelle Kohler for help with the SOP for entering specimen data into the INHS Insect Collection Database using TaxonWorks (https://www.gbif.org/dataset/68513375-3aa5-4f6f-9975-d97d56c21d61).

References

1. Trivellone V, Wei W, Filippin L et al (2021) Screening potential insect vectors in a museum biorepository reveals undiscovered diversity of plant pathogens in natural areas. Ecol Evol 11: 6493–6503. https://doi.org/10.1002/ece3.7502
2. Lee I-M, Davis RE, Gundersen-Rindal DE (2000) Phytoplasma: phytopathogenic mollicutes. Ann Rev Microbiol 54:221–255. https://doi.org/10.1146/annurev.micro.54.1.221
3. Berges R, Rott M, Seemüller E (2000) Range of phytoplasma concentrations in various plant hosts as determined by competitive polymerase chain reaction. Phytopathology 90:1145–1152. https://doi.org/10.1094/PHYTO.2000.90.10.1145

4. Marzachì C (2006) Molecular diagnosis of phytoplasmas. Arab J Pl Prot 24:139–142. https://doi.org/10.14601/Phytopathol_Mediterr-1750
5. Galetto L, Nardi M, Saracco P et al (2009) Variation in vector competency depends on chrysanthemum yellows phytoplasma distribution within Euscelidius variegatus. Entomol Exp Appl 131:200–207. https://doi.org/10.1111/j.1570-7458.2009.00845.x
6. Galetto L, Marzachì C (2009) Real-time PCR diagnosis and quantification of phytoplasmas. In: Weintraub PG, Jones P (eds) Phytoplasmas: genomes, plant hosts and vectors. CABI, Wallingford, pp 1–18
7. Trivellone V, Cao Y, Dietrich CH (2022) Comparison of traditional and next-generation approaches for uncovering phytoplasma diversity, with discovery of new groups, subgroups and potential vectors. Biology 11:977. https://doi.org/10.3390/biology11070977
8. Wei W, Trivellone V, Dietrich CH et al (2021) Identification of phytoplasmas representing multiple new genetic lineages from phloem-feeding leafhoppers highlights the diversity of phytoplasmas and their potential vectors. Pathogens 10:352. https://doi.org/10.3390/pathogens10030352
9. Brooks DR (1993) Extending the symbiotype concept to host voucher specimens. J Parasitol 79:631
10. Thompson CW, Phelps KL, Allard MW et al (2021) Preserve a voucher specimen! the critical need for integrating natural history collections in infectious disease studies. 12:1–20. https://doi.org/10.1128/mbio.02698-2011
11. Buckner JC, Sanders RC, Faircloth BC et al (2021) The critical importance of vouchers in genomics. elife 10:e68264. https://doi.org/10.7554/eLife.68264
12. Hassler HB, Probert B, Moore C et al (2022) Phylogenies of the 16S rRNA gene and its hypervariable regions lack concordance with core genome phylogenies. Microbiome 10:104. https://doi.org/10.1186/s40168-022-01295-y
13. Clarridge JE (2004) Impact of 16S rRNA gene sequence analysis for identification of bacteria on clinical microbiology and infectious diseases. Clin Microbiol Rev 17:840–862. https://doi.org/10.1128/CMR.17.4.840-862.2004
14. Bartoš O, Chmel M, Swierczková I (2024) The overlooked evolutionary dynamics of 16S rRNA revises its role as the "gold standard" for bacterial species identification. Sci Rep 14:9067. https://doi.org/10.1038/s41598-024-59667-3
15. Glaeser SP, Kämpfer P (2015) Multilocus sequence analysis (MLSA) in prokaryotic taxonomy. Syst Appl Microbiol 38:237–245. https://doi.org/10.1016/j.syapm.2015.03.007
16. Kuo C-H, Ochman H (2009) Inferring clocks when lacking rocks: the variable rates of molecular evolution in bacteria. Biol Direct 4:35. https://doi.org/10.1186/1745-6150-4-3517
17. Martini M, Quaglino F, Bertaccini A (2019) Multilocus genetic characterization of phytoplasmas. In: Bertaccini A, Oshima K, Kube M et al (eds) Phytoplasmas: plant pathogenic bacteria - III. Springer Singapore, Singapore, pp 161–200. https://doi.org/10.1007/978-981-13-9632-8_9
18. Bertaccini A, Arocha-Rosete Y, Contaldo N et al (2022) Revision of the "*Candidatus* Phytoplasma" species description guidelines. Int J Syst Evol Microbiol 72(4). https://doi.org/10.1099/ijsem.0.005353
19. Lemmon AR, Emme SA, Lemmon EM (2012) Anchored hybrid enrichment for massively high-throughput phylogenomics. Syst Biol 61:727–744. https://doi.org/10.1093/sysbio/sys049
20. Cao Y, Dietrich CH (2021) Identification of potential host plants of sap-sucking insects (Hemiptera: Cicadellidae) using anchored hybrid by-catch data. Insects 12:964. https://doi.org/10.3390/insects12110964
21. Inaba J, Shao J, Trivellone V et al (2023) Guilt by association: DNA barcoding-based identification of potential plant hosts of phytoplasmas from their insect carriers. Phytopathology 113:413–422. https://doi.org/10.1094/PHYTO-09-22-0323-R
22. Benítez-Villaseñor A, Granados Mendoza C, Wanke S et al (2023) The use of anchored hybrid enrichment data to resolve higher-level phylogenetic relationships: a proof-of-concept applied to Asterales (Eudicotyledoneae; angiosperms). Mol Phylogenet Evol 181:107714. https://doi.org/10.1016/j.ympev.2023.107714
23. Bogarín D, Pérez-Escobar OA, Groenenberg D et al (2018) Anchored hybrid enrichment generated nuclear, plastid and mitochondrial markers resolve the *Lepanthes horrida* (Orchidaceae: Pleurothallidinae) species complex. Mol Phylogenet Evol 129:27–47. https://doi.org/10.1016/j.ympev.2018.07.014

24. Favre A, Pringle JS, Heckenhauer J et al (2020) Phylogenetic relationships and sectional delineation within Gentiana (Gentianaceae). Taxon 69:1221–1238. https://doi.org/10.1002/tax.12405
25. Fragoso-Martínez I, Salazar GA, Martínez-Gordillo M et al (2017) A pilot study applying the plant anchored hybrid enrichment method to New World sages (*Salvia* subgenus *Calosphace*; Lamiaceae). Mol Phylogenet Evol 117: 124–134. https://doi.org/10.1016/j.ympev.2017.02.006
26. Granados Mendoza C, Jost M, Hágsater E et al (2020) Target nuclear and off-target plastid hybrid enrichment data inform a range of evolutionary depths in the orchid genus epidendrum. Front Plant Sci 10. https://doi.org/10.3389/fpls.2019.01761
27. Mitchell N, Lewis PO, Lemmon EM et al (2017) Anchored phylogenomics improves the resolution of evolutionary relationships in the rapid radiation of *Protea* L. Am J Bot 104: 102–115. https://doi.org/10.3732/ajb.1600227
28. Wanke S, Granados Mendoza C, Mueller S et al (2017) Recalcitrant deep and shallow nodes in Aristolochia (Aristolochiaceae) illuminated using anchored hybrid enrichment. Mol Phylogenet Evol:111–123. https://doi.org/10.1016/j.ympev.2017.05.014
29. Banker SE, Lemmon AR, Hassinger AB et al (2020) Hierarchical hybrid enrichment: multitiered genomic data collection across evolutionary scales, with application to chorus frogs (Pseudacris). Syst Biol 69:756–773. https://doi.org/10.1093/sysbio/syz074
30. Berv JS, Singhal S, Field DJ et al (2024) Genome and life-history evolution link bird diversification to the end-cretaceous mass extinction. Sci Adv 10:eadp0114. https://doi.org/10.1126/sciadv.adp011
31. Brandley MC, Bragg JG, Singhal S et al (2015) Evaluating the performance of anchored hybrid enrichment at the tips of the tree of life: a phylogenetic analysis of Australian Eugongylus group scincid lizards. BMC Evol Biol 15:62. https://doi.org/10.1186/s12862-015-0318-0
32. Breinholt JW, Earl C, Lemmon AR et al (2018) Resolving relationships among the megadiverse butterflies and moths with a novel pipeline for anchored phylogenomics. Syst Biol 67: 78–93. https://doi.org/10.1093/sysbio/syx048
33. Buenaventura E, Szpila K, Cassel BK et al (2020) Anchored hybrid enrichment challenges the traditional classification of flesh flies (Diptera: Sarcophagidae). Syst Entomol 45: 281–301. https://doi.org/10.1111/syen.12395
34. Cao Y, Dietrich CH, Zahniser JN et al (2022) Dense sampling of taxa and characters improves phylogenetic resolution among deltocephaline leafhoppers (Hemiptera: Cicadellidae: Deltocephalinae). Syst Entomol 47:430–444. https://doi.org/10.1111/syen.12540
35. Dornburg A, Townsend JP, Brooks W et al (2017) New insights on the sister lineage of percomorph fishes with an anchored hybrid enrichment dataset. Mol Phylogenet Evol 110:27–38. https://doi.org/10.1016/j.ympev.2017.02.017
36. Douglas HB, Kundrata R, Brunke AJ et al (2021) Anchored phylogenomics, evolution and systematics of elateridae: are all bioluminescent Elateroidea derived click beetles? Biology 10:451. https://doi.org/10.3390/biology10060451
37. Dowdy NJ, Keating S, Lemmon AR et al (2020) A deeper meaning for shallow-level phylogenomic studies: nested anchored hybrid enrichment offers great promise for resolving the tiger moth tree of life (Lepidoptera: Erebidae: Arctiinae). Syst Entomol 45:874–893. https://doi.org/10.1111/syen.12433
38. Espeland M, Breinholt J, Willmott KR et al (2018) A comprehensive and dated phylogenomic analysis of butterflies. Curr Biol 28:770–778.e5. https://doi.org/10.1016/j.cub.2018.01.061
39. Eytan RI, Evans BR, Dornburg A et al (2015) Are 100 enough? Inferring acanthomorph teleost phylogeny using anchored hybrid enrichment. BMC Evol Biol 15:113. https://doi.org/10.1186/s12862-015-0415-0
40. Glon MG, Broe MB, Crandall KA et al (2022) Anchored hybrid enrichment resolves the phylogeny of Lacunicambarus Hobbs, 1969 (Decapoda: Astacidea: Cambaridae). J Crustac Biol 42:ruab073. https://doi.org/10.1093/jcbiol/ruab073
41. Haddad S, Shin S, Lemmon AR et al (2018) Anchored hybrid enrichment provides new insights into the phylogeny and evolution of longhorned beetles (Cerambycidae). Syst Entomol 43:68–89. https://doi.org/10.1111/syen.12257
42. Hamilton CA, St Laurent RA, Dexter K et al (2019) Phylogenomics resolves major relationships and reveals significant diversification rate shifts in the evolution of silk moths and relatives. BMC Evol Biol 19:182. https://doi.org/10.1186/s12862-019-1505-1
43. Hamilton CA, Lemmon AR, Lemmon EM et al (2016) Expanding anchored hybrid

enrichment to resolve both deep and shallow relationships within the spider tree of life. BMC Evol Biol 16:212. https://doi.org/10.1186/s12862-016-0769-y

44. Johnson KP, Dietrich CH, Friedrich F et al (2018) Phylogenomics and the evolution of hemipteroid insects. Proc Natl Acad Sci USA 115:12775–12780. https://doi.org/10.1073/pnas.1815820115
45. Maddison WP, Evans SC, Hamilton CA et al (2017) A genome-wide phylogeny of jumping spiders (Araneae, Salticidae), using anchored hybrid enrichment. ZooKeys 695:89–101. https://doi.org/10.3897/zookeys.695.1385246
46. Marchán DF, James SW, Lemmon AR et al (2022) A strong backbone for an invertebrate group: anchored phylogenomics improves the resolution of genus-level relationships within the Lumbricidae (Annelida, Crassiclitellata). Org Divers Evol 22:915–924. https://doi.org/10.1007/s13127-022-00570-y
47. Mayer C, Dietz L, Call E et al (2021) Adding leaves to the Lepidoptera tree: capturing hundreds of nuclear genes from old museum specimens. Syst Entomol 46:649–671. https://doi.org/10.1111/syen.12481
48. Mossop KD, Lemmon AR, Moriarty Lemmon E et al (2023) Phylogenomics and biogeography of arid-adapted Chlamydogobius goby fishes. Mol Phylogenet Evol 182:107757. https://doi.org/10.1016/j.ympev.2023.107757
49. Ogden TH, Breinholt JW, Bybee SM et al (2019) Mayfly phylogenomics: initial evaluation of anchored hybrid enrichment data for the order Ephemeroptera. Zoosymposia 16:167–181. https://doi.org/10.11646/zoosymposia.16.1.14
50. Piller KR, Parker E, Lemmon AR et al (2022) Investigating the utility of anchored hybrid enrichment data to investigate the relationships among the killifishes (Actinopterygii: Cyprinodontiformes), a globally distributed group of fishes. Mol Phylogenet Evol 173:107482. https://doi.org/10.1016/j.ympev.2022.107482
51. Prum RO, Berv JS, Dornburg A et al (2015) A comprehensive phylogeny of birds (Aves) using targeted next-generation DNA sequencing. Nature 526:569–573. https://doi.org/10.1038/nature15697
52. Rawlinson KC (2021) Phylogenetic relationships of genera in the caddisfly family Limnephilidae using anchored hybrid enrichment-based Phylogenomic analysis (Insecta:Trichoptera). Thesis, Brigham Young University, 9766
53. Stout CC, Tan M, Lemmon AR et al (2016) Resolving Cypriniformes relationships using an anchored enrichment approach. BMC Evol Biol 16:1–13. https://doi.org/10.1186/s12862-016-0819-5
54. Young AD, Lemmon AR, Skevington JH et al (2016) Anchored enrichment dataset for true flies (order Diptera) reveals insights into the phylogeny of flower flies (family Syrphidae). BMC Evol Biol 16:143. https://doi.org/10.1186/s12862-016-0714-0
55. Nguyen HDT, McCormick W, Eyres J et al (2021) Development and evaluation of a target enrichment bait set for phylogenetic analysis of oomycetes. Mycologia 113:856–867. https://doi.org/10.1080/00275514.2021.1889276
56. OpenRefine (2024). https://openrefine.org/. Accessed 21 July 2025
57. R Core Team (2024) R: a language and environment for statistical computing. https://www.R-project.org/. Accessed 21 July 2025
58. Christensen NM, Nicolaisen M, Hansen M et al (2004) Distribution of phytoplasmas in infected plants as revealed by real-time PCR and bioimaging. Mol Plant-Microbe Interact 17:1175–1184. https://doi.org/10.1094/MPMI.2004.17.11.1175
59. Angelini E, Bianchi LG, Filippin L et al (2007) A new TaqMan method for the identification of phytoplasmas associated with grapevine yellows by real-time PCR assay. J Microbiol Methods 68:613–622. https://doi.org/10.1016/j.mimet.2006.11.015
60. Katoh K, Standley DM (2013) MAFFT multiple sequence alignment software version 7: improvements in performance and usability. Mol Biol Evol 30:772–780. https://doi.org/10.1093/molbev/mst010
61. Emms DM, Kelly S (2019) OrthoFinder: phylogenetic orthology inference for comparative genomics. Genome Biol 20:238. https://doi.org/10.1186/s13059-019-1832-y
62. Hollingsworth PM, Forrest LL, Spouge JL et al (2009) A DNA barcode for land plants. PNAS 106:12794–12797. https://doi.org/10.1073/pnas.090584510
63. Ettinger CL, Byrne FJ, Collin MA et al (2021) Improved draft reference genome for the glassy-winged sharpshooter (Homalodisca vitripennis), a vector for Pierce's disease. G3 Genes|Genomes|Genetics 11:jkab255. https://doi.org/10.1093/g3journal/jkab255
64. Bennett GM, Abbà S, Kube M et al (2016) Complete genome sequences of the obligate symbionts "*Candidatus* Sulcia muelleri" and "ca. Nasuia deltocephalinicola" from the

pestiferous leafhopper macrosteles quadripunctulatus (Hemiptera: Cicadellidae). Genome Announc 4:e01604–15. https://doi.org/10.1128/genomeA.01604-15

65. Bolger AM, Lohse M, Usadel B (2014) Trimmomatic: a flexible trimmer for Illumina sequence data. Bioinformatics 30:2114–2120. https://doi.org/10.1093/bioinformatics/btu170

66. Andrews S. FastQC: a quality control tool for high-throughput sequence data. https://www.bioinformatics.babraham.ac.uk/projects/fastqc/. Accessed 21 July 2025

67. Prjibelski A, Antipov D, Meleshko D, Lapidus A, Korobeynikov A (2020) Using SPAdes de novo assembler. Curr Protoc Bioinformatics 70(1):e102. https://doi.org/10.1002/cpbi.102

68. Altschul SF, Madden TL, Schäffer AA et al (1997) Gapped BLAST and PSI-BLAST: a new generation of protein database search programs. Nucleic Acids Res 25:3389–3402. https://doi.org/10.1093/nar/25.17.3389

69. Ratnasingham S, Herbert PDN (2007) Bold: the barcode of life data system (http://www.barcodinglife.org). Mol Ecol Notes 7:355–364. https://doi.org/10.1111/j.1471-8286.2007.01678.x

70. Zhao Y, Wei W, Lee I-M et al (2009) Construction of an interactive online phytoplasma classification tool, iPhyClassifier, and its application in analysis of the peach X-disease phytoplasma group (16SrIII). Int J Syst Evol Microbiol 59:2582–2593. https://doi.org/10.1099/ijs.0.010249-0

71. Brown ME, Ottati S, Trivellone V (2025) A non-destructive, fast, inexpensive, non-toxic chelating resin-based DNA extraction protocol for insect voucher specimens and associated microbiomes. J Insect Sci 25(3):17. https://doi.org/10.1093/jisesa/ieaf062

Chapter 10

Insect Cell Cultures as Platforms for the Study of Insect Vectors' Physiology and Vector–Phytoplasma Interactions

Rosemarie Tedeschi

Abstract

Insect cell cultures are a useful tool to study cell biology and physiology, metabolic pathways, pathogen infection, and their transmission by insect vectors, as well as cellular immune responses. Specifically, cell cultures of phytoplasma vectors offer a controlled environment to investigate vector–pathogen interactions at the cellular level. Furthermore, the effects of phytoplasmas on vector physiology and biology can be analyzed, promoting understanding of vector competence and thus supporting the development of targeted strategies for controlling these vectors. This chapter describes optimized protocols for the establishment and maintenance of both primary cell culture and cell lines.

Key words Primary cell cultures, Cell lines, Immunocytes, Vector–phytoplasma interactions, Immune responses, Insecticide screening

1 Introduction

Cell culture is the process by which cells are isolated from living eukaryotic organisms and grown in vitro under controlled conditions, outside their natural environment. In vitro cultivation of insect cells dates back more than a century, with the first attempts by Goldschmidt and Wilhelm [1], who cultivated sperm cells from *Samia cecropia* L. (Lepidoptera: Saturniidae), directly in drops of hemolymph of the moth. In this way, the authors were able to observe the entire process of spermatogenesis until spermatozoa were obtained. Over the years, the technique has evolved and has been applied to investigate various entomological aspects. On the one hand, the development of several new culture media based on hemolymph composition of different insect species allowed maintaining insect cells alive for a longer period [2–5]. These advances have made it possible to move from primary cell cultures to cell lines. The former consist of cells derived from a tissue or organ explant; they are only able to duplicate for a limited number of

Katrin Janik and Mattia Tabarelli (eds.), *Phytoplasma: Methods and Protocols*, Methods in Molecular Biology, vol. 3008, https://doi.org/10.1007/978-1-0716-5104-9_10,

passages, after which they undergo senescence and die. On the contrary, cell lines, which are obtained from a stabilized primary culture that has undergone the process of immortalization by chemical or viral agents, consist of cells that have acquired a degree of genotypic and phenotypic uniformity in the population and are able to replicate for an unlimited number of passages. Although immortalized cells are suitable for conducting research in controlled environments, proteome profiling reveals that immortalized cell lines can have significant rearrangement of metabolic pathways and up-regulation in cell cycle-associated genes [6]. In any case, despite their shorter lifespan, primary cell cultures can provide significant results for several types of experiments, since they tend to retain in vivo characteristics to a greater extent than continuous cell lines [7–9].

As mentioned before, primary cell culture refers to the initial stage of the culture after cells are isolated from the tissue. At this stage, proliferating cells can be subcultured (i.e., passaged) by transferring them to a new flask/dish with fresh medium. If the cell culture continues to grow and is passaged at least ten times, it is considered a continuously replicating cell line. As these cultures are passaged, cells with the highest growth capacity predominate, resulting in a degree of genotypic and phenotypic uniformity in the population. If a subpopulation of a cell line is positively selected from the culture, by cloning or some other method, this cell line becomes a cell strain, characterized often by additional genetic changes [10].

The applications of cell culture in entomological research are many. Insect cell cultures are used to study cell biology and physiology, metabolic pathways, virus infection, and their transmission by insect vectors, as well as cellular immune responses [6, 11, 12]. Most interest has focused on mechanisms underlying cellular and humoral immunity and the interaction with microorganisms [13–15].

Even in the context of insect vectors of phytoplasmas, the opportunities are numerous. Insect cell cultures offer a controlled environment to investigate interactions between insect vectors and phytoplasmas at the cellular level. They are a valuable way to study the intracellular behavior of phytoplasmas, how they invade insect cells, and how they manage to evade insect immune responses. Moreover, insect cell cultures may help analyze the effects of phytoplasmas on vector physiology and biology, such as changes in gene expression, immune modulation, and impacts on reproduction. All this information is critical for understanding vector competence (i.e., the ability of an insect to acquire, harbor, and transmit phytoplasmas), as well as for developing targeted control strategies for these vectors, potentially leading to innovative pest management solutions that mitigate the spread of phytoplasma diseases in crops.

For instance, primary cultures of insect vector immunocytes allow detailed analyses of the interactions between phytoplasmas and the cellular immune responses of the host. Primary cell cultures of immunocytes have been developed from the three psyllid species *Cacopsylla melanoneura* (Förster), *Cacopsylla pyri* (L.) (vectors of '*Candidatus* Phytoplasma mali' and '*Candidatus* Phytoplasma pyri,' respectively) and *Cacopsylla crataegi* (Schrank) (Hemiptera, Psyllidae), whose role in transmitting phytoplasmas has never been demonstrated, although it may be capable of harboring them [16]. Three different cell types were observed, and phagocytosis activity was confirmed for all psyllid cell cultures through functionality tests [15], setting the basis for a better understanding of the interplay between symbionts and immunocytes. For instance, primary immunocyte cultures may help explain why some species are phytoplasma vectors while others are not, and may resolve contradictions concerning the vector ability of some species. Phytoplasmas are known to alter gene expression and immune signaling in their vectors, likely as a strategy to enhance their own survival and transmission [17, 18]. By using insect cell cultures, it is possible to dissect these molecular interactions more precisely, identifying potential targets for disrupting the phytoplasma-vector relationship and preventing disease transmission.

In vitro cultivation of immunocytes from *Euscelidius variegatus* (Kirschbaum) (Cicadellidae, Deltocephalinae), a leafhopper frequently used as a laboratory experimental model of insect vector of phytoplasmas, made it possible to prove their ability to synthesize antimicrobial peptides belonging to the defensin family. On the contrary, genes coding for cecropins were not found [19]. Furthermore, Gonella et al. [18] studied the expression of four genes related to major components of immunity in *E. variegatus*-cultured immunocytes from healthy individuals, as well as from specimens undergoing early- and late-phytoplasma infection, treated or not treated with different Asaia strains, a symbiotic bacterium previously demonstrated to affect phytoplasma acquisition by leafhoppers [20].

Moreover, cell lines of *E. variegatus* were established to study colonization of insect cells by phytoplasmas and, in particular, to study the cellular interactions between FD phytoplasma proteins and insect cells, to clarify the mechanisms of phytoplasma invasion in insect cells [21–24].

Finally, since insect cell lines have been used in insecticide screenings and toxicology studies [25, 26], they could help develop new strategies of disease control by blocking the phytoplasma–insect cell interaction or assist in finding new mechanisms of action for innovative insecticidal molecules against the insect vectors.

2 Materials

2.1 Basic Cell Culture Laboratory Equipment

1. Room ultraviolet air purifier (i.e., germodepurator) to sterilize the working environment.
2. Laminar flow hood or biological safety cabinet (to manage reagents and cells in sterile conditions).
3. Stereo microscope (for insect dissection).
4. Inverted microscope with 10×, 20×, and 40× objectives (for cell observation).
5. Mechanical pipetting devices.
6. Refrigerated incubator capable of maintaining cells at a stable temperature of 24–28 °C.
7. A small brush for insect manipulation and forceps, sleeved needles and tweezers for insect dissections.
8. Sterilizer (e.g., glass beads) for forceps, sleeved needles and tweezers.
9. Small Petri dishes for insect washing.
10. Filter paper for drying insects after washing.
11. Tissue culture dishes or multi-well plates for cell growing.
12. Controlled-rate freezer or specialized test tube containers able to achieve a rate of cooling very close to −1 °C/min.

2.2 Reagents

1. 70% Ethanol.
2. 1.15% and 0.115% Sodium hypochlorite solutions.
3. Ultrapure, sterile water.
4. Culturing media (*see* **Note 1**).
5. 0.05–0.25% trypsin solution.
6. Liquid nitrogen and dimethyl sulfoxide (DMSO) or glycerol for cultured cell cryopreservation.

2.3 Supplements

1. 200 mM L-Glutamine solution 200 mM (*see* **Note 2**).
2. Antibiotics: gentamicin (at a final concentration of 50 μg/mL) + penicillin/streptomycin (at a final concentration of 50 U/mL and 50 μg/mL, respectively) + nystatin (100 U/mL) or amphotericin B (0.5 μg/mL) (*see* **Note 3**).

3 Methods

While specific protocols have been developed for each study, the fundamental procedures outlined below can be applied broadly to all cases.

3.1 Hood Preparation and Sterile Handling

1. Wear a lab coat and gloves.
2. Turn on the environmental UV.
3. Sweep the internal surfaces of the hood with 70% ethanol. Start from the back and proceed forward.
4. Close the hood and switch on the UV light for at least 30 min.
5. After 30 min, switch the UV light off and start the laminar airflow. Wait for 30–40 min (*see* **Note 4**).
6. Spray 70% ethanol onto all the instruments before putting them inside the hood (pipettes, reagent bottles, racks, etc., even the bits of parafilm that are used to seal bottles) and let them dry under the hood.
7. Spray down your gloves regularly with 70% ethanol.
8. Keep containers closed unless adding to or taking anything from them.
9. Open sterile equipment or culture dishes only inside the hood.
10. Bring only the items you need for a particular procedure into the hood, to prevent cluttering your working space (*see* **Note 5**).
11. Always seal bottles before leaving the hood and store them in the refrigerator or freezer wrapped in foil.
12. Do not cover the hood grate to prevent disrupting the airflow.
13. Always keep liquids tilted, both when opening and closing them and when withdrawing the desired volume. Take the exact volumes, do not reload the excess into the stock solutions.
14. Be careful not to run your arms or various objects over open bottles, caps, plates, etc.
15. Whenever you have left the hood, wash your gloves with EtOH 70% before re-entering.
16. When working under the laminar flow hood, avoid leaving the hood frequently to discard pipettes, parafilm, filters, or other waste.

3.2 Insect Dissection

1. Turn on the sterilizer, clean with 70% ethanol, and sterilize the sleeved needles and tweezers (at least two to have spares) for 30 s.
2. Clean a small brush first with 1.15% sodium hypochlorite and then with 70% ethanol.
3. Disinfect the insect surface by applying the following serial washing procedure: 10 s in 75% ethanol, then 10 s in 0.115% sodium hypochlorite, followed by 30 s in 75% ethanol, and finally rinse in ultrapure, sterile water for 20 s. Perform the

washings in small Petri dishes, completely submerging a single individual (*see* **Note 6**).

4. Dry on a filter paper for some seconds.
5. Dissect the insect with sterile sleeved needles, forceps or squeezers, directly into a tissue culture dish or a multi-well plate (*see* **Note 7**).

3.3 Initiation and Maintenance of Primary Cell Cultures

1. Regardless of the tissue you want to start from, remove all remaining body fragments from the dissection and incubate the plates at 24–28 °C (generally 27 °C is near optimum for many insects).
2. After 16–24 h, observe the cells at the inverted microscope and repeat regularly (daily or every other day, *see* **Note 8**).
3. After approximately 1 week, remove a part of spent culture and replace it by new fresh medium.
4. Continue to observe the cells regularly, daily or every other day (*see* **Note 9**).

3.4 Subculture of the Cells

When cells have covered most of the available surface (i.e., ca. 80%), a subculture may be attempted. Passaging/subculturing refers to diluting cells back to a density that maintains log phase growth and maximum viability. Different approaches can be followed depending on the force with which the cells are attached to the surface.

1. *Non-attached or loosely attached cells:* draw an appropriate volume of medium with a sterile pipette and spray it across the cell surface to dislodge them, then transfer the cell suspension to a new cell culture plate (or flask) with fresh medium.
2. *Strongly attached cells:* cool the culture at 4 °C for 20 min to depolymerize microtubules that are important for cell attachment, then give the culture a couple of taps with the palm of your hand to encourage the cells to loosen, then proceed as described in the previous point.
3. *Very strongly attached cells:* an enzymatic treatment should be used (trypsinization).

3.5 Establishment and Maintenance of Cell Lines

Cell lines can be established from primary cells or can be bought from commercial or non-profit suppliers (i.e., cell banks, *see* **Note 10**). For their maintenance, the same instructions given for primary cultures apply, with the need to dilute them to maintain log phase growth and maximum viability (*see* **Note 11**).

1. Regularly check cell viability by staining the cells with trypan blue, which enters non-viable cells only, and thus, dead cells become blue. Cell viability should be at least 95% for healthy log-phase cultures (*see* **Note 12**).

2. Replace or refresh medium every 2–3 days or as needed, especially after passaging.

   Once the cell lines are established and are doubling regularly, they can be deep frozen (below −70 °C) for future studies (*see* **Note 13**).

4 Notes

1. Culture media for insect cells must contain the same nutrients found in media for mammalian cell cultures (carbohydrates, amino acids, and salts), but at concentrations adapted to insect cell metabolism [27, 28]. Furthermore, insect media are generally more acidic (ranging in pH from 6.2 to 6.9) and buffered with sodium phosphate, so that a CO_2 incubator is not required for insect cell culture. A further difference is that the osmotic pressure is generally more than twice as high in insect cell culture media than the typical osmolarity of media for mammalian cells. Several commercial media for insect cells are available. They were optimized for some orders of insects, but they can be adapted, individually or in combination, also for other orders. However, developing new cell culture protocols for insect species often involves multiple trials and errors before an optimal culture medium is established. For Hemiptera species, several different kinds of cell culture media were developed, such as the Mitsuhashi and Maramorosch's medium (MM's medium) [29] for leafhoppers, Kimura's medium [30] for planthoppers, Ex-Cell 401 (JRH Bioscience) [31], and Ex-Cell 405 (JRH Bioscience) [32] for sharpshooters, the HH70 medium and the CLG#2 supplemented with heat-inactivated FBS [15, 33, 34] for psyllids. Some other media, originally designed for other insects, resulted suitable for leafhoppers, such as Schneider's Drosophila medium and Grace's medium [23]. In the last decade, serum-free media have become dominant in insect cell culture in order to overcome the drawbacks of serum (cost, lot-to-lot variability, potential adventitious agents or contaminants, interference with product purification, etc.). Moreover, in response to increasingly stringent regulations, there is a strong demand to reduce the use of animal-derived components. Supplements designed to replace serum and other animal-derived ingredients are protein hydrolysates (peptones) of plant or microbial origin.
2. L-Glutamine solution is added as a major energy source for better growth of insect cells.
3. Antibiotics are required for primary insect cell cultures. On the contrary, it is not appropriate to use antibiotics in continuous cell lines to avoid the development of resistance to them. In any

case, if any form of resistance is suspected during a cell culture, it is necessary to discard the culture immediately.

4. In the meantime, thaw reagents and prepare the material needed for the next steps.
5. Having a clear working space will significantly reduce the chance of contamination. Ensure easy access to items in the hood and maintain plenty of free space in the center of the hood to work in.
6. The volume of each reagent depends on the size of the insect. It should be enough to fully immerse the specimens (e.g., 4–5 mL in a 60 mm Petri dish is sufficient for an adult psyllid or leafhopper).
7. The choice of tissue to start with depends on the size of the insect, which affects the difficulty of dissecting certain organs, and on the type of study to be conducted. For example, cell cultures can start from embryos, especially if you want to start new cell lines (after decontamination of the egg surface and egg disruption directly in the medium), but also from the fat body, imaginal discs, midguts, and nerve tissue. To study the cellular immune system, cell cultures can start isolating hemocytes, releasing the insect hemolymph directly into the medium after a small cut in the insect body (without affecting the gut), and shaking it in the medium.
8. The culture medium should always appear clear. Turbidity or mold that prevent cells from being observed are signs of bacterial and fungal contamination, respectively. In such cases, the entire culture must be autoclaved and discarded. Cell growth may be visible immediately or after some weeks; normally, mitosis requires a few days to start, as cells adapt to the new environment.
9. Regular cell observation is of fundamental importance not only to follow the regular development of the cells of interest, but also to detect any abnormalities, such as contaminations or granularity around the nucleus, detachment of the cells from the substrate, or cytoplasmic vacuolation, which are generally signs of cell deterioration. These can be consequences of senescence, the presence of toxic substances, or the need for a medium change.
10. In order to simplify the work of researchers, an online database describes the currently known cell lines from different insect taxonomic orders developed from agricultural pests and includes insect cell lines from insects that feed on crops (https://entomology.ca.uky.edu/aginsectcellsdatabase). The designation, tissue and species of origin, cell line developer, unique characteristics, its use in various applications, publications, and patents, and, when known, insect virus susceptibility

are provided in that database. Each cell line is described there with its distinct characteristics, including growth rate and protein expression capabilities.

11. Always use a different bottle of culture medium for each cell line, in order to avoid cross-cell contamination. Moreover, only handle one cell line at a time. So, at any given time in the hood, you should have only one parent culture, one new flask (already labeled with the cell identity), one bottle of medium (and a container with the enzyme, if using one), along with a pipettor and a pipette.
12. Trypan blue molecules are excluded from viable cell membranes but enter non-viable cells. Consequently, viable cells will remain unstained, while non-viable cells will take up the blue dye.
13. The most effective way to cryopreserve cultured cells is by storing them in liquid nitrogen using complete medium supplemented with a cryoprotectant like dimethyl sulfoxide (DMSO) or glycerol. Cryoprotective agents lower the medium's freezing point and enable a more gradual cooling process, significantly minimizing the formation of ice crystals that can harm cells and lead to cell death. To ensure optimal preservation, freeze-cultured cells at high density, before freezing, verify that cell viability is at least 90%, as freezing less viable cultures can compromise recovery. Slow cooling is critical: reduce the temperature gradually at a rate of approximately 1 °C per min. This can be achieved using a controlled-rate freezer or specific freezing containers.

References

1. Goldschmidt R (1915) Some experiments on spermatogenesis in vitro. Proc Natl Acad Sci USA 1:220–222
2. Trager W (1935) Cultivation of the virus of grasserie in silkworm tissue cultures. J Exp Medic 61(4):501
3. Wyatt G, Loughheed T, Wyatt SS (1956) The chemistry of insect hemolymph; organic components of the hemolymph of the silkworm, *Bombyx mori*, and two other species. J Gen Physiol 39:853–868. https://doi.org/10.1085/jgp.39.6.853
4. Grace TDC (1962) Establishment of four strains of cells from insect tissue grown in vitro. Nature 195:788–789. https://doi.org/10.1038/195788a0
5. Schneider I (1964) Differentiation of larval drosophila eye-antennal discs in vitro. J Exp Zool 156:91–104. https://doi.org/10.1002/jez.1401560108
6. Arunkarthick S, Asokan R, Aravintharaj R et al (2017) A review of insect cell culture: establishment, maintenance and applications in entomological research. J Entomol Sci 52(3): 261–273. https://doi.org/10.18474/0749-8004-52.3.261
7. Craven RA, Stanley AJ, Hanrahan S et al (2006) Proteomic analysis of primary cell lines identifies protein changes present in renal cell carcinoma. Proteomics 6:2853–2864. https://doi.org/10.1002/pmic.200500574
8. Merkley MA, Hildebrandt E, Podolsky RH et al (2009) Large-scale analysis of protein expression changes in human keratinocytes immortalized by human papilloma virus type 16 E6 and E7 oncogenes. Proteome Sci 7:29. https://doi.org/10.1186/1477-5956-7-29
9. Pan C, Kumar C, Bohl S et al (2009) Comparative proteomic phenotyping of cell lines and primary cells to assess preservation of cell type-

specific functions. Mol Cell Proteomics 8:443–450. https://doi.org/10.1074/mcp.M800333-MCP200
10. Mandrioli M, Monti M, Tedeschi R (2015) A practical guide to insect cell cultures: establishment and maintenance of primary cell cultures. Halteres 6:132–141
11. Pinto CPG, Rickes LN, Zotti MJ et al (2019) Compared activity of agonist molecules towards ecdysone receptor in insect cell-based screening system. Arq Inst Biol 86. https://doi.org/10.1590/1808-1657000692018
12. Ghosh A, Dhall H, Dietzgen RG et al (2020) Insect cell culture as a tool in plant virus research: a historical overview. Phytoparasitica 48(2):287–303. https://doi.org/10.1007/s12600-019-00729-x
13. Fallon AM, Sun D (2001) Exploration of mosquito immunity using cells in culture. Insect Biochem Mol Biol 31(3):263–278. https://doi.org/10.1016/S0965-1748(00)00102-9
14. Smagghe G, Goodman CL, Stanley D (2009) Insect cell culture and applications to research and pest management. In Vitro Cell Dev Biol Anim 45(3):93–105. https://doi.org/10.1007/s11626-009-9194-0
15. Monti M, Mandrioli M, Bextine B et al (2014) Maintenance of primary cell cultures of immunocytes from *Cacopsylla* spp. psyllids: a new in vitro tool for the study of crop pest insects. In Vitro Cell Dev Biol Anim 50(9):797–801. https://doi.org/10.1007/s11626-014-9756-9
16. Tedeschi R, Lauterer P, Bertignono L et al (2008) Hawthorn psyllid fauna in northwestern Italy. Bull Insectol 61(1):143–144
17. Galetto L, Abbà S, Rossi M et al (2018) Two phytoplasmas elicit different responses in the insect vector *Euscelidius variegatus* Kirschbaum. Infect Immun 86(5):10–1128. https://doi.org/10.1128/IAI.001128-18
18. Gonella E, Mandrioli M, Tedeschi R et al (2019) Activation of immune genes in leafhoppers by phytoplasmas and symbiotic bacteria. Front Physiol 10:795. https://doi.org/10.3389/fphys.2019.00795
19. Tedeschi R, Monti M, Gonella E et al (2017) Molecular and cellular analysis of immunity in the phytoplasma vector *Euscelidius variegatus*: exploiting immunity to improve biological control strategies. Invertebr Surviv J 14(1):63–72
20. Gonella E, Crotti E, Mandrioli M et al (2018) *Asaia* symbionts interfere with infection by Flavescence dorée phytoplasma in leafhoppers. J Pest Sci 91:1033–1046. https://doi.org/10.1007/s10340-018-1014-0
21. Arricau-Bouvery N, Duret S, Dubrana MP et al (2018) Variable membrane protein A of flavescence dorée phytoplasma binds the midgut perimicrovillar membrane of *Euscelidius variegatus* and promotes adhesion to its epithelial cells. Appl Environ Microbiol 84(8):e02487–17. https://doi.org/10.1128/AEM.02487-17
22. Arricau-Bouvery N, Duret S, Dubrana MP et al (2021) Interactions between the flavescence dorée phytoplasma and its insect vector indicate lectin-type adhesion mediated by the adhesin VmpA. Sci Rep 11(1):11222. https://doi.org/10.1038/s41598-021-90679-5
23. Arricau-Bouvery N, Dubrana MP, Canuto F et al (2023) Flavescence dorée phytoplasma enters insect cells by a clathrin-mediated endocytosis allowing infection of its insect vector. Sci Rep 13(1):2211. https://doi.org/10.1038/s41598-023-29458-5
24. Canuto F, Duret S, Dubrana MP et al (2023) A knockdown gene approach identifies an insect vector membrane protein with leucine-rich repeats as one of the receptors for the VmpA adhesin of flavescence dorée phytoplasma. Front Cell Infect Microbiol 13:1289100. https://doi.org/10.3389/fcimb.2023.1289100
25. Mak M, Beattie KD, Basta A et al (2021) Triangulation of methods using insect cell lines to investigate insecticidal mode-of-action. Pest Manag Sci 77(1):492–501. https://doi.org/10.1002/ps.6045
26. He X, Lu L, Huang P et al (2023) Insect cell-based models: cell line establishment and application in insecticide screening and toxicology research. Insects 14(2):104. https://doi.org/10.3390/insects14020104
27. Lynn DE (2002) Methods for maintaining insect cell cultures. J Insect Sci 2(1):9. https://doi.org/10.1673/031.002.0901
28. Lynn DE (2007) Routine maintenance and storage of lepidopteran insect cell lines and baculoviruses. In: Murhammer DW (ed) Baculovirus and insect cell expression protocols. Methods in molecular biology, vol 388. Humana Press. https://doi.org/10.1007/978-1-59745-457-5_9
29. Mitsuhashi J, Maramorosch K (1964) Leafhopper tissue culture: embryonic, nymphal, and imaginal tissues from aseptic insects. Contr Boyce Thompson Inst 22:435–460
30. Kimura I (1984) Establishment of new cell lines from leafhopper vector and inoculation of its cell monolayers with rice dwarf virus. Proc Jpn Acad Ser B 60(6):198–201. https://doi.org/10.2183/pjab1984.60.198

31. Kamita SG, Nagasaka K, Chua JW et al (2005) A baculovirus-encoded protein tyrosine phosphatase gene induces enhanced locomotory activity in a lepidopteran host. PNAS 102(7): 2584–2589. https://doi.org/10.1073/pnas.0409672102
32. Boyapalle S, Pal N, Miller WA et al (2007) A glassy-winged sharpshooter cell line supports replication of *Rhopalosiphum padi* virus (Dicistroviridae). J Invertebr Pathol 94(2):130–139. https://doi.org/10.1016/j.jip.2006.11.002
33. Marutani-Hert M, Hunter WB, Hall DG (2009) Establishment of Asian citrus psyllid (*Diaphorina citri*) primary cultures. In Vitro Cell Dev Biol Anim 45:317–320. https://doi.org/10.1007/s11626-009-9196-y
34. Wu K, Ortgiesen GJ, Goodman CL et al (2023) Optimized conditions for the long-term growth of primary cell cultures derived from the Asian citrus psyllid, *Diaphorina citri* (Liviidae: Hemiptera). In Vitro Cell Dev Biol Anim 59(4):235–240. https://doi.org/10.1007/s11626-023-00772-3

Part III

Functional Characterization and Experimental Systems: Experimental Approaches to Understand Phytoplasma Biology

Chapter 11

Protocols for Grapevine Transformation: Tissue Preparation, Protoplast Isolation, and Cultivation

Rebeka Strah, Edoardo Bertini, Sara Zenoni, Maruša Pompe Novak, Mario Pezzotti, and Marina Dermastia

Abstract

Grapevine is among the most important economic plants that are susceptible to phytoplasma disease. Tools and procedures for genetic manipulation of grapevine are cumbersome and sometimes difficult to replicate; therefore, most of the research on plant–phytoplasma interactions focuses on species with more established transformation procedures instead, even though they might not be hosts of the studied phytoplasma. Studying the mechanisms of plant–phytoplasma interactions directly in the host plants would bring new and more accurate insights into the disease mechanisms. In this chapter, we present the protocols to obtain transformation-competent grapevine material from different tissues and the procedures for protoplast isolation and cultivation.

Key words Grapevine, Transformation, Transfection, Embryogenic, Callus, Protoplast

1 Introduction

Phytoplasmas are notoriously difficult to maintain outside the host tissue, limiting experimental approaches to study their biology. Consequently, most research on phytoplasmas is conducted in host plants. This research has been extended through the production of transgenic plants, which are currently the most viable option. Plants can be genetically modified to either alter their susceptibility to phytoplasma infection [1–3] or to express different phytoplasma proteins to study their possible role in plant–phytoplasma interaction [4].

While procedures for transient and stable transformation of various leafy plants are well established, woody species like grapevine (*Vitis vinifera* L.) still present numerous challenges. Despite these challenges, improving the transformation of the grapevine is

Rebeka Strah and Edoardo Bertini contributed equally with all other contributors.

Katrin Janik and Mattia Tabarelli (eds.), *Phytoplasma: Methods and Protocols*, Methods in Molecular Biology, vol. 3008, https://doi.org/10.1007/978-1-0716-5104-9_11,

highly desirable. Experimentally, working directly with the target species provides more reliable results, while economically, new genetic technologies offer the potential to develop resistance traits without compromising the oenological properties of a grapevine variety. Although producing stably transformed grapevine by co-cultivating embryogenic callus with *Agrobacterium tumefaciens* is laborious and time-consuming, this method yields consistent results [5, 6]. However, obtaining embryogenic callus is neither efficient nor possible for all varieties, and usually depends on collecting grapevine flowers—a process that is both time-consuming and seasonal.

Several studies have tested various grapevine varieties for their ability to generate embryogenic callus, revealing that success rates depend heavily on the combination of cultivar and induction medium used [7–9]. In addition, maintaining callus cultures and propagating the material for transformation experiments is laborious and slow.

In an effort to achieve faster results, grapevine researchers have turned to protoplast isolation methods. Protoplasts have been successfully obtained from leaves [10, 11], berries [12], and embryogenic callus [13–15], with the choice of tissue depending mainly on the main purpose of the experiment. Leaves are often used for transient expression experiments due to their relative ease of acquisition compared to embryogenic callus, which remains the preferred tissue for generating transgenic plants, as it is currently the only type of grapevine tissue from which whole plants can reliably be regenerated from a single cell.

This chapter provides details and procedures for the extraction of transformation- and regeneration-competent tissues from grapevine, protoplast isolation, transfection, and regeneration procedures that have been successfully used by our research team. These techniques are valuable not only for molecular biology research on grapevine but also for precision breeding efforts.

2 Materials

2.1 Explant Sources

1. Dormant 1-year-old canes.
2. In vitro grapevine tissue culture.

2.2 Equipment

1. Plant growth chamber set to 22 °C and a 16 h photoperiod.
2. Forceps.
3. Sterile and non-sterile needles (21G, 40 mm for work with callus and 50 mm for work with flower explants).
4. Scalpels (No. 11).
5. Stereomicroscope (magnification about 1.5×).

6. Microscope (magnification 20× and 40×).
7. Fluorescent microscope (magnification 20× and 40×).
8. Inverted microscope (magnification 20× and 40×).
9. Autoclave.
10. pH meter.
11. Laminar airflow sterile culture hood.
12. Bead sterilizer or ethanol burner.
13. Gyrating shaker.
14. Micropipettes (sizes 10–100 μL and 100–1000 μL).
15. Cell counting chambers.
16. Plate incubator (27 °C).
17. Centrifuge (with holders to centrifuge 15 mL tubes and option to adjust acceleration and deceleration).

2.3 Chemicals

1. Acetocarmine.
2. Sterile water.
3. 50 mg/mL Fluorescin diacetate (FDA) diluted in acetone.
4. 70% ethanol.
5. 3 g/L Indoleacetic acid.
6. 0.55 M sucrose.
7. 3% sodium hypochlorite.
8. Tween 20.

2.4 Consumables

1. Sterile 50- and 15-mL tubes.
2. Pots for planting grapevine canes.
3. Plant substrate (universal for potted leafy plants).
4. Microscope slides and cover glass.
5. Sterile 90 mm and 35 mm petri dishes.
6. 50 mL syringe.
7. Sterile tissue culture pots or tubes (e.g., the Magenta™ GA-7 plant culture box).
8. 0.2 μm filter.
9. Sterile micropipette tips, intact and cut (sizes 10–100 μL and 100–1000 μL).
10. Sterile nylon filter (mesh size between 50 and 100 μm).
11. Sterile 2 mL tubes.

2.5 Culture Media

1. Embryogenic callus induction medium MSII: mix in distilled water: 4.43 g/L Murashige and Skoog macroelements, 0.103 g/L Murashige and Skoog microelements, 0.103 g/L Murashige and Skoog vitamins, 20 g/L sucrose, 5 μM 6-benzylaminopiruine (BAP), 2.5 μM 2,4-dichlorophenoxyacetic acid (2,4-D) 2.5 μM naphthoxyacetic acid (NOA), 7 g/L tissue culture agar [8] (*see* **Notes 1**–**4**). Sterilize by autoclaving at 121 °C for 15–20 min.
2. Embryogenic callus induction medium PIV: mix in distilled water: 1.99 g/L Nitsch macroelements, 0.103 g/L Murashige and Skoog microelements (see medium MSII), 0.112 g/L B5 vitamins, 60 g/L sucrose, 4.5 μM 2,4-D, 8.9 μM BAP, and 3 g/L phytagel [9, 16] (*see* **Notes 5** and **6**). Sterilize by autoclaving at 121 °C for 15–20 min.
3. Embryogenic callus induction medium NB2: mix in distilled water: 1.99 g/L Nitsch macroelements (see medium PIV), 0,063 g/L Nitsch microelements, 0.107 g/L Nitsch vitamins, 60 g/L sucrose, 5 μM 2,4-D, 1 μM BAP, and 7 g/L tissue culture agar [16] (*see* **Notes** 7 and **8**). Sterilize by autoclaving at 121 °C for 15–20 min.
4. Embryogenic callus propagation medium C1P: mix in distilled water: 2.22 g/L Murashige and Skoog macroelements, 51 mg/L Murashige and Skoog microelements (see medium MSII), 0.122 g/L T vitamins (*see* **Note 9**), 0.112 g/L amino acid mix (*see* **Note 10**), 22 mM NaFeEDTA, 1 g/L casein hydrolysate, 30 g/L sucrose, 5 μM 2,4-D, 1 μM BAP, and 5 g/L phytagel [17]. Sterilize by autoclaving at 121 °C for 15–20 min.
5. Embryo development medium ×6: mix in distilled water: 6.1 g/L modified MS macroelements (*see* **Note 11**), 0.103 g/L Murashige and Skoog microelements (see MSII medium), 0.101 g/L modified Murashige and Skoog vitamins (*see* **Note 12**), 60 g/L sucrose, 1 g/L caseine hydrolysate, 0.05 g/L activated charcoal, and 7 g/L tissue culture agar [18]. Sterilize by autoclaving at 121 °C for 15–20 min.
6. Protoplast solid medium: mix in distilled water: Nitsch medium (see medium NB2) with 30 g/L sucrose, 54 g/L glucose (*see* **Note 13**), 2 mg/L 1-naphtaleneacetic acid (NAA), 0.5 mg/L BAP, 2 g/L gellan gum. Sterilize by autoclaving at 121 °C for 15–20 min.
7. Protoplast liquid medium: this medium has the same composition as the protoplast solid medium, except for foregoing the gellan gum, and the addition of 3 g/L activated charcoal [19]. Sterilize by autoclaving at 121 °C for 15–20 min.

8. Germination medium: mix in distilled water: Nitsch medium (see medium NB2), 30 g/L sucrose, and 2 g/L gellan gum. Sterilize by autoclaving at 121 °C for 15–20 min.
9. Shooting medium MG1: mix in distilled water: 1.99 g/L Nitsch macroelements (see medium PIV), 0.103 g/L Murashige and Skoog microelements (see medium MSII), 22 mM NaFeEDTA, 0.112 g/L vitamins B5 (see medium PIV), 30 g/L sucrose, 2.5 g/L activated charcoal, and 7 g/L tissue culture agar. Add BAP according to the cultivar of choice [20]. Sterilize by autoclaving at 121 °C for 15–20 min.
10. Shooting medium C2D: mix in distilled water: 4.9 g/L C2D macroelements, 80 mg/L C2D microelements, 13 mg/L C2D vitamins (*see* **Notes 14–16**), 30 g/L sucrose, and 7 g/L tissue culture agar. Add BAP according to the cultivar of choice [21]. Sterilize by autoclaving at 121 °C for 15–20 min.

2.6 Protoplast Isolation and Transfection

1. Digestion solution for callus: combine in distilled water: 2% (w/v) Cellulase Onozuka RS, 1% (w/v) Macerozyme R-10, 0.05% (w/v) Pectolyase Y-23, 1470.1 mg/L $CaCl_2$ $2H_2O$, 97.2 mg/L 2-(*N*-morpholino) ethanesulfonic acid (MES), and 91.09 g/L mannitol [15, 19]. Filter-sterilize through a 0.22 μm filter.
2. Digestion solution for leaves: 1.5% (w/v) Cellulase R-10, 0.4% (w/v) Macerozyme R-10, 0.1% (w/v) bovine serum albumin (BSA), 147.01 mg/L $CaCl_2 \cdot 2H_2O$, 9304 mg/L MES, 72.83 g/L mannitol, and 1491.03 mg/L KCl [10]. Filter-sterilize through a 0.22 μm filter.
3. Protoplast wash solution: 147.01 mg/L $CaCl_2 \cdot 2H_2O$ and 91.09 mg/L mannitol. Filter-sterilize through a 0.22 μm filter.
4. W5 solution: 390.4 mg/L MES, 9 g/L NaCl, 54 g/L glucose, 372.76 mg/L KCl, and 55.88 g/L $CaCl_2 \cdot 2H_2O$ [15]. Filter-sterilize through a 0.22 μm filter.
5. MMG solution: 780.8 mg/L MES, 72.83 g/L mannitol, and 1428.16 mg/L $MgCl_2$.
6. PEG solution: 40% (w/v) PEG 4000, 72.83 g/L mannitol, and 14.701 g/L $CaCl_2 \cdot 2H_2O$ (*see* **Note 17**). Filter-sterilize through a 0.22 μm filter.

3 Methods

Here, we describe the procedures for obtaining transfection-competent material, isolating protoplasts, and performing transfection. All steps, including surface sterilization, preparation of explants, propagation, and protoplast isolation, should be performed using sterile tools and within a sterile environment under a laminar flow hood.

Regardless of the starting material, protoplast isolation begins by cutting the tissue into smaller pieces and incubating it in an enzyme solution for several hours to break down the cell walls and release the protoplasts. The solution is then filtered, usually through a nylon mesh with a pore size between 100 and 50 μm, depending on the desired protoplast size, and subsequently washed in a washing buffer. Optionally, a sucrose flotation step can be performed to remove cell debris. After washing, the protoplasts are transfected with a vector of choice, typically using polyethylene glycol (PEG), followed by further washing and culturing in either a liquid or solid medium, depending on the intended application.

3.1 Embryogenic Callus from Grapevine Flowers

This section describes how to obtain grapevine flowers from cuttings and use them as explants to generate embryogenic callus. Alternatively, the flowers can also be harvested from flowering field-grown plants.

3.1.1 Obtaining Grapevine Flowers

1. After grapevines shed their leaves, obtain fully lignified 1-year-old canes. Keep them at 4 °C for at least 2 weeks to ensure complete dormancy.
2. Cut the canes into cuttings, long approximately 30 cm, with fresh cuts at the top and bottom. Ensure each cut is made at least 2 cm away from the dormant bud.
3. Remove all buds except the top one. Trim 2 cm of wood from the bottom of each cutting.
4. Dip the bottoms of the cuttings into a solution of 3 g/L indoleacetic acid for 5–6 s and immediately transfer them into pots filled with the growth substrate of choice (*see* **Note 18**). Maintain the potted cuttings under light conditions with a 16-h photoperiod for 3–5 weeks to develop the inflorescence.
5. As the bud begins to open, use small tweezers to remove the first young leaves. If possible, also remove all but one inflorescence. Later in the growth process, preserve some leaves to provide the plant with enough energy for the developing inflorescence.
6. Monitor the development of the inflorescence. When it reaches stage III or IV (refer to [22] for detailed stages of development), the flowers should no longer be wrapped in hairs, with lengthened pedicels and be separate from each other (Fig. 1a). At this stage, use small tweezers to collect a few flowers from the inflorescence to confirm the pollen developmental stage (*see* **Note 19**).
7. Place a droplet of water on a Petri dish or another transparent surface and put the inflorescence in the water droplet to

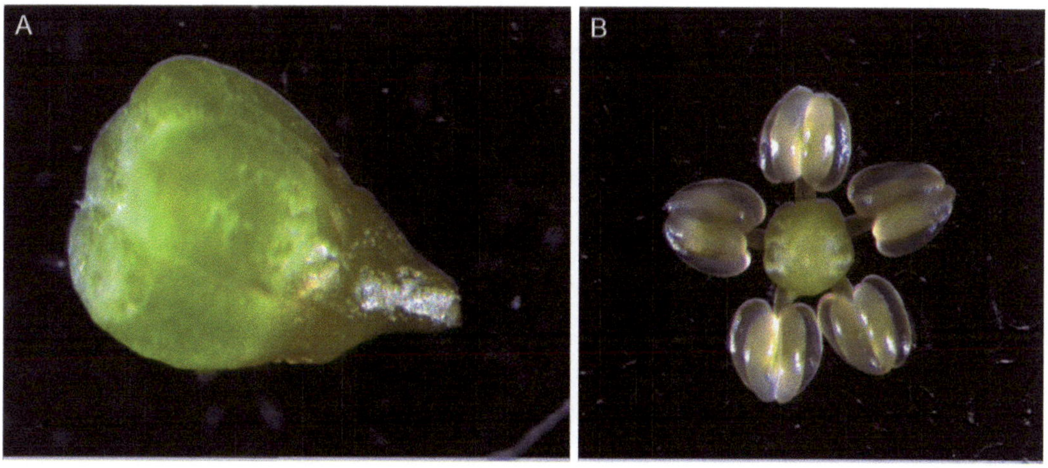

Fig. 1 Closed and opened grapevine flowers in the fourth pollen developmental stage. (**a**) An intact flower; (**b**) An opened flower. (Images were made at 15× magnification)

prevent desiccation. Under a stereomicroscope, use needles or scalpels to open the flower. Start by cutting away the pedicel, then peel away the calyptra. Add more water if the sample begins to dry. At the developmental stage VI, the anthers should appear translucent, with the two inner pollen sacks turning yellow and slightly opaque (Fig. 1b).

8. To confirm the pollen developmental stage, place the anthers on a microscope slide and crush them thoroughly. Add a drop of acetocarmine to stain the chromosomes and examine the slide under a microscope. Stage IV of pollen development is characterized by the presence of tetrads, identifiable by the occurrence of microspores separated within the anther sacs (Fig. 2).

3.1.2 Explant Preparation

1. Once the inflorescence is confirmed to be in the ideal stage of pollen development, collect it into a 50 mL sterile tube. Add approximately 40 mL of a 3% sodium hypochlorite solution and 40 μL of Tween 20 to sterilize the tissue.
2. Gently shake the tube for 10 min, then wash the inflorescence two to three times with sterile water. Place it in a sterile Petri dish and chill at 4 °C in the dark for 2 or 3 days.
3. Place a stereomicroscope inside the laminar airflow hood and surface sterilize it with 70% ethanol. Open the flowers as described in Subheading 3.1.1. Transfer the opened flowers aseptically to the induction medium MSII or PIV (*see* **Note 20**). Ensure the entire basal part of the flower is in contact with the induction medium (*see* **Note 21**).

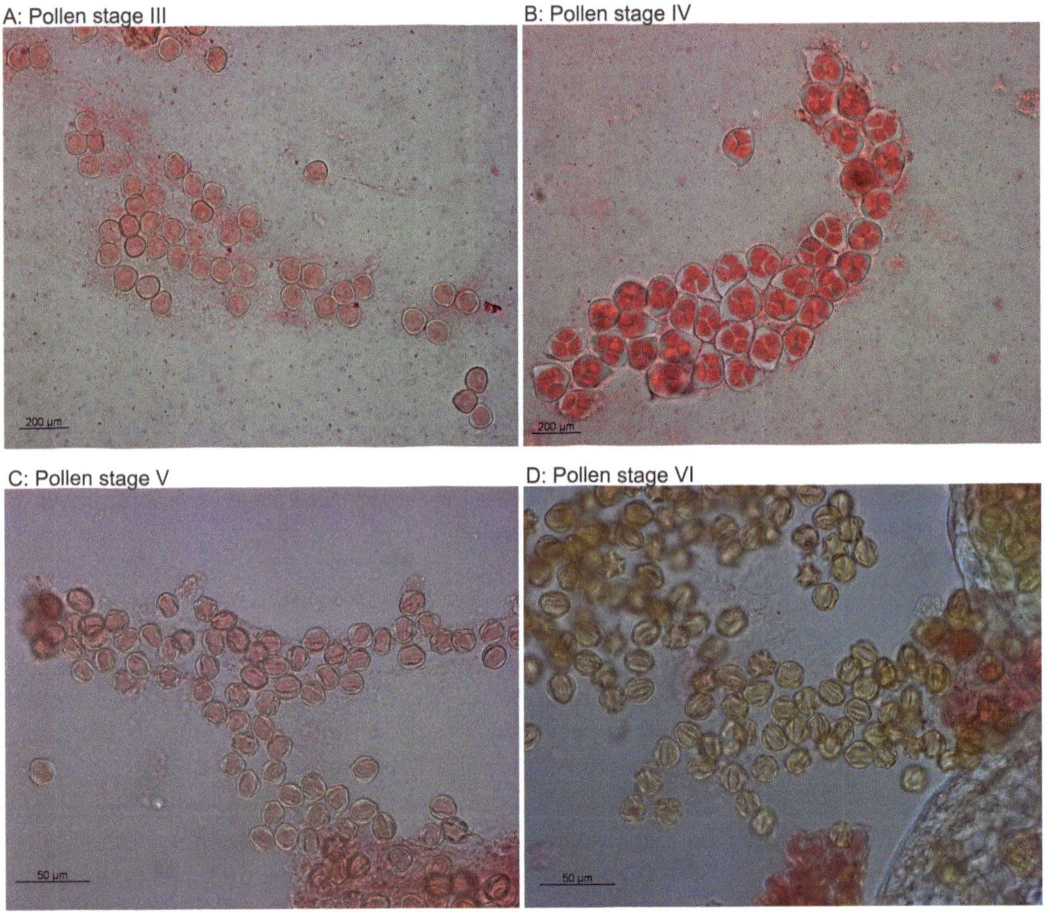

Fig. 2 Grapevine pollen developmental stages, stained with acetocarmine. (**a**) Pollen in stage III; (**b**) Pollen in stage IV with fully formed tetrads; (**c**) Pollen in stage V, in this stage, the penetration of acetocarmine into the cells diminishes; (**d**) Pollen in stage VI is mature, with no acetocarmine penetration

4. Incubate the transferred explants in darkness at 27 °C. The embryogenic callus may start appearing after 5 weeks (*see* **Note 22**).
5. Collect the newly formed callus and transfer it to the callus propagation medium C1P (Fig. 3a).

3.2 Embryogenic Callus from Leaves

Inducing embryogenic callus from tissue culture leaves can have very low success rates, depending on the grapevine variety. Therefore, prepare a large number of tissue culture plantlets in advance using standard procedures for sterile tissue culture work.

1. Using forceps and a scalpel, collect newly formed, still closed leaves from the apex of the plantlet.

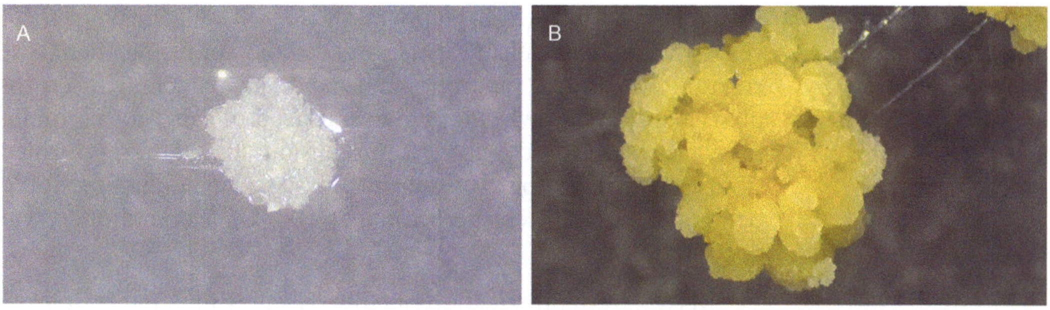

Fig. 3 Callus propagation. (**a**) Propagated embryogenic callus; (**b**) 5-week-old callus maintained on medium. (Images were made at 15× magnification)

2. Sterilely transfer the leaves onto the NB2 induction medium. Ensure that one of the adaxial sides of each leaf is fully in contact with the medium.
3. Proceed as described in **steps 4** and **5** of Subheading 3.1.2.

3.3 Maintaining and Propagating Embryogenic Callus

To maintain the embryogenic potential of the callus culture, it needs to be kept in darkness at 27 °C and propagated at regular intervals. The propagation medium should be poured in thick layers in sterile Petri dishes.

1. Place a stereomicroscope into the laminar flow hood and sterilize it using 70% ethanol.
2. Using a scalpel (preferably No. 11) and either small forceps or a sterile needle to stabilize the tissue, excise the proembryonic masses from the rest of the callus. These should be microscopic in size.
3. Collect several of the proembryonic masses together and transfer them to fresh C1P medium to create a new culture spot (Fig. 3a).
4. Manipulate the new culture spot gently with a scalpel so that the proembryonic masses remain close together while maintaining the layer as thin as possible, maximizing the contact of each individual mass with the medium underneath.
5. Repeat the procedure once every 4–6 weeks. Five-week-old propagated callus is depicted in Fig. 3b.
6. Once a larger amount of the embryogenic callus is obtained, test its embryogenic potential by propagating a portion of the material onto the ×6 embryo induction medium in the same manner as for propagation purposes.
7. Maintain the callus on ×6 medium under the same conditions as for propagation cultures. After 5 weeks, somatic embryos should form from the transferred proembryonic masses.

3.4 Protoplast Isolation from Embryogenic Callus

Protoplasts are plant cells deprived of the cell walls, making them very sensitive to mechanical stress. To avoid rupturing them, use cut tips, avoid rapid pipetting, and handle containers with protoplasts gently and slowly. Placing several layers of paper towels under tube racks is also advisable. Resuspension should be performed slowly, taking care not to create any bubbles. Centrifugation should be gentle, with all steps performed at 100× *g*, acceleration 1, and deceleration 0.

1. Make a fresh propagation of embryogenic callus and let it grow for 8 days.
2. One day before isolation, prepare the protoplast digestion and wash solutions and keep them at 4 °C until use.
3. Aseptically transfer the 8-day-old callus into a 35 mm Petri dish and add 10 mL of digestion solution per 1 g of callus tissue. Place the Petri on a gyrating shaker and agitate at about 1× *g*.
4. After 1 h of shaking, break up the larger callus pieces with a sterile needle and syringe by slowly aspirating and expelling them.
5. Continue shaking for a total of 3–6 h. Periodically check the degree of tissue degradation by sampling 10–20 μL of the digesting callus and observing it under a microscope (*see* **Note 23**).
6. Place a sterile nylon 60 μm pore size filter over a 90 mm Petri dish and tape it to the working surface on one side (*see* **Note 24**). Put 1 mL of the digestion solution on top of the filter and gently touch it to the bottom of the Petri dish to allow the liquid to pass through (*see* **Note 25**). Tape the other side of the filter so that it is stretched taut over the Petri dish.
7. When digestion is sufficient, gently pass the protoplast solution through the filter. Wash the bottom of the small Petri dish several times with the digestion solution and pass the liquid through the filter as well. Most of the protoplasts will be collected once the solution is visibly clear. Carefully remove and discard the filter.
8. Transfer the filtered solution into a 15 mL tube and wash the Petri dish as described in the previous step.
9. To wash the protoplasts, centrifuge the tube for 5 min at 100× *g*, with the acceleration set to 1 and deceleration to 0. Carefully remove the supernatant without disturbing the pellet. Resuspend the protoplasts in 5 mL of the wash solution gradually, first adding about 1 mL.
10. Repeat the wash step at least one more time.
11. Prepare a 0.55 M sucrose solution and place 5–10 mL in a tube. Gently add an overlay of protoplasts resuspended in the

wash solution on top of the sucrose by pipetting along the sides of the tube. After centrifugation, a band of viable protoplasts should appear at the interface between the sucrose and wash solution. Collect them carefully with a 200 μL tip (from which the upper, tapered part was cut off to prevent shearing) (*see* **Note 26**).

12. Centrifuge the protoplasts again and resuspend them in 2–10 mL of the wash solution. If implementing **step 11**, centrifugation is not necessary. At this point, take an aliquot to count the number of protoplasts (*see* **Note 27**) and for viability assays.
13. Count the total number of protoplasts using your preferred counting chamber according to the instructions (*see* **Note 28**). Calculate the concentration of protoplasts and the volume containing the desired number of protoplasts for your experiment (*see* **Note 29**).
14. To assess protoplast viability, dilute the 50 mg/mL FDA stock by adding 20 μL of FDA to 1 mL of protoplast wash solution. Mix 20 μL of the protoplast suspension with the diluted FDA. Incubate in the dark for 10 min, then observe using a fluorescent microscope. Cells exhibiting strong fluorescence are considered viable.
15. Distribute the required volume of protoplasts into new, separate tubes (*see* **Note 30**), centrifuge them, and remove the excess liquid.

3.5 Protoplast Isolation from Leaves

Many researchers have reported success in isolating protoplasts from grapevine leaves, which are primarily used for transient expression studies [10, 11, 23]. While we have achieved significant success with protoplast isolation and cultivation from grapevine embryogenic callus as described above, our attempts to isolate protoplasts from grapevine leaves yielded mixed results. Nevertheless, leaves are an attractive source material, because they are much easier to obtain and maintain than embryogenic callus. Here, we present our current method of protoplast isolation from leaves, with the caveat that it may produce variable results from one isolation attempt to another.

1. Aseptically collect the first opened leaves from approximately 2-month-old grapevine tissue culture plantlets and place them in a sterile petri dish.
2. Cut the leaves into thin strips using a very sharp scalpel, ensuring that the blade cuts cleanly through the tissue without tearing it.

3. Add 10 mL of protoplast digestion solution for leaves per 1 g of leaves and shake at 50 rpm for approximately 3–6 h (*see* **Note 31**).
4. Continue as described in Subheading 3.4, starting from **step 6** onwards.

3.6 Protoplast Transfection and Cultivation

3.6.1 Protoplast Transfection

1. Add 200 μL of MMG solution to each protoplast aliquot and gently resuspend them.
2. Add 20–60 μg of the selected transfection vector (*see* **Note 32**) and resuspend again.
3. Add an equal volume of PEG and mix gently but thoroughly.
4. Incubate the mixture in the dark at room temperature for 5–10 min.
5. To stop the reaction, add 5 mL of W5 solution. Centrifuge and discard the supernatant.
6. Repeat **step 5** at least one more time.

3.6.2 Protoplast Cultivation

1. Centrifuge the protoplasts and remove as much supernatant as possible.
2. Add 1 mL of warm solid protoplast medium and resuspend the protoplasts (*see* **Notes 33** and **34**).
3. Transfer the protoplast suspension to a 35 mm Petri dish, depositing it in a drop at the center. Allow the medium to solidify and cool completely.
4. Once the medium is solidified, add 1 mL of liquid protoplast medium containing glucose.
5. Incubate in darkness at 27 °C.
6. For protoplasts transfected with a fluorescent proteins expression system, observe results on the second or third day of cultivation.

3.6.3 Long-Term Protoplast Cultivation and Regeneration

Protoplasts can be cultivated for an extended period depending on the experiment goals. Protoplasts from embryogenic callus can be reliably regenerated into embryos, although this is challenging with protoplasts regenerated from somatic tissues.

1. Continue the cultivation of protoplasts as described in the previous section.
2. Weekly, replace the liquid medium of the protoplast culture. After the first 2 weeks of cultivation, switch to liquid medium without added glucose.
3. For protoplast regeneration into embryos, monitor the cell growth and division over time using an inverted microscope. Embryos may become visible to the naked eye after approximately 1–2 months of cultivation.

4 Notes

1. The recipes for macroelements, microelements, and vitamins of different plant tissue culture media refer to premade formulations that can be bought or prepared as stock solutions. Detailed compositions are listed in Notes below.
2. Murashige and Skoog macroelements: 1650 mg/L NH_4NO_3, 440 mg/L $CaCl_2 \times \cdot 2H_2O$, 180.7 mg/L $MgSO_4 \cdot \times 7H_2O$, 170 mg/L KH_2PO_4, 1900 mg/L KNO_3.
3. Murashige and Skoog microelements: 6.2 mg/L H_3BO_3, 0.025 mg/L $CoCl_2 \cdot \times 6H_2O$, 27.8 mg/L $FeSO_4 \cdot \times 7H_2O$, 22.3 mg/L $MnSO_4 \cdot \times 4H_2O$, 0.83 mg/L KI, 0.25 mg/L $Na_2MoO_4 \cdot \times 2H_2O$, 8.6 mg/L $ZnSO_4 \cdot \times 7H_2O$, 36.70 mg/L FeNa-EDTA, 0.025 mg/L $CuSO_4 \cdot \times 5H_2O$.
4. Murashige and Skoog vitamins: 100 mg/L myo-Inositol, 0.5 mg/l nicotinic acid, 0.5 mg/L pyridoxine HCl, 0.1 mg/L thiamine HCl, 2 mg/L glycine.
5. Nitsch macroelements: 68 mg/L KH_2PO_4, 166 mg/L $CaCl_2$, 720 mg/L NH_4NO_3, 950 mg/L KNO_3, 90.27 mg/L $MgSO_4$).
6. B5 vitamins: 100 mg/L myo-Inositol, 1 mg/L nicotinic acid, 1 mg/L pyridoxine HCl, 10 mg/L thiamine HCl.
7. Nitsch microelements: 0.025 mg/L $CuSO_4 \cdot 5H_2O$, 36.7 mg/L FeNa-EDTA, 6.2 mg/L H_3BO_3, 18.94 mg/L $MnSO_4 \cdot H_2O$.
8. Nitsch vitamins: 0.05 mg/L biotin, 0.5 mg/L folic acid, 2 mg/L glycine, 100 mg/L myo-Inositol, 5 mg/L nicotinic acid, 0.5 mg/L pyridoxin HCl, 0.5 mg/L thiamin HCl.
9. T vitamins: 100 mg/L myo-Inositol, 10 mg/L nicotinic acid, 10 mg/L thiamine HCl, 1 mg/L pyridoxine HCl, 1 mg/L D-panthotenic acid, 0.1 mg/L biotin.
10. Amino acid mix: 100 mg/L L-glutamic acid, 10 mg/L L-phenylalanine, 2 mg/L glycine.
11. Modified Murashige and Skoog macroelements: 440 mg/L $CaCl_2 \cdot 2H_2O$, 180.7 mg/L $MgSO_4 \cdot 7H_2O$, 170 mg/L KH_2PO_4, 1900 mg/L KNO_3, 3033 mg/L KNO_3, 364 mg/L NH_4Cl.
12. Modified Murashige and Skoog vitamins: 100 mg/L myo-Inositol, 0.5 mg/L nicotinic acid, 0.5 mg/L pyridoxine HCl, 0.1 mg/L thiamine HCl.
13. Glucose is sensitive to heat degradation and must not be autoclaved. Instead, prepare a liquid stock solution, sterilize it using a 0.2 μm filter, and add it to the autoclaved medium in a sterile environment.

14. C2D macroelements: 1650 mg/L NH_4NO_3, 370 mg/L $MgSO_4 \cdot 7H_2O$, 170 mg/L KH_2PO_4, 1900 mg/L KNO_3, 709 mg/L $Ca(NO_3)_2 \cdot 4H_2O$, 27.8 mg/L $FeSO_4 \cdot 7H_2O$, 37.3 mg/L Na_2EDTA.
15. C2D microelements: 6.2 mg/L H_3BO_3, 0.025 mg/L $CoCl_2 \cdot 6H_2O$, 27.8 mg/L $FeSO_4 \cdot 7H_2O$, 0.845 mg/L $MnSO_4 \cdot 4H_2O$, 0.25 mg/L $Na_2MoO_4 \cdot 2H_2O$, 8.6 mg/L $ZnSO_4 \cdot 7H_2O$, 36.70 mg/L FeNa-EDTA, 0.025 mg/L $CuSO_4 \cdot 5H_2O$.
16. C2D vitamins: 1 mg/L thiamine HCl, 10 mg/L myo-Inositol, 1 mg/L nicotinic acid, 1 mg/L pyridoxine HCl.
17. Always prepare the PEG solution fresh on the day of transfection.
18. To induce root development, place cuttings in flasks with distilled water as an alternative to using growth regulators.
19. In a grapevine inflorescence, flowers develop in a gradient, with basal flowers at a later stage than those at the apex. Typically, flowers are sampled from the bottom third of the inflorescence, and only this portion is used to prepare explants.
20. The efficiency of an embryogenic callus induction depends on the specific medium and grapevine variety combination. If information for your variety is unavailable, we recommend testing both induction media and exploring other media reported in the literature.
21. During flower opening, some stamens will separate from the flower. Embryogenic callus is most often induced from filaments attached to the anther. Only transfer the flowers and stamens with intact anthers, as seen in Fig. 4.
22. Examine the explants for the appearance of embryogenic callus every 2 weeks, as callus formation can occur up to 5 months after initial cultivation. Embryogenic callus consists of proembryonic masses, which appear as clusters of small globules on the surface of the callus. These are white or light yellow, compact, and slightly translucent (Fig. 5).
23. Prolonged exposure to the protoplast digestion solution can damage the protoplasts. Aim for a balance between digestion and protoplast viability, ensuring a high number of perfectly spherical cells with minimal visibly burst cells and cell debris.
24. Instead of nylon filter cloth, filters that fit into 50 mL tubes can be used. Place the filter directly on top of the tube and tilt the tube to allow the filtered solution to run down the sides instead of dripping from the top, which helps maintain protoplast viability. Although spherical bottom tubes are reported to be better for maintaining protoplast viability during subsequent

Fig. 4 An opened grapevine flower with two detached stamens. (The image was made at 15× magnification)

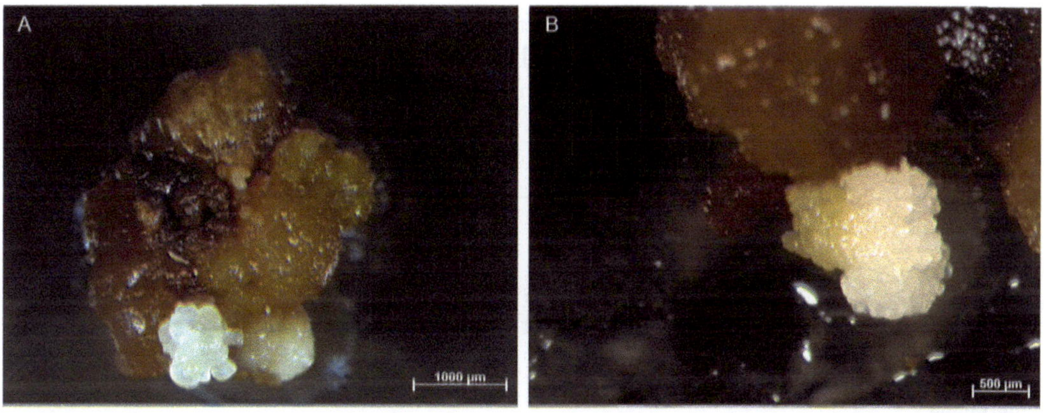

Fig. 5 Callus induction. (**a**, **b**) Embryogenic callus induced from grapevine flowers at different magnification

centrifugation steps, we did not encounter issues with conical-bottom tubes.

25. Wetting the nylon mesh beforehand helps the solution with protoplasts pass through more easily. Gently touch the mesh; it may take several attempts to get the first drop of fluid through.
26. This step is optional depending on the purpose of protoplast isolation and the amount of cell debris generated.
27. Protoplasts settle very quickly. Prepare aliquots immediately after resuspension or resuspend again if the solution has stood for a while.

28. Use a Neubauer hemocytometer or other counting chambers to count cells. Due to varying levels of cell debris and clumps, which depend on the previous steps and the size of the filter mesh, ensure only single, perfectly spherical cells are counted.
29. For simple protoplast cultivation, use 1×10^5 cells per culture. For transfection, use 1.5 or 2×10^5 cells to account for potential cell mortality. You may adjust these numbers based on specific experimental needs.
30. Use either 15 mL or 2 mL tubes. If using 2 mL tubes, avoid safe-lock types and open them carefully to prevent damaging the protoplasts.
31. Choose a Petri dish size that prevents overcrowding of leaf material, as excessive crowding can reduce digestion efficiency.
32. Add a minimal volume of transfection vector to avoid excessive dilution of the MMG solution. The optimal volume is 50 μL or less.
33. For protoplast cultivation, we tested various liquid media, including MMG, W5, and protoplast wash solution. Due to inconsistent results with these media, we prefer using a solid medium for protoplast culture.
34. Ensure the solid medium is not hot to the touch. Because small volumes of media cool down and solidify rapidly, this step should be performed swiftly, but still gently to avoid rupturing the protoplasts.

Acknowledgement

This chapter includes data from the project J4-2544, funded by the Slovenian Agency for Research and Innovation (ARIS).

References

1. Sung Y-C, Lin C-P, Hsu H-J et al (2019) Silencing of CrNPR1 and CrNPR3 alters plant susceptibility to periwinkle leaf yellowing phytoplasma. Front Plant Sci. https://doi.org/10.3389/fpls.2019.01183
2. Thomazella DPDT, Seong K, Mackelprang R et al (2021) Loss of function of a DMR6 ortholog in tomato confers broad-spectrum disease resistance. Proc Natl Acad Sci USA 118. https://doi.org/10.1073/pnas.2026152118
3. Pirrello C, Zeilmaker T, Bianco L et al (2021) Mining grapevine downy mildew susceptibility genes: a resource for genomics-based breeding and tailored gene editing. Biomolecules 11. https://doi.org/10.3390/biom11020181
4. Dermastia M, Tomaž Š, Strah R et al (2023) Candidate pathogenicity factor/effector proteins of '*Candidatus* Phytoplasma solani' modulate plant carbohydrate metabolism, accelerate the ascorbate–glutathione cycle, and induce autophagosomes. Front Plant Sci. https://doi.org/10.3389/fpls.2023.1232367
5. Fujita K, Matsuoka T, Suzuki S, Takayanagi T (2009) In planta transformation technique for grapevines (*Vitis vinifera* L) using dormant buds. J Plant Biochem Biotechnol 18:161–167. https://doi.org/10.1007/BF03263314

6. Zhang X, Wu Y, Li Z et al (2021) Advancements in plant regeneration and genetic transformation of grapevine (Vitis spp.). J Integr Agric 20:1407–1434. https://doi.org/10.1016/S2095-3119(20)63586-9
7. Gambino G, Ruffa P, Vallania R, Gribaudo I (2007) Somatic embryogenesis from whole flowers, anthers and ovaries of grapevine (*Vitis* spp.). Plant Cell Tissue Organ Cult 90:79–83. https://doi.org/10.1007/s11240-007-9256-x
8. Dhekney S, Li Z, Compton M (2009) Optimizing initiation and maintenance of Vitis embryogenic cultures. HortScience 44. https://doi.org/10.21273/HORTSCI.44.5.1400
9. Franks T, Gang He D, Thomas M (1998) Regeneration of transgenic shape *Vitis vinifera* L. Sultana plants: genotypic and phenotypic analysis. Mol Breed 4:321–333. https://doi.org/10.1023/A:1009673619456
10. Zhao F-L, Li Y-J, Hu Y et al (2016) A highly efficient grapevine mesophyll protoplast system for transient gene expression and the study of disease resistance proteins. Plant Cell Tissue Organ Cult 125:43–57. https://doi.org/10.1007/s11240-015-0928-7
11. Katsirdakis KC, Roubelakis-Angelakis KA (1992) Modified culture conditions for increased viability and cell wall synthesis in grapevine (*Vitis vinifera* L. cv. Sultanina) leaf protoplasts. Plant Cell Tissue Organ Cult 28:255–260. https://doi.org/10.1007/BF00036121
12. Wang H, Wang W, Zhan J et al (2015) An efficient PEG-mediated transient gene expression system in grape protoplasts and its application in subcellular localization studies of flavonoids biosynthesis enzymes. Sci Hortic 191:82–89. https://doi.org/10.1016/j.scienta.2015.04.039
13. Bertini E, Tornielli GB, Pezzotti M et al (2019) Regeneration of plants from embryogenic callus-derived protoplasts of Garganega and Sangiovese grapevine (*Vitis vinifera* L.) cultivars. Plant Cell Tissue Organ Cult 138:239–246. https://doi.org/10.1007/s11240-019-01619-1
14. Scintilla S, Salvagnin U, Giacomelli L et al (2022) Regeneration of non-chimeric plants from DNA-free edited grapevine protoplasts. Front Plant Sci 13. https://doi.org/10.3389/fpls.2022.1078931
15. Najafi S, Bertini E, D'Incà E et al (2023) DNA-free genome editing in grapevine using CRISPR/Cas9 ribonucleoprotein complexes followed by protoplast regeneration. Hortic Res 10. https://doi.org/10.1093/hr/uhac240
16. Nitsch JP, Nitsch C (1969) Haploid plants from pollen grains. Science 163:85–87. https://doi.org/10.1126/science.163.3862.85
17. Torregrosa L (1998) A simple and efficient method to obtain stable embryogenic cultures from anthers of *Vitis vinifera* L. Vitis 37(2):91–92
18. Li J, Wang X, Wang X, Wang Y (2014) Embryo rescue technique and its applications for seedless breeding in grape. Plant Cell Tissue Organ Cult 120:861–880. https://doi.org/10.1007/s11240-014-0656-4
19. Zhu Y-M, Hoshino Y, Nakano M et al (1997) Highly efficient system of plant regeneration from protoplasts of grapevine (*Vitis vinifera* L.) through somatic embryogenesis by using embryogenic callus culture and activated charcoal. Plant Sci 123:151–157. https://doi.org/10.1016/S0168-9452(96)04557-8
20. Iocco P, Franks T, Thomas MR (2001) Genetic transformation of major wine grape cultivars of *Vitis Vinifera* L. Transgenic Res 10:105–112. https://doi.org/10.1023/A:1008989610340
21. Chee R, Pool R, Bucher D (1984) A method for large scale in vitro propagation of vitis. N Y Food Life Sci Bull 109. https://hdl.handle.net/1813/5140
22. Gribaudo I, Gambino G, Vallania R (2004) Somatic embryogenesis from grapevine anthers: the optimal developmental stage for collecting explants. Am J Enol Vitic 55:427–430. https://doi.org/10.5344/ajev.2004.55.4.427
23. Brehm I, Preisig-Müller R, Kindl H (1999) Grapevine protoplasts as a transient expression system for comparison of stilbene synthase genes containing cGMP-responsive promoter elements. Z Naturforsch C J Biosci 54:220–229. https://doi.org/10.1515/znc-1999-3-412

Chapter 12

Expression of Recombinant Phytoplasma Proteins in *E. coli*

Kajohn Boonrod

Abstract

Since phytoplasmas cannot be cultured in vitro, it is impossible to study a protein of interest in vivo by manipulating its corresponding gene. It is therefore necessary to recombinantly express a protein of interest for use in an in vitro study. Although recombinant expression of proteins in *E. coli* is well established, there are some issues to consider when expressing recombinant phytoplasma proteins. Here, we describe a detailed method for the recombinant expression of an effector protein, SAP11, from '*Candidatus* Phytoplasma mali' in *E. coli*. Some important points are highlighted.

Key words Protein expression, Recombinant protein, *E. coli*, Phytoplasma protein, Effector

1 Introduction

As phytoplasma cannot be cultivated in vitro, to study the function of phytoplasma proteins, it is necessary to express the target protein in plant models or in other microorganisms to obtain the recombinant protein for further studies. *Escherichia coli* (*E. coli*) has long been a workhorse for recombinant protein production. Despite huge advances over other expression systems, the production of heterologous recombinant proteins in microbial expression systems remains simpler and less expensive than in alternative systems such as mammalian cell culture [1, 2]. *E. coli* offers several advantages. These include the ease of genetic manipulation, the use of a simple growth medium, and rapid cell growth. It can also be up-scaled using a simple fermentation process. In addition, recombinant protein from *E. coli* is a virus-free product, has high product yields, and is cost-effective to produce [3].

Many factors need to be considered before a recombinant protein can be expressed in *E. coli*. These include promoter, fusion tag for detection and purification, bacterial strain, temperature after induction, agitation, and codon usage [3]. If the protein is to be used for biochemical analysis, it is necessary to keep the protein in

Katrin Janik and Mattia Tabarelli (eds.), *Phytoplasma: Methods and Protocols*, Methods in Molecular Biology, vol. 3008,
https://doi.org/10.1007/978-1-0716-5104-9_12,

its natural form to maintain its activity. Therefore, expressing the target protein in the periplasmic space of *E. coli* by fusing the target gene with a periplasmic leader sequence (e.g., pelB) can improve its solubility and function [4, 5]. Alternatively, fusing the target protein to a tag such as glutathione S-transferase (GST), maltose-binding protein (MBP), N-utilization substance A (NusA), thioredoxin (TRX), or adding an additive during induction can improve protein solubility [6, 7]. However, a soluble protein may lack the desired activity if it requires specific conditions, a particular pH, a reducing environment, or certain ion concentrations. Therefore, these factors must be optimized for each protein to achieve its optimal activity. After expression, the recombinant protein needs to be purified. Therefore, the purification process should be considered prior to cloning the target gene into a desired expression vector. His-tags are commonly used to facilitate purification of the target protein in both native and denatured forms. GST and MBP-fused proteins can only be purified under native conditions. The MBP tag is used to increase the solubility of the target protein [8, 9] while the GST tag is widely used to prepare a protein for later use in pull-down assays [10]. The details of using these tags are well described [11].

For expressing phytoplasma proteins in *E. coli*, the most concern is correct codon usage. It was shown that the codon content modulates a kinetic competition between protein elongation and mRNA degradation that is a central feature of the physiology and possibly the regulation of translation in *E. coli* [12]. Moreover, it was shown that the G + C content of the genes that are sought to be expressed should be between 30% and 70%. Any value outside this range could adversely affect transcriptional and translational efficiency [13]. Although phytoplasma is a cell-wall-less bacteria, the genome analysis of four different phytoplasmas indicates that they are very different from *E. coli* [14]. Phytoplasmas have a G + C content of less than 28% [14] and possibly use different codons than *E. coli*. Thus, it is worth to redesign the gene of interest using optimized codon usage and G + C content for recombinant expression of a phytoplasma protein in *E. coli*. This strategy has been successfully applied to express an immunodominant protein (Imp) and an effector protein (SAP11) of '*Candidatus* Phytoplasma mali' in our laboratory [15–17]. Several phytoplasma proteins contain transmembrane domain(s). For most studies, the expression of a transmembrane protein in *E. coli* often leads to the formation of protein aggregates or inclusion bodies and the use of denatured proteins is limited to certain downstream applications [18]. To overcome this obstacle, many strategies have been developed [18]. Strategies to express transmembrane proteins, such as immunodominant proteins, were well described [19, 20]. In this chapter, we describe a general protocol to express phytoplasma proteins and address different factors that may affect the expression of the target protein.

2 Materials

2.1 Bacterial Strain and Plasmid

1. *E. coli* strain BL21(DE3): Any *E. coli* strain that carries the T7 polymerase gene can be used when the gene will be expressed under the control of a T7 promoter.
2. pMAL-c2X expression plasmid: Different expression plasmids can be used for the expression of a protein in *E. coli*. For this work, we used the pMAL-c2X plasmid.

2.2 Bacteria Culture, Induction, and Protein Extraction

1. 2× YT liquid medium: 16 g tryptone, 10 g yeast extract, 5 g NaCl. Mix all ingredients and add water to 1 L. Sterilize by autoclaving. Store at room temperature. The medium is stable for several months.
2. 2× YT agar medium: 16 g tryptone, 10 g yeast extract, 5 g NaCl, 15 g agar. Mix all ingredients and add water to 1 L. Sterilize by autoclaving. Store at room temperature. The medium is stable for several months.
3. Antibiotic solution: Dissolve the corresponding antibiotic (i.e., for the selection of bacteria successfully transformed with the expression plasmid) as recommended by the manufacturer. It is recommended to prepare a 100× concentrated stock solution.
4. 1 M Isopropylthio-β-galactoside (IPTG): Weigh 2.38 g of IPTG. Add 10 mL sterile water. Dissolve completely. Store at −20 °C. The solution is stable for several months.
5. Soluble protein extracting buffer: 1% (w/v) n-octyl-β-D-thio-glucopyranoside (OTG) in 50 mM Tris-HCl pH 7.5. Add 0.05% (w/v) sodium azide (NaN_3). Store at room temperature. The solution can be stored for several months.
6. DNase I: Dissolve lyophilized DNase I (1500 units) in 550 μL RNase-free water. Mix gently by inverting. Do not vortex. Divide it into 1 mL single-use aliquots, and store at −20 °C. The solution can be stored for at least 1 year.
7. RNase: Dissolve 100 mg RNase in 100 μL 10 mM Tris–HCl pH 7.6 and add 10 mL water. Boil for 10 min. Cool down the RNase solution, make 1 mL aliquots, and store at −20 °C. The RNase solution can be stored for at least 1 year.
8. 8 M Urea: Dissolve 480.48 g of urea in 1 L of distilled water.

2.3 SDS–PAGE

1. 30% Acrylamide gel solution: Aqueous solution containing 29.22% (w/v) acrylamide and 0.78% (w/v) bis-acrylamide for a total concentration of 30% (w/v) of (bis-)acrylamide and a

monomer to crosslinker ratio of 37.5:1. Store at 4 °C. The solution can be stored for at least 1 year.

2. 3.5× Gel buffer: 1.25 M Bis-Tris. Adjust pH to 6.5–6.8 with HCl. Store at room temperature. The buffer is stable for several months.
3. N,N,N′,N′-Tetramethylethylendiamin (TEMED): Store at 4 °C.
4. Ammonium persulfate (APS): 10% (w/v) solution in water, Store at 4 °C. The solution can be stored for several months.
5. 4× Laemmli buffer: Weigh 1.54 g Tris-base and 4 g sodium dodecyl sulfate (SDS) and dissolve both in 20 mL glycerol and 10 mL ß-mercaptoethanol. Add 200 mg bromophenol blue. Mix all components and add water to 50 mL. Prepare 2 mL aliquots. Leave one aliquot at 4 °C for current use and store the remaining aliquots at −20 °C. The frozen aliquots are stable for several months.
6. Isopropanol: Stored at room temperature.
7. 5× Protein loading buffer: 10% (w/v) SDS, 50% (v/v) glycerol, 250 mM Tris-HCl, 0.5% (w/v) bromophenol blue. Mix all components and adjust pH to 6.8. Store at room temperature. The buffer can be stored for several months.
8. 1 M Sodium bisulfite: Dissolve 5.2 g sodium bisulfite in 50 mL distilled water. Store at room temperature. The solution can be stored for several months.
9. 5× Running buffer for low-molecular-weight (MW) proteins: use for separating small proteins with a molecular weight between 2 and 50 kDa. 250 mM 2-(*N*-morpholino) ethanesulfonic acid (MES), 250 mM Tris-HCl, 5 mM ethylenediaminetetraacetic acid (EDTA), 0.5% (w/v) sodium dodecyl sulfate (SDS). Store at room temperature. The buffer can be stored for several months. Use 1× running buffer for electrophoresis by diluting 200 mL 5× running buffer with 800 mL deionized water and add 5 mL of 1 M sodium bisulfite into the 1× running buffer directly before use.
10. 5× Running buffer for high-MW proteins: use for separating proteins having a molecular weight over 20 kDa. 250 mM 3-(*N*-morpholino) propanesulfonic acid (MOPS), 250 mM Tris-HCl, 5 mM EDTA, 0.5% (w/v) SDS. Store at room temperature. The buffer is stable for several months. Use 1× running buffer for electrophoresis by diluting 200 mL 5× running buffer with 800 mL deionized water and add 5 mL of 1 M sodium bisulfite into the 1× running buffer directly before use.
11. Coomassie brilliant blue staining solution: Mix 0.2–0.3% (w/v) Coomassie R-250, 10% (v/v) methanol, 10% (v/v)

acetic acid. Add water to 1 L. Store at room temperature. The solution can be stored for several months.

12. Destaining solution: 0.5 M NaCl. Store at room temperature. The solution can be stored for several months.
13. Ponceau S solution: 5% (v/v) glacial acetic acid and 0.1% (w/v) Ponceau S. Store at room temperature. The solution can be stored for several months.
14. Pre-stained protein marker: Store at −20 °C.

2.4 Equipment

1. Incubator with shaker and cooling temperature (14–37 °C).
2. Mini protein gel electrophoresis apparatus.
3. Power supply for electrophoresis chamber.
4. Centrifuge: Benchtop centrifuge with fixed rotor maximum speed up to 10,000× *g* for 1.5–2 mL reaction tubes.
5. Heat block: For reaction tubes 1.5 mL and operating temperature up to 100 °C.
6. Low-speed rotary shaker.

2.5 Consumables

1. Reaction tubes: 1.5 mL.
2. Petri dishes: 90 mm × 15 mm.
3. Erlenmeyer flasks: 50 mL.
4. Conical Sterile Polypropylene Centrifuge Tubes: 50 mL.

3 Methods

3.1 Protein Expression

1. Optimize codons of the target gene for the expression in *E. coli* and clone the gene in the destination vector (*see* **Note 1**).
2. Transform the plasmid into *E. coli* expressing strain (e.g., BL21 (DE3)+ or T7 express). Plate the transformed bacteria on 2× YT agar supplemented with the corresponding antibiotic.
3. Inoculate a colony of *E. coli* carrying the expression plasmid from the 2× YT agar plate in 10 mL 2× YT medium supplemented with the corresponding antibiotic overnight at 37 °C, 300 rpm (*see* **Note 2**).
4. Dilute the overnight culture 1:100 in 2× YT medium supplemented with the appropriate antibiotic to a final volume of 20 mL (e.g., add 20 μL bacterial overnight culture to 20 mL 2× YT medium). Incubate the cell mixture at 37 °C, 300 rpm until an OD_{600} of 0.4–0.5 is reached (*see* **Note 3**). Induce the cells with IPTG by adding 20 μL 1 M IPTG to get a final concentration of 1 mM IPTG. Divide the cell mixture into two shaker flasks with 10 mL culture broth in each (*see* **Note 4**). Incubate one of the two cultures overnight at 14 °C,

160 rpm, while the other for 3 h at 37 °C, 300 rpm (*see* **Note 5**). Collect 500 μL of each induced bacterial culture and centrifuge at 10,000× *g* for 5 min. Discard the supernatant and store the pellets at −20 °C or proceed with the protein extraction from the pellet.

3.2 Protein Extraction

1. Add 100 μL soluble protein extracting buffer, 10 μL DNase, and 10 μL RNase to the bacterial pellet to digest the DNA and RNA (*see* **Note 6**). Vortex the mixture vigorously. Incubate the mixture at 37 °C for 30 min and then centrifuge at 10,000× *g* for 5 min. Collect the supernatant for soluble protein analysis and extract the aggregated proteins (inclusion bodies) by adding 100 μL of 4× Laemmli buffer to the pellet (*see* **Note 7**).
2. Aliquot 20 μL of each extract for SDS–PAGE analysis.

3.3 SDS–PAGE

1. Assemble the protein gel apparatus.
2. Prepare a 10% separating gel mixture as indicated in Table 1.
3. Pour (or pipette) the gel mixture into the space between two glass plates.
4. Pipette 1 mL of isopropanol to even out the gel surface and allow the gel to polymerise for 15 min.
5. Turn the gel cassette upside-down to remove the isopropanol.
6. Prepare the collecting gel mixture as indicated in Table 2.
7. Pipette the mixture into the space, place a comb, and allow the gel to polymerize for 15 min.
8. Put the gel in the running chamber and add the running buffer.
9. Pull the comb out slowly and wash the gel slot using a 1 mL pipette to get rid of the non-polymerized gel mixture (optional).

Table 1
Ingredients of the separating gel

Reagent	10%	12.50%	15%	17.50%
30% Acrylamide gel solution	2.1 mL	2.6 mL	3.2 mL	3.7 mL
Gel buffer	1.6 mL	1.6 mL	1.6 mL	1.6 mL
Water	1.95 mL	2 mL	1.5 mL	1 mL
TEMED	8 μL	8 μL	8 μL	8 μL
10% APS	60 μL	60 μL	60 μL	60 μL

For separation of high molecular weight proteins (over 80 kDa), 10% gel is recommended, while a 12.5–15% is suitable for proteins with a molecular weight of 80–20 kDa and a 17.50% is suitable for very low molecular weight proteins (<20 kDa)

Table 2
Ingredients of the collecting gel

30% Acylamine gel solution	310 μL
Gel buffer	310 μL
Water	1.8 mL
TEMED	5 μL
10% APS	25 μL

10. Prepare the samples by adding 5 μL of 5× protein loading buffer to each protein sample (20 μL) and boiling at 95 °C in a heat block for 5 min. Add 2 μL of ß-mercaptoethanol (*see* **Note 8**) and load the sample into a gel slot.
11. Electrophorese the gel by applying 170 V for 45–60 min.

3.4 Staining and Destaining the Gel with Coomassie Brilliant Blue

1. After the SDS–PAGE, carefully remove the gel and put it into a container. Add the Coomassie Brilliant Blue staining solution to cover the gel (*see* **Note 9**). Stain the gel for at least 30 min, then remove the staining solution and rinse the stained gel with water.
2. Add destaining solution to cover the gel. Destain the gel on a low-speed rotary shaker until defined protein bands are clearly visible (Fig. 1, *see* **Note 10**).

4 Notes

1. To optimize codon usage, e.g., the following program can be used: https://www.novoprolabs.com/tools/codon-optimization. If the gene is to be synthesized, most manufacturers offer an "optimize codon usage" service.
2. The addition of an antibiotic (e.g., chloramphenicol) is suggested to select bacteria carrying a lambda DE3 prophage carrying the gene for T7 RNA polymerase under the control of a LacUV5 promoter, e.g., in the bacterial strain BL21(DE3). If the bacterial strain carries the T7 RNA polymerase in the lac operon (no prophage) (e.g., T-Express, New England Biolabs), no additional antibiotic is required to select the bacterial strain.
3. Bacteria grown to an $OD_{600} \geq 0.5$ are not recommended to be used for induction.
4. A high concentration of IPTG could reduce the expression of soluble protein. Low concentrations of IPTG could increase the protein solubility, since the protein will be expressed slowly.

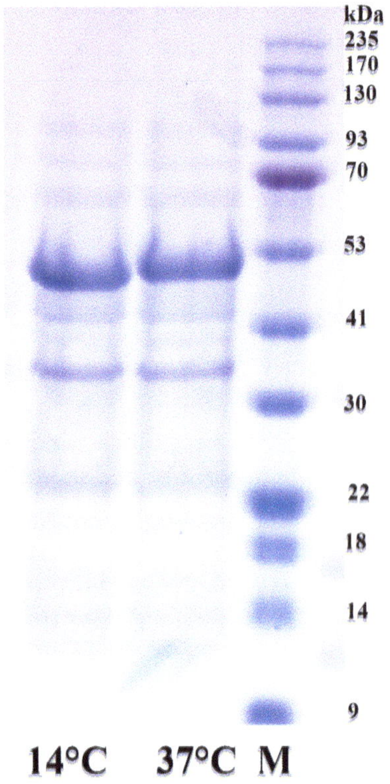

Fig. 1 Recombinant expression of SAP11 fused with MBP in *E. coli*. SAP11 with optimised codon usage for expression in *E. coli* BL21(DE3)+ was synthesised and cloned into the pMal-c2X vector. The expression of SAP11 from this expression vector leads to an MBP–SAP11 fusion protein. Bacterial cells were induced with IPTG and incubated at 37 °C and 14 °C for 3 h and overnight, respectively. The cells were harvested, and the fraction containing soluble proteins was analysed on SDS–PAGE and stained with Coomassie Brilliant Blue. The dominant band at approx. 52 kDa is MBP–SAP11 (left lane: after induction at 14 °C overnight; middle lane: after induction at 37 °C for 3 h; right lane: size marker)

5. The temperature during induction can affect the solubility of the expressed protein. At low temperatures, bacterial cells grow and express the target protein slowly, which could reduce the aggregation of the protein. Therefore, the optimal temperature for soluble protein expression should be carefully determined.
6. Any nuclease can be used instead of DNase/RNase. Ultrasound can also be used to destroy the nucleic acid–protein complex.
7. As the Laemmli Buffer contains a high concentration of SDS, most proteins will be denatured and solubilised. If proteins from inclusion bodies will be used for further examination, it is advisable to resolubilize them at this stage using 8 M urea.

8. The addition of a reducing agent such as ß-mercaptoethanol will destroy the protein dimer. To detect dimerization or multimerization of the target protein, the protein sample should be electrophoresed in parallel with and without the addition of a reducing agent.
9. If the gel will be further used for blotting, staining with a removable dye such as Ponceau S is recommended.
10. The use of destaining buffers containing acetic acid and methanol is faster than 0.5 M NaCl, but 0.5 M NaCl is much cheaper, non-toxic, and more environmentally friendly.

References

1. Tripathi NK, Shrivastava A (2019) Recent developments in bioprocessing of recombinant proteins: expression hosts and process development. Front Bioeng Biotechnol 7:420. https://doi.org/10.3389/fbioe.2019.00420
2. Pouresmaeil M, Azizi-Dargahlou S (2023) Factors involved in heterologous expression of proteins in *E. coli* host. Arch Microbiol 205: 212. https://doi.org/10.1007/s00203-023-03541-9
3. Tungekar AA, Castillo-Corujo A, Ruddock LW (2021) So you want to express your protein in *Escherichia coli*? Essays Biochem 65:247–260. https://doi.org/10.1042/EBC20200170
4. Lundell D, Greenberg R, Alroy Y et al (1990) Cytoplasmic and periplasmic expression of a highly basic protein, human interleukin 4, in *Escherichia coli*. J Ind Microbiol 5:215–227. https://doi.org/10.1007/BF01569679
5. Gao B, Zhangsun D, Hu Y, Wu Y et al (2013) Expression and secretion of functional recombinant μO-conotoxin MrVIB-His-tag in *Escherichia coli*. Toxicon 72:81–89. https://doi.org/10.1016/j.toxicon.2013.06.012
6. Nag N, Khan H, Tripathi T (2022) Strategies to improve the expression and solubility of recombinant proteins in *E. coli*. In: Advances in protein molecular and structural biology methods. Academic Press, pp 1–12
7. Atroshenko DL, Sergeev EP, Golovina DI et al (2024) Additivities for soluble recombinant protein expression in cytoplasm of *Escherichia coli*. Fermentation 10:120. https://doi.org/10.3390/fermentation10030120
8. Raran-Kurussi S, Waugh DS (2012) The ability to enhance the solubility of its fusion partners is an intrinsic property of maltose-binding protein but their folding is either spontaneous or chaperone-mediated. PLoS One 7:e49589. https://doi.org/10.1371/journal.pone.0049589
9. Waugh DS (2016) The remarkable solubility-enhancing power of *Escherichia coli* maltose-binding protein. Postepy Biochem 62:377–382. https://doi.org/10.18388/pb.2016_41
10. Gnanasekaran P, Pappu HR (2023) Detection of protein-protein interactions using glutathione-S-transferase (GST) pull-down assay technique. Methods Mol Biol 2690: 111–115. https://doi.org/10.1007/978-1-0716-3327-4_10
11. Terpe K (2003) Overview of tag protein fusions: from molecular and biochemical fundamentals to commercial systems. Appl Microbiol Biotechnol 60:523–533. https://doi.org/10.1007/s00253-002-1158-6
12. Boël G, Letso R, Neely H et al (2016) Codon influence on protein expression in *E. coli* correlates with mRNA levels. Nature 529:358–363. https://doi.org/10.1038/nature16509
13. Puigbò P, Guzmán E, Romeu A et al (2007) OPTIMIZER: a web server for optimizing the codon usage of DNA sequences. Nucleic Acids Res 35(Web Server issue):W126–W131. https://doi.org/10.1093/nar/gkm219
14. Kube M, Mitrovic J, Duduk B et al (2012) Current view on phytoplasma genomes and encoded metabolism. Sci World J 2012: 185942. https://doi.org/10.1100/2012/185942
15. Boonrod K, Munteanu B, Jarausch B et al (2012) An immunodominant membrane protein (Imp) of "*Candidatus* Phytoplasma mali" binds to plant actin. Mol Plant Microbe Interact 25:889–895. https://doi.org/10.1094/MPMI-11-11-0303
16. Boonrod K, Kuaguim L, Braun M et al (2023) Identification of the actin-binding region and binding to host plant apple actin of immunodominant transmembrane protein of "*Candidatus* Phytoplasma mali". Int J Mol Sci 24:968. https://doi.org/10.3390/ijms24020968

17. Strohmayer A, Moser M, Si-Ammour A et al (2019) "Candidatus Phytoplasma mali" genome encodes a protein that functions as a E3 ubiquitin ligase and could inhibit plant basal defense. Mol Plant Microbe Interact 32: 1487–1495. https://doi.org/10.1094/MPMI-04-19-0107-R
18. Geertsma ER, Groeneveld M, Slotboom DJ et al (2008) Quality control of overexpressed membrane proteins. Proc Natl Acad Sci USA 105(15):5722–5727. https://doi.org/10.1073/pnas.0802190105
19. Depping P, Román Lara MM, Kesidis A et al (2022) Heterologous expression of membrane proteins in *E. coli*. Methods Mol Biol 2507:59–78. https://doi.org/10.1007/978-1-0716-2368-8_4
20. Galetto L, Siampour M, Marzachì C (2013) Preparation of phytoplasma membrane recombinant proteins. In: Dickinson M, Hodgetts J (eds) Phytoplasma, Methods in molecular biology, vol 938. Humana Press, Totowa. https://doi.org/10.1007/978-1-62703-089-2_30

Chapter 13

Transient Expression of Phytoplasma Proteins *in Planta* Using an Agro-Infiltration Method

Kajohn Boonrod

Abstract

Agrobacterium infiltration is a very versatile technique for the transient expression of a protein in plants. This technique allows the researcher to study a protein of interest *in planta* in a very short time. Although the method is well established, there are some limitations to the transient expression of a phytoplasma protein. In this chapter, we describe the method in detail and address some of the issues necessary to improve the efficiency of the method for the transient expression of phytoplasma proteins *in planta*.

Key words Agrobacterium, Agro-infiltration, Transient expression

1 Introduction

Agrobacterium tumefaciens (*A. tumefaciens*) is an Alphaproteobacterium of the family Rhizobiaceae. It is a rod-shaped, gram-negative soil bacterium that causes crown gall disease in over 140 species of eudicots [1]. As part of the bacterial life cycle, the bacterium can deliver DNA fragments (T-DNA) encoding genes required for the formation of crown gall tumours from tumour-inducing plasmids into plant nuclei, which are incorporated into the plant genome at semi-random locations [2]. Using a constructed binary vector system in which the T-DNA is free of tumour-inducing genes, any desired gene or sequence fragment can be expressed under the control of a eukaryotic promoter, such as the cauliflower mosaic virus 35S promoter. The T-DNA molecules introduced into a plant cell can either be stably integrated into the plant genome by recombination and thus become heritable, which is the basis of current plant transformation techniques [3], or non-integrated T-DNAs that are later translated and the transcribed protein is stable for a certain time [4, 5]. Transient bacterial expression of a target protein has become a widely used tool in plant pathology and cell biology studies. This technique is commonly known as agroinfiltration or

Katrin Janik and Mattia Tabarelli (eds.), *Phytoplasma: Methods and Protocols*, Methods in Molecular Biology, vol. 3008, https://doi.org/10.1007/978-1-0716-5104-9_13,

agroinjection. Agroinfiltration is influenced by several factors such as the foreign gene to be introduced, the type of plant host and the plant growth conditions after agroinfiltration [6].

In phytoplasma research, agroinfiltration is widely used to study the localization and function of phytoplasma proteins, such as effectors or immunodominant proteins, in model plants [7–9]. So far, two model plants are commonly used for this purpose, *Arabidopsis thaliana* (*A. thaliana*) and *Nicotiana benthamiana* (*N. benthamiana*). It is known that agroinfiltration of *A. thaliana* is suboptimal due to plant immune responses triggered upon the perception of *Agrobacterium* [10–12]. Therefore, if *A. thaliana* is to be used for studies, the use of protoplasts is recommended [13]. However, Zhang and co-workers successfully agroinfiltrated *A. thaliana* leaves using *A. tumefaciens* strain *AGL1* in combination with the surfactant Silwet L-77 and by keeping the plants in the dark for 24 h [14]. This method greatly improved the efficacy of transient transformation of *A. thaliana* leaves.

In this chapter, we will use *N. benthamiana* as a model plant for agroinfiltration since it expresses much higher levels of transcript mRNAs and their products than *A. thaliana* [15]. This may be since in *N. benthamiana*, RNA-dependent RNA polymerase 1 (RdRP1), which is involved in antiviral defence, is naturally truncated resulting in reduced hypersusceptibility to many different plant viruses and *Agrobacterium* [16, 17]. It should be noted that to achieve high levels of transgene expression, the codon usage of the phytoplasma gene needs to be optimized for each model plant. We have found that transient expression of the original phytoplasma gene (sequence) can result in low or no expression. In this work, we have developed and optimized a method of agroinfiltration to achieve transient expression of phytoplasma genes from T-DNAs. Using this procedure, we transiently expressed the plant codon usage optimized genes SAP11-like effector of '*Candidatus* Phytoplasma mali' fused with GFP *in planta*.

2 Materials

2.1 Bacteria and Plasmid

1. *A. tumefaciens* strain *ATHV*.
2. Binary vector: e.g., pPZP200 and pCambia carrying a target gene under control of the cauliflower mosaic virus 35S promoter.

2.2 Solutions

1. 100 mM $MgCl_2$: weigh 20.33 g of magnesium chloride hexahydrate ($MgCl_2 \cdot 6H_2O$; 100% purity) and dissolve in deionized or distilled water. Adjust volume to 1000 mL. Sterile filtrate and store at 4 °C. The solution should be ice-cold when used for the procedure.

2. 20 mM $CaCl_2$: Weigh 4.38 g of calcium chloride hexahydrate ($CaCl_2{\cdot}6H_2O$; 100% purity). And dissolve in deionized or distilled water. Adjust volume to 1000 mL. Sterile filtrate and store at 4 °C. The solution can be stored for up to several months. The solution should be ice-cold when used for the procedure.
3. 86% Glycerol: Dilute 88.07 mL of 95% (v/v) glycerol to a final volume of 100 mL with deionized (distilled) water and autoclave. Store at room temperature. The diluted glycerol can be stored for several months.
4. Rifampicin solution: Weigh 0.20 g rifampicin and dissolve in 10 mL 100% ethanol. Aliquot in 2 mL and store at −20 °C. The solution can be stored for several months.
5. Antibiotic solution: Dissolve the corresponding antibiotic in solution as recommend by manufacture at 100× of used final concentration. Aliquot in 2 mL and store at −20 °C. Antibiotic solutions can be stored for several months.
6. Liquid nitrogen.

2.3 Media

1. SOB: weigh 5 g yeast extract, 20 g tryptone, 0.58 g NaCl, 0.19 g KCl, and 2.4 g $MgSO_4$. Dissolve all substances in 800 mL of distilled water. Bring the final volume to 1000 mL with distilled water. Sterilize the medium by autoclaving. Aliquot in 2 mL and store at −20 °C for at least 1 year.
2. 2× YT medium: weigh 16 g tryptone, 10 g yeast extract and 5 g NaCl. Dissolve all substances in 900 mL of distilled water. Adjust pH to 7.0 with 5 N NaOH. Adjust to 1000 mL with distilled water. Sterilize by autoclaving. Store at room temperature. The medium can be stored for several months.
3. 2× YT agar: 16 g tryptone, 10 g yeast extract, 5 g NaCl, 15 g agar. Mix all ingredients and add water to 1 L. Sterilize by autoclaving. Store at room temperature. The medium can be stored for several months.
4. Infection Medium: weigh 86 mg Murashige and Skoog Media (MS), 0.4 g sucrose, 200 μL of 1 M 2-morpholinoethanesulphonic acid (MES), 10 μL of 500 mM acetosyringone. Add water to 20 mL. Prepare the infection medium directly before use.

2.4 Plants

1. *Nicotiana benthamiana* in "6 leaves"-stage (*see* Fig. 1a).

2.5 Consumables

1. Reaction tubes: 2 mL.
2. Petri dishes.
3. Erlenmeyer flasks: 50 mL.

Fig. 1 Infiltration of *Agrobacterium* into *N. benthamiana* leaves. A suspension of the Agrobacterium strain *ATHV* transformed with a binary plasmid expressing SAP11 fused to GFP and expressed under the control of a 35S promoter was transformed into *N. benthamiana* leaves. (**a**) The 6-leaves stage is ideal for agroinfiltration. (**b**) Agrobacterium is induced in the induction medium for 2 h. (**c**) The bacterial cell suspension is gently infiltrated into the abaxial surface of the plant leaves. (**d**) The cell suspension should be infiltrated and dispersed smoothly (right), not with tension (left)

4. Conical Sterile Polypropylene Centrifuge Tubes: 15 mL.
5. Insulin Syringe (1 mL).

2.6 Equipment

1. Growth chamber: The growth condition for *N. benthamiana* is "short day" (8 h light/16 h dark) with a light intensity of 80–100 μmol m^{-2} s^{-1}, 21 °C during the day and 16 °C during the night period with a humidity of about 60%.
2. Incubator for 28 °C.
3. Tabletop centrifuge with fixed rotor and maximum speed is 1750× *g* for 50 mL conical tubes.
4. Benchtop centrifuge with fixed rotor maximum speed up to 7000× *g* for 1.5–2 mL reaction tubes.

2.7 Optional Equipment

1. Confocal microscope with the corresponding laser and filter sets to detect the signal of the fluorescence tag.

3 Methods

This method is a modification of a technique described by Jyothishwaran et al. [18].

3.1 Sequence Optimization and Cloning

1. Optimize the codon usage of the target gene for expression in *N. benthamiana* using an appropriate software (*see* **Note 1**).
2. Clone the gene in a binary expression vector (e.g. pZP2000) in which expression is under the control of a 35S promoter. In this work, we cloned *SAP11* as a fusion protein with the green fluorescence protein (GFP) using an expression vector with a 35S promoter.

3.2 Preparation of Agrobacterium Competent Cells

1. Inoculate a colony of *A. tumefaciens* strain *ATHV* (*see* **Note 2**) in 10 mL 2× YT medium supplemented with rifampicin (20 μg/mL) for selecting *Agrobacterium* and incubate at 28 °C with 200 rpm shaking overnight.
2. Resuspend 1 mL of the *A. tumefaciens* overnight culture in 50 mL 2× YT media supplemented with rifampicin (20 μg/mL) and incubate the cell suspension at 28 °C with 200 rpm shaking. Let the bacteria grow to an OD_{600} of 0.3 (*see* **Note 3**).
3. Chill the bacteria cell suspension on ice for 15 min and centrifuge at 7000× *g* at 4 °C.
4. Resuspend the cell pellet in 10 mL ice-cold 100 mM $MgCl_2$ and incubate on ice for 30 min.
5. Centrifuge the cells at 7000× *g*, 4 °C for 5 min and resuspend the pellet in 2 mL sterile, ice-cold 20 mM $CaCl_2$.
6. Incubate the cell mixture for 90 min on ice and then add 500 μL of 85% glycerol. Aliquot the competent cells in 50 μL and snap freeze in liquid nitrogen (*see* **Note 4**). Aliquots kept at −80 °C are stable for several months.

3.3 Agrobacterium Transformation

1. Thaw an aliquot of *Agrobacterium* competent cells on ice for 5 min and add 1 μg of the desired expression plasmid. For this work we used pPZP200:*35S:SAP11-GFP*.
2. Incubate the *Agrobacterium*/plasmid mixture for 10 min on ice. Immerse the cell mixture in liquid nitrogen for shock freezing.
3. Incubate the frozen mixture immediately at 37 °C for 5 min and then add 200 μL of SOB.
4. Incubate the cell mixture for 1 h at 28 °C with shaking at 200 rpm. Then plate the cells on a selective 2× YT agar media and incubate the plate at 28 °C for 2 days (*see* **Note 5**). Light reddish transformed colonies should be observed on the plates.

3.4 Induction of Agrobacterium Virulence Gene Expression

1. Inoculate a colony of *Agrobacterium* transformed with the desired plasmid (pPZP200-35S:*SAP11-GFP*) in 10 mL 2× YT media supplemented with rifampicin (20 μg/mL) and a selective antibiotic (selecting the bacterium carrying the desired plasmid) at 28 °C with 200 rpm for 1–2 days (*see* **Note 6**).
2. Transfer 2 mL of the bacteria culture into a reagent tube and centrifuge the cell suspension at 7000× *g* for 5 min and resuspend the pellet with 2 mL infection medium.
3. Incubate the cell mixture at room temperature for at least 2 h.

3.5 Agroinfiltration

1. Prior to agroinfiltration, *N. benthamiana* (6 leaves stage) should not be watered for 12 h (*see* **Note** 7) or the plant should be removed from the growth chamber and left at room temperature for at least 30 min. In this condition, the leaves lose water and it is easier to infiltrate the cell suspension into a leaf (Fig. 1a, b).
2. Infiltrate the induced *Agrobacterium* cells into the abaxial surface of several plant leaves using an insulin syringe (Fig. 1c, d) by gently applying the infiltration force to press the cell suspension into the leaf without damaging the cells (*see* **Note 8**).
3. Keep the infiltrated plants in the growth chamber for 2 days (*see* **Note 9**). The expressed protein can be directly visualized under a confocal microscope for expression control and protein localization (Fig. 2) or be subjected to further analyses by methods such as co-immunoprecipitation, destabilization assays, protein purification, etc.

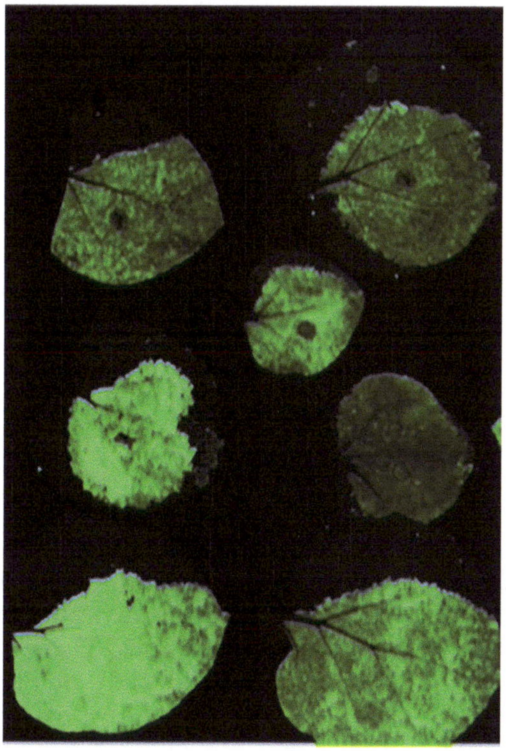

Fig. 2 *N. benthamiana* leaves expressing the recombinant GFP-tagged phytoplasmal protein SAP11. After 2 days post-infiltration, the expression of the recombinant phytoplasma protein was visualised by confocal microscopy using the appropriate laser and filter set for GFP visualization

4 Notes

1. To optimize codon usage, e.g. the following program can be used: https://www.novoprolabs.com/tools/codon-optimization. If the gene will be synthesized, most manufactures offer an "optimize codon usage" service.
2. Using a different strain may result in low protein expression. If *A. thaliana* is used for the study, the strain *GV1301* is recommended. The strain *GV1301* is also suitable for the generation of transgenic *A. thaliana* plant lines by floral dip.
3. Overgrowth of the bacterium $OD_{600} > 0.3$ will result in inefficient competent cells and consequently reduced transformation efficiency due to thick cell wall formation.
4. To maintain competent cells at −80 °C, rapid freezing of bacteria is strongly recommended since it results in higher transformation efficiency after prolonged storage than slow freezing.

5. If the cells appear before or after 2 days, it is likely that a contamination with other bacteria or yeast occurred. Confirmation of successful transformation by colony PCR using specific primers (for the amplification of the cloned gene of interest) is recommended.
6. The bacterial cells can also be directly induced at this stage by adding acetosyringone to final concentration 100 μM into the culture medium.
7. The plant's growth stage is very important. Plants older than the 6-leaves stage (*see* Fig. 1a) could result in low expression of the target protein, and it is difficult to infiltrate the bacterial suspension into the plant leaf.
8. The infiltration should be smooth and without much force. The *Agrobacterium* cell suspension should be distributed within the infiltrated leaf.
9. The optimal incubation time for post-infiltration is 2 days. At this time, the protein is highly expressed, and after this time, the recombinant protein may be degraded, and gene silencing could take place. Therefore, infiltrated plants should be used for further analysis after about 2 days of infiltration.

References

1. Smith EF, Townsend CO (1907) A plant-tumor of bacterial origin. Science 26:671–673. https://doi.org/10.1126/science.25.643.671
2. Chilton MD, Drummond MH, Merio DJ et al (1977) Stable incorporation of plasmid DNA into higher plant cells: the molecular basis of crown gall tumorigenesis. Cell 11:263–271. https://doi.org/10.1016/0092-8674(77)90043-5
3. Gelvin SB (2003) *Agrobacterium*-mediated plant transformation: the biology behind the "gene-jockeying" tool. Microbiol Mol Biol Rev 67:16–37. https://doi.org/10.1128/MMBR.67.1.16-37.2003
4. Janssen BJ, Gardner RC (1990) Localized transient expression of GUS in leaf discs following cocultivation with *Agrobacterium*. Plant Mol Biol 14:61–72. https://doi.org/10.1007/BF00015655
5. Krenek P, Samajova O, Luptovciak I et al (2015) Transient plant transformation mediated by *Agrobacterium* tumefaciens: principles, methods and applications. Biotechnol Adv 1:1024–1042. https://doi.org/10.1016/j.biotechadv.2015.03.012
6. Del Toro F, Tenllado F, Chung BN et al (2014) Expression by agroinfiltration at high temperature. Mol Plant Pathol 15:848–857. https://doi.org/10.1111/mpp.12136
7. Boonrod K, Kuaguim L, Braun M et al (2023) Identification of the actin-binding region and binding to host plant apple actin of immunodominant transmembrane protein of "*Candidatus* Phytoplasma mali". Int J Mol Sci 24:968. https://doi.org/10.3390/ijms24020968
8. Strohmayer A, Moser M, Si-Ammour A et al (2019) "*Candidatus* Phytoplasma mali" genome encodes a protein that functions as a E3 ubiquitin ligase and could inhibit plant basal defense. Mol Plant Microbe Interact 32:1487–1495. https://doi.org/10.1094/MPMI-04-19-0107-R
9. Boonrod K, Strohmayer A, Schwarz T et al (2022) Beyond destabilizing activity of SAP11-like effector of *Candidatus* Phytoplasma mali strain PM19. Microorganisms 10:1406. https://doi.org/10.3390/microorganisms10071406
10. Mangano S, Gonzalez CD, Petruccelli S (2014) *Agrobacterium* tumefaciens-mediated transient transformation of *Arabidopsis thaliana* leaves. In: Sanchez-Serrano J, Salinas J (eds) *Arabidopsis* protocols, Methods in molecular biology, vol 1062. Humana Press, Totowa.

https://doi.org/10.1007/978-1-62703-580-4_8

11. Zhu W, Hu X, Wang C et al (2017) NPR1-dependent salicylic acid signaling inhibits *Agrobacterium*-mediated transient expression in *Arabidopsis*. Sci China Life Sci 60:668–670. https://doi.org/10.1007/s11427-016-0383-y
12. Rosas-Díaz T, Cana-Quijada P, Amorim-Silva V et al (2017) *Arabidopsis* NahG plants as a suitable and efficient system for transient expression using *Agrobacterium* tumefaciens. Mol Plant 10:353–356
13. Sugio A, Kingdom HN, MacLean AM et al (2011) Phytoplasma protein effector SAP11 enhances insect vector reproduction by manipulating plant development and defense hormone biosynthesis. Proc Natl Acad Sci USA 108:1254–1263. https://doi.org/10.1073/pnas.1105664108
14. Zhang Y, Chen M, Siemiatkowska B et al (2020) A highly efficient *Agrobacterium*-mediated method for transient gene expression and functional studies in multiple plant species. Plant Commun 1(5):100028. https://doi.org/10.1016/j.xplc.2020.100028
15. Andrews LB, Curtis WR (2005) Comparison of transient protein expression in tobacco leaves and plant suspension culture. Biotechnol Prog 21:946–952. https://doi.org/10.1021/bp049569k
16. Yang SJ, Carter SA, Cole AB et al (2004) A natural variant of a host RNA-dependent RNA polymerase is associated with increased susceptibility to viruses by Nicotiana benthamiana. Proc Natl Acad Sci USA 20:6297–6302. https://doi.org/10.1073/pnas.0304346101
17. Ying XB, Dong L, Zhu H et al (2010) RNA-dependent RNA polymerase 1 from Nicotiana tabacum suppresses RNA silencing and enhances viral infection in Nicotiana benthamiana. Plant Cell 22:1358–1372. https://doi.org/10.1105/tpc.109.072058
18. Jyothishwaran G, Kotresha D, Selvaraj T et al (2007) A modified freeze–thaw method for efficient transformation of *Agrobacterium* tumefaciens. Curr Sci 93:770–772

Chapter 14

An *In Vitro* Ubiquitination Assay to Study Phytoplasmal Effector Function

Margot Raffeiner

Abstract

The attachment of one or more ubiquitin molecules onto a target substrate, ubiquitination for short, is a posttranslational modification that determines its correct further processing within a eukaryotic cell. Ubiquitination was shown to be a central player in the regulation of many cellular processes from protein degradation to protein sorting (e.g., endocytosis), DNA repair, or other non-proteolytic pathways. Because of its importance in maintaining cellular homeostasis, it is not surprising that, in the course of evolution, pathogens, including plant pathogens such as phytoplasma, have developed strategies to manipulate ubiquitination of host targets to their own benefit. Manipulation includes the translocation of so-called effector proteins that themselves ubiquitinate host targets, thereby mediating their subsequent degradation to dampen host immune responses.

The identification and characterization of phytoplasmal effector proteins that ubiquitinate target substrates, as well as the detection of a target ubiquitination is crucial to understand phytoplasma-associated pathogenesis.

Here, we provide an easy-to-follow protocol adapted from a method described by Furlan and Trujillo (2017) to study the *in vitro* E3 ligase activity potential of phytoplasma effector proteins. Additionally, this protocol allows to study the *in vitro* ubiquitination of putative target substrates by phytoplasma effectors.

Key words Posttranslational modification, Ubiquitin, *In vitro* ubiquitination, Phytoplasma, Effector proteins, Effector targets

1 Introduction

Ubiquitination is a central posttranslational modification (PTM) employed by eukaryotes to regulate a vast variety, if not almost all, cellular processes. This includes non-proteolytic processes such as DNA repair, transcription, or translation, as well as proteolytic pathways such as protein degradation via the ubiquitin-26S proteasome system (UPS) or autophagy-related vacuolar degradation [1]. Ubiquitination is defined by the attachment of the signal molecule ubiquitin onto a target substrate via a three-step enzymatic cascade. The attachment of this 8.5 kDa polypeptide that

Katrin Janik and Mattia Tabarelli (eds.), *Phytoplasma: Methods and Protocols*, Methods in Molecular Biology, vol. 3008, https://doi.org/10.1007/978-1-0716-5104-9_14,

consists of 76 amino acids and is highly conserved among eukaryotes involves three distinct enzyme classes, E1-activating enzymes (E1), E2-conjugating enzymes (E2), and E3 ligases (E3). The genomes of different organisms encode different numbers of E1, E2, and E3 enzymes. In Arabidopsis, for instance, only two E1 enzymes have been identified so far, while there appear to be at least 38 E2 enzymes [2–4]. The largest enzyme group with approximately 1.500 representatives in Arabidopsis is the family of E3 ligases, which consists of four subgroups [5, 6].

During the first ATP-dependent step, ubiquitin is adenylated at its C-terminal glycine residue and bound to the cysteine sulfhydryl group of the E1 via a thioester linkage upon cleavage of AMP. Secondly, E2s can transfer ubiquitin molecules via transthioesterification from the E1 to a cysteine in their own active site. In the last step of the ubiquitination cascade, E3 ligases play the major role. E3s are responsible for the specific selection of substrates and build a scaffold between the right E2 and the selected substrate. The actual ubiquitin transfer is then established by the formation of an isopeptide bond between the C-terminal glycine of the ubiquitin and the ε-amino group of a lysine in the substrate protein [7]. Another layer of specificity is added to this process by additional ubiquitin-modifying enzymes such as deubiquitinating enzymes (DUBs), even though they are less studied than the aforementioned canonical enzyme groups [8]. A simplified schematic of the ubiquitination cascade is shown in Fig. 1.

To understand how this posttranslational modification can regulate such a diverse array of cellular processes, it must be noted that ubiquitination is much more than a simple three-step enzymatic cascade where only one ubiquitin molecule is attached to a target substrate. Even if the so-called monoubiquitination, the transfer of a single ubiquitin molecule to a substrate, is a possibility, many proteins are marked with longer polyubiquitin chains. Those chains can be generated because ubiquitin itself contains seven lysins (K6, K11, K27, K29, K33, K48, and K63) where other ubiquitin molecules can bind to. The length and topology (linkage type) of the ubiquitin chain attached to a target substrate determine which path the substrate protein will take. It can be, for example, either directed towards degradation, can be stabilized, or can be trafficked to another location in the cell. All different forms of ubiquitin conjugation to a specific protein are summarized by the ubiquitin code [1]. Here, E2 enzymes play a particularly important role. They not only transfer ubiquitin to the corresponding E3 or the target substrate, but also the formation of different ubiquitin linkages has been shown to depend on E2 enzymes to some extent [7]. It must be noted that, although many E3 ligases can work together with "generic" E2 enzymes such as *At*UBC8, others might require a more specific E2 enzyme (e.g., one that is only able to catalyze the production of ubiquitin chains of a certain linkage type). This

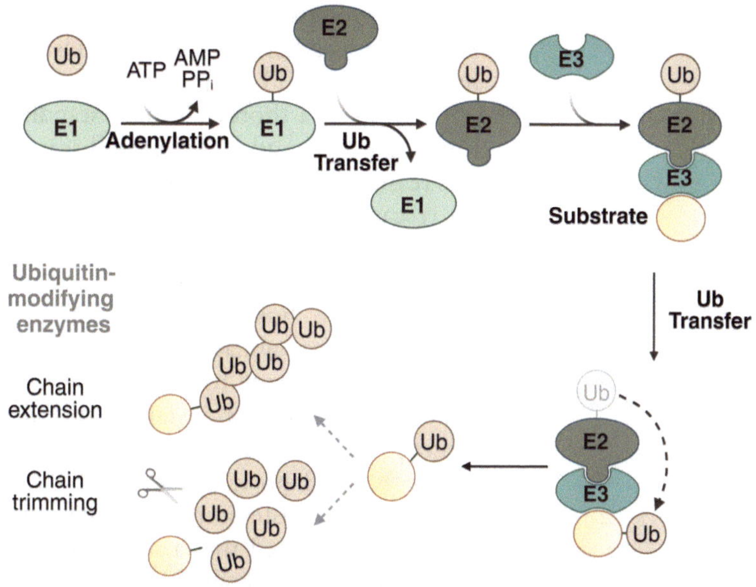

Fig. 1 Schematic representation of the ubiquitination cascade. The conventional ubiquitin labeling of substrate proteins takes place via a three-step enzymatic cascade. In the first step, ubiquitin molecules (Ub) are bound to an E1-activating enzyme (E1) via adenylation. This is followed by Ub transfer from the E1- to an E2-conjugating enzyme (E2). In the last step, the E3 ligase (E3) binds to the corresponding substrate and, via binding to the E2, the Ub is transferred from the E2 to the substrate. Other ubiquitin-modifying enzymes can have a supporting function in the polyubiquitination of substrate proteins (chain extension). Deubiquitinating enzymes (DUBs), on the other hand, can catalyze the cleavage of ubiquitin molecules from the substrate (chain trimming). This figure was modified after the model "Protein Ubiquitylation Pathway" from BioRender.com. (Source: https:/app.biorender.com/biorender-templates. Adapted from Figure 1 in [1]. The figure was created with Biorender.com)

should always be considered when studying components of the ubiquitination cascade because it can be important for the interpretation of obtained experimental results. Testing different E2–E3 pairings in an *in vitro* ubiquitination assay might be informative in some cases.

Phytoplasma-associated diseases, characterized by symptoms associated with developmental alterations such as stunted growth, phyllody (flowers are converted to leaves), virescence (non-green parts of flowers turn green), proliferation of axillary buds resulting in witches' brooms or flower sterility, lead to severe losses in food crop production worldwide [9]. As many other bacterial phytopathogens, phytoplasmas produce effector proteins to manipulate host cellular processes for their own benefit [10]. Despite the fact that a number of candidate effector proteins were identified in different phytoplasma species, so far only a few phytoplasma effector proteins have been characterized in depth to provide important insights into the mechanisms of host manipulation on a molecular level [11–16].

However, a thorough understanding of phytoplasma effector function is key to addressing the threat posed by this pathogen. For this reason, it is extremely important to keep developing new tools for researching effector functions.

In the past years, it has been shown that some phytoplasma species manipulate the stability of target proteins, potentially involving the host ubiquitin-26S proteasome system (UPS), which mediates regulated protein turnover in all eukaryotic organisms. SAP54 secreted by Aster Yellows phytoplasma strain Witches' Broom (AY-WB), for instance, mediates the degradation of its target proteins via the UPS of the host plant by recruiting members of the RAD23 protein family, which shuttle ubiquitinated substrates to the proteasomal degradation machinery [17]. Another study demonstrated that a secreted effector protein from '*Candidatus* Phytoplasma mali' PM19_00185, possesses *in vitro* E3 ubiquitin ligase activity, potentially marking host proteins for their degradation to interfere with immune responses [16]. Interestingly, another effector secreted by the AY-WB phytoplasma, SAP05, was more recently shown to mediate ubiquitin-independent degradation of its target proteins via the UPS by binding to the 26S proteasome subunit RPN10 [18]. Those examples suggest that manipulation of protein stability involving the host UPS is a major virulence strategy of phytoplasma. Thus, a thorough understanding of the molecular mechanisms behind this virulence strategy could represent a major step forward in understanding the pathosystem, holding great potential to devise novel control strategies.

Here, a step-by-step protocol, adapted from a method described by Furlan and Trujillo (2017), is described [7]. The described method has been applied to study the *in vitro* E3 ligase activity potential of phytoplasma effector proteins and can be used to verify whether a candidate phytoplasma effector protein possesses *in vitro* E3 ubiquitin ligase activity. Furthermore, this protocol can be used to verify whether an E3 ligase effector protein ubiquitinates its putative target substrates *in vitro*. It also provides the basis for further downstream analyses to elucidate effector protein functions related to ubiquitination.

The work from Stromeyer and colleagues [16] on the PM19_00185 effector protein has already demonstrated that an *in vitro* ubiquitination assay is a powerful tool to unravel phytoplasma effector function. The protocol described here has previously allowed to prove the *in vitro* E3 ligase activity of the type-III effector protein XopL from the Gram-negative phytopathogen *Xanthomonas campestris* pv. *vesicatoria* on its target substrate *At*SH3P2 [19] It is likewise suitable to study phytoplasma effector functions associated with ubiquitination and host UPS manipulation.

2 Materials

All purchased reagents are to be stored according to the manufacturer's instructions. All other solutions/reagents are to be prepared in ultrapure water (resistivity of 18 MΩ.cm at 25 °C produced from tap water) and stored at the indicated temperatures for the indicated time periods. Important: In some steps of the following sections, product examples are mentioned to facilitate the understanding of the protocol. These products are possible examples and not exclusive recommendations. Similar products from other companies or suppliers might be as suitable for the method.

2.1 Solutions

1. Luria-Bertani (LB) growth medium: 10 g/L tryptone (microbiology grade), 10 g/L NaCl, 5 g/L yeast extract, adjust pH to 7.0 with NaOH. For the production of solid growth medium, add 15 g/L agar-agar. Autoclave at 121 °C for 20 min and store at room temperature for up to one year (*see* **Note 1**).
2. Antibiotics for resistance selection of *E. coli* strains during cloning and recombinant protein expression.
3. Tris–HCl, pH 8.0: 1 M stock solution, store at room temperature for up to one year (*see* **Note 1**).
4. NaCl: 5 M stock solution, store at room temperature for up to one year (*see* **Note 1**).
5. EDTA pH 8.0: 0.5 M stock solution, store at room temperature for up to one year (*see* **Note 1**).
6. 1,4-dithiothreitol (DTT): 1 M stock solution. Aliquots of the stock solution can be stored at −20 °C for up to one year (*see* **Note 1**).
7. Protease inhibitor cocktail (PIC, e.g., cOmplete™ Ultra Tablets, Merck, *see* **Note 2**). To be used according to the manufacturer's instructions.
8. 4x Laemmli loading buffer: 400 mM Tris–HCl pH 6.8, 0.4 M DTT, 40% (v/v) glycerol, 6 mM bromophenol blue, 8% (w/v) sodium dodecyl sulfate (SDS). Prepare 1 mL aliquots. The loading buffer can be stored at −20 °C for up to one year.

2.2 Cloning

1. Polymerase chain reaction (PCR) reagents to amplify the genes of interest (GOIs).
2. Arabidopsis cDNA to amplify the E1 activating and E2 conjugating enzymes required for the *in vitro* ubiquitination reaction (e.g., E1 = *At*UBA1, AT2G30110; E2 = the generic

*At*UBC8, AT5G41700 or others, e.g., *At*UBC35, AT1G78870; *see* Subheading 4.2.2).

3. Phytoplasma cDNA to amplify GOIs (e.g., effector protein with putative E3 ligase function).
4. Phytoplasma host cDNA (e.g., *Malus domestica*) to amplify GOIs (i.e., potential phytoplasma effector targets), usually identified because they are binding to the protein with putative E3 ligase function).
5. Cloning vectors to generate *Escherichia coli* (*E. coli*) expression constructs of the aforementioned GOIs (*see* **Notes 3** and **4**). Depending on the applied cloning strategy, an "entry" vector for each GOI might be required as well (e.g., when using Gateway™ cloning technology, Thermo Fisher Scientific).
6. *E. coli* DH5α and BL21 (DE3) competent cells.

2.3 Expression of Recombinant Proteins and Protein Purification

1. Isopropyl ß-D-1-thiogalactopyranoside (IPTG): 0.5 M stock solution. Aliquots of the stock solution can be stored at −20 °C for up to one year (*see* **Note 1**).
2. Coomassie protein stain (e.g., Instant*Blue*™ Protein Stain, Merck, **Note 2**).
3. Bovine serum albumin (BSA): Prepare aliquots of a 1 mg/mL stock solution. The stock solution can be stored at −20 °C for up to one year (*see* **Note 1**).
4. Lysis and wash buffer for His_{6x}-tagged protein purification: 20 mM sodium phosphate (a 0.1 M stock solution can be prepared and stored at room temperature for up to two months), 0.5 M NaCl, 20 mM imidazole (a 5 M stock solution can be prepared and stored at 4 °C for several weeks); adjust buffer to pH 7.4 (*see* **Note 5**).
5. Elution buffer for His_{6x}-tagged protein purification: 20 mM sodium phosphate (see above), 0.5 M NaCl, 500 mM imidazole; adjust buffer to pH 7.4 (*see* **Note 5**). The elution buffer must be freshly prepared for each protein purification.
6. Lysis buffer for MBP-tagged protein purification: 20 mM Tris–HCl pH 8.0, 500 mM NaCl, 20 mM DTT, 1 mM sodium fluoride (NaF, aliquots of a 1 M stock solution can be prepared and stored at −20 °C for up to one year (*see* **Note 1**), 0.1% (v/v) Triton-X-100, 1x PIC.
7. Wash buffer for MBP-tagged protein purification: 20 mM Tris–HCl pH 8.0, 150 mM NaCl, 1 mM DTT.
8. Elution buffer for MBP-tagged protein purification: 20 mM Tris–HCl pH 8.0, 150 mM NaCl, 1 mM DTT, 50 mM maltose (sterile filtered 0.5 M stock solution which can be stored at 4 °C for up to one year (*see* **Note 1**), Polyethersulfurone (PES)

syringe filters with a filter pore size of 0.45 μm can be used for sterile filtration).

9. Lysis buffer for GST-tagged protein purification: 50 mM Tris–HCl pH 8.0, 250 mM NaCl, 1 mM EDTA, 0.2% (v/v) Triton-X-100, 1 mM DTT, 1x PIC.
10. Wash buffer for GST-tagged protein purification: 100 mM Tris–HCl pH 8.0, 150 mM NaCl, 0.2% (v/v) NP-40, 1x PIC.
11. Elution buffer for GST-tagged protein purification: 100 mM Tris–HCl pH 8.0, 20 mM reduced glutathione.
12. Affinity chromatography columns for the purification of His-tagged proteins (e.g., HisTrap™ FF, Merck, *see* **Notes 2** and **5**).
13. Amylose resin for the purification of MBP-tagged proteins (e.g., from NEB, *see* **Note 2**).
14. Affinity chromatography resin for the purification of GST-tagged proteins purification (e.g., Glutathione Sepharose™ 4B, Merck, *see* **Note 2**).
15. Chromatography Columns (e.g., Econo-Pac®, BioRad; *see* **Note 2**).
16. Ultra Centrifugal Filters: MWCO 10 kDa, MWCO 50 kDa, and MWCO 100 kDa (e.g., Amicon®, Merck; *see* **Note 2**).
17. Exchange buffer: 100 mM HEPES-NaOH pH 7.5 (a sterile filtered 1 M stock solution can be prepared and stored at room temperature for up to one year (*see* **Note 1**), PES syringe filters with a filter pore size of 0.45 μm can be used), 2 mM EDTA, 5 mM dithioerythritol (DTE, aliquots of a 1 M stock solution can be prepared and stored at −20 °C for up to one year; *see* **Note 1**), 0.4 mM phenylmethylsulfonyl fluoride (PMSF, aliquots of a 1 M stock solution can be prepared and stored at −20 °C for up to one year (*see* **Note 1**); powder is soluble in pure ethanol), 10% (v/v) glycerol.

2.4 In vitro Ubiquitination Assay

1. Adenosine triphosphate (ATP): prepare aliquots of a 0.1 M stock solution and store them at −20 °C for up to one year (*see* **Note 1**).
2. Ubiquitin: Different versions of ubiquitin are commercially available. In this protocol, His_{6x}-tagged recombinant human ubiquitin (R&D Systems, discontinued product) and HA-tagged recombinant human ubiquitin (R&D Systems) were used, but also untagged or another His_{6x}- or HA-tagged ubiquitin from human or other organisms might be suitable; this can be experimentally tested. Dissolve the ubiquitin according to manufacturer's instructions at a 2.5 mg/mL concentration and prepare 20–50 μL aliquots. Several freeze-thaw

cycles are not harmful to ubiquitin stability. Ubiquitin aliquots can be stored at −20 °C for up to one year (*see* **Note 1**).

3. Buffer stock solutions: 1 M Tris–HCl pH 7.5; 0.1 M $MgCl_2$; 0.5 M KCl; 0.1 M DTT. Prepare 1 mL aliquots; several freeze-thaw cycles are not harmful to buffer stability. The buffer stocks can be stored at −20 °C for up to one year (*see* **Note 1**). The stock solutions are used to prepare the reaction mix/ubiquitination buffer.
4. Purified E1 ubiquitin activating enzyme (UBA): His_{6x}-tagged *At*UBA1 can be expressed in *E. coli* and purified using the Äkta™ Start system (*see* **Note 6**). Commercially available UBA proteins might be suitable as well, but this needs to be experimentally tested (*see* **Note 7**).
5. Purified E2 ubiquitin conjugating enzyme (UBC): His_{6x}-tagged *At*UBC8 was expressed in *E. coli* and purified using Äkta™ Start system. Commercially available UBC proteins might be suitable as well but need to be experimentally tested (*see* **Note 7**).
6. Purified putative E3 ligase and purified putative target substrate (*see* **Note 7**).

2.5 SDS–PAGE and Immunoblotting

1. Polyacrylamide gels: 12% and 4–20% gradient gels (e.g., 4–20% Mini-PROTEAN® TGX™ Precast Protein Gels, BioRad, **Note 2**). Gels with different gel percentages might be suitable as well, but this needs to be experimentally tested.
2. SDS–PAGE running buffer and western blot transfer buffer according to the system used.
3. Nitrocellulose or PVDF membrane (if a PVDF membrane is used, use pure ethanol or methanol to activate the membrane).
4. TBS-T buffer (20 mM Tris, 500 mM NaCl, 0.1% (v/v) Tween-20) and TBS buffer (TBS-T but without Tween-20). Stock solutions can be prepared and stored at room temperature for up to one year (*see* **Note 1**).
5. Blocking solution: 5% (w/v) skimmed milk in TBS-T buffer.
6. Antibodies: Ubiquitin antibody (P4D1) HRP; mouse monoclonal (e.g., catalogue no. sc-8017 HRP, SantaCruz; **Note 2**), GST antibody (B-14) HRP; mouse monoclonal (e.g., catalogue no. sc-138 HRP, SantaCruz; **Note 2**), MBP antibody; mouse monoclonal (e.g., catalogue no. E8032S, NEB; **Note 2**), Mouse IgG (H+L) HRP secondary antibody; goat polyclonal (e.g., catalogue no. 10158113, Fisher Scientific; **Note 2**).

2.6 Equipment

1. Thermal Cycler for PCR and *in vitro* ubiquitination reaction.
2. Benchtop thermomixer for *E. coli* heat shock transformation and sample denaturation.
3. 37 °C incubators and 16 °C/37 °C shaking incubators to grow *E. coli*.
4. Spectrophotometer and cuvettes to measure the optical density of bacteria at 600 nm (OD_{600}).
5. Ultrasonicator for cell lysis.
6. Microcentrifuge (with a rotor for 1.5 mL and for 2 mL PCR microcentrifuge tubes, reaching 13,000× *g*) and centrifuge (with a rotor for 15 mL and for 50 mL centrifuge tubes, reaching 7000× *g*).
7. Tube rotator utilized during recombinant protein expression and protein purification.
8. Protein electrophoresis and blotting equipment.
9. Western blot imaging apparatus for chemiluminescent imaging of the chemiluminescent detection reagent.
10. Optional: protein purification system (e.g., ÄKTA™ start; used in this protocol to purify His_{6x}-tagged proteins; *see* **Note 2**).

3 Methods

All methods described in the following paragraph have been experimentally proven to be suitable for carrying out an *in vitro* ubiquitination assay. However, also other cloning, protein expression or protein purification methods and the use of commercially available reagents might be applicable to efficiently validate E3 ligase activity of POIs, their autoubiquitination as well as ubiquitination of putative E3 ligase target substrates. The use of material other than described in this protocol might require individual adjustments of the given *in vitro* ubiquitination assay protocol.

3.1 Cloning

In the following paragraph, the method described involves Gateway cloning. However, other cloning methods can be used to generate appropriate expression vectors for the described E1, E2, and E3. Primer sequences for AtUBA1 (E1), AtUBC8 (E2), and HRD1a (as an E3 positive control) with the respective primer overhangs for TOPO directional cloning into pENTR™/D-TOPO™ Gateway™ entry vectors are given in Table 1.

1. Use Arabidopsis cDNA as a template and gene-specific primers (with specific overhangs suitable for the used cloning strategy) for the amplification of the Arabidopsis E1 enzyme *At*UBA1

Table 1
Gateway cloning primers for selected loci

Enzyme	Locus tag	Primer name	Primer sequence (5′–3′)
E1	AT2G30110	AtUBA1_fwd AtUBA1_rev	**CACC**CTTCACAAGCGAGCTAGTG TCACCTGAAGTAGATAGAGACGAG
E2	AT5G41700	AtUBC8_fwd AtUBC8_rev	**CACC**GCTTCGAAACGGATCTTGAAG TTAGCCCATGGCATACTTCTG
E3	AT3G16090.1	HRD1a_fwd HRD1a_rev	**CACC**ATTCGACTAAGAACATACGC TCACTCTGCTGCATCAGCAAC

Primer sequences designed for Gateway Cloning of selected Arabidopsis genes. Forward primers include a CACC overhang (indicated in bold in the primer sequence) for directional cloning into pENTR™/D-TOPO™ vectors

(AT2G30110) and the E2 enzyme *At*UBC8 (AT5G41700). Clone the amplicon into Gateway®-compatible "entry" vectors (or other suitable vectors with the respective tags).

2. For the production of Gateway®-compatible "entry" clones, use the pENTR™/D-TOPO® Kit (Thermo Fisher Scientific). The cloning must be performed according to the manufacturer's instructions. Perform LR reactions for cloning the desired genes into the respective destination vectors (*see* **Note 3**) using the Gateway® LR Clonase™ II Enzyme Mix (ThermoFisher Scientific) and following the manufacturer's instructions.
3. Clone the putative E3 ligases as well as putative E3 ligase target substrates into respective destination vectors using the above-described cloning method (*see* **Note 8**).

3.2 Expression of Recombinant Proteins and Protein Purification

3.2.1 Expression of Recombinant Proteins

1. Transform all plasmids containing the respective expression constructs (His_{6x}-*At*UBA1, His_{6x}-*At*UBC8, as well as MBP-putative E3, GST-putative E3 target substrate, or vice versa) into the *E. coli* BL21 (DE3) expression strain.
2. Inoculate one colony in 10 mL LB growth medium containing the construct-specific antibiotics for resistance selection (LB_{AB}) and grow the preculture at 37 °C overnight (ON) at 180 rpm.
3. Transfer the preculture to 500 mL of fresh LB_{AB} and grow the culture to an OD_{600} of approximately 0.5 (37 °C, 180 rpm, 3–4 h).
4. Once an OD_{600} of approximately 0.5 is reached, aliquot 1 mL of culture, transfer it to a microcentrifuge tube, and spin it down for 1 min at 13,000× *g*. Remove the supernatant and resuspend the pellet in 4x Laemmli buffer (per OD_{600} = 0.1, use 8 μL 4x Laemmli; e.g., if the culture has an OD_{600} of 0.5, resuspend the pellet in 40 μL 4x Laemmli)—This will be the t_0

sample to verify induction of the POI later on. Store sample at −20 °C until further use.

5. Keep the rest of the culture on ice for 30 min.
6. Induce protein expression by adding 0.5 mM IPTG to the culture.
7. Induce proteins at either 37 °C for 3–4 h at 180 rpm or at 16 °C ON at 180 rpm (*see* **Note 9**).
8. After induction of protein expression with IPTG (3–4h or ON), measure the OD_{600} of the culture again, transfer 500 μL of the culture to a microcentrifuge tube and spin it down at 13,000× *g* for 1 min. Remove the supernatant and resuspend the pellet in 4x Laemmli buffer (per $OD_{600} = 0.1$, use 4 μL 4x Laemmli; e.g., If the culture has an OD_{600} of 2, resuspend the pellet in 80 μL 4x Laemmli)—This will be the t_{IND} sample to verify induction of the POI later on. Store sample at −20 °C until further use.
9. The remaining culture is transferred to 50 mL centrifuge tubes. Centrifuge tubes at max. speed for 5 min at 4 °C.
10. Remove supernatant and store the pellet at −20 °C until further use. The pellet can be stored for several weeks to months.
11. To verify induction of the POI, denature t_0 and t_{IND} samples for 10 min at 95 °C and load 10 μL of each sample on a 12% polyacrylamide gel. After SDS–PAGE, stain the gel with ready-to-use Coomassie protein stain (e.g., Instant*Blue*™, **Note 2**). An example of a successfully stained protein gel containing recombinant proteins expressed as described above is provided in Fig. 2).

3.2.2 Protein Purification

All steps of the protein purification need to be carried out on ice or at 4 °C to avoid protein degradation.

1. The sample preparation procedure is the same for all POIs that need to be purified, independently of the affinity tag they are fused to. Only the lysis buffers vary accordingly, as described below.
2. Use 1 mL of the respective lysis buffer per 50 mL of pellet (e.g., 10 mL lysis buffer for a 500 mL pellet). For His_{6x}-tagged proteins, use the buffer listed under **point 5** in Subheading 2.3. For MBP-tagged proteins, use the buffer listed under **point** 7 in Subheading 2.3. For GST-tagged proteins, use the buffer listed under **point 10** in Subheading 2.3.
3. Resuspend the pellets in the respective lysis buffer and sonicate the cell lysates on ice 8–10 times for 30 s with an amplitude of 29–31%. Lysates should become clearer after sonication.

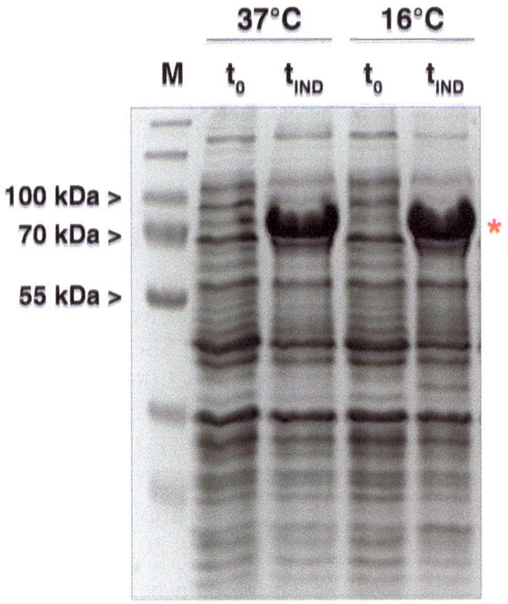

Fig. 2 Example of recombinant protein expression in *E. coli*. Recombinant protein expression of a protein of interest (POI) was induced as described in the Methods section. Induction was tested under two different conditions (37 °C for 4 h and 16 °C overnight) to find optimal induction conditions for this specific POI. To verify protein induction of the POI, 10 μL of the sample taken before addition of IPTG (t_0) was loaded on a 12% polyacrylamide gel next to 10 μL of sample taken after addition of IPTG (t_{IND}). After resolving proteins on the gel, the gel was stained with Coomassie Instant*Blue*™ Protein Stain (Merck). The appearance of an additional band at the size of the POI (including the size of the fused MBP tag) in the t_{IND} samples, marked by the red asterisk, indicates that the POI is efficiently induced at both tested conditions. M = PageRuler™ Plus Prestained Protein Ladder (ThermoFisher Scientific)

4. Centrifuge the sonicated cell lysates in a benchtop centrifuge for 20 min at max. speed and 4 °C to pellet insoluble debris.
5. Purification of His_{6x}-tagged proteins can be carried out using the ÄKTA™ start protein purification system (*see* **Note 5**) following the manufacturer's instructions. Buffers listed under **points 5** and **6** in Subheading 2.3 should be used.
6. Purification of MBP-tagged proteins can be carried out as follows:
7. Equilibrate 50 μL of amylose resin per 1 mL cell lysate in a microcentrifuge tube (e.g., use 500 μL amylose resin for 10 mL of lysate). To do so, centrifuge the resin for 1 min at 800× *g* and 4 °C. Remove the supernatant and add 750 μL of MBP lysis buffer (*see* **point 7**, Subheading 2.3), invert the tube several times, and centrifuge for 1 min at 800× *g* and 4 °C. Repeat this step twice. After the last removal of the

supernatant, add 500 μL of the lysis buffer to the resin and keep the tube on ice until further usage.

8. Take the supernatant from **point 4** (this Subheading) and transfer it to a 20 mL Econo-Pac® Chromatography Column (*see* **Note 2**).
9. Add the equilibrated amylose resin and incubate the mixture for 2 h on a rotator at 4 °C.
10. After 2 h, open the cap of the column to allow the lysate to flow through while the amylose resin (where your POI is now bound to) is retained by the filter.
11. Wash the amylose resin with 5 mL of wash buffer (*see* **point 8**, Subheading 2.3) for 5 times by letting the buffer flowing through the column.
12. Carefully close the cap again and add 1 mL of elution buffer (*see* **point 9**, Subheading 2.3) to the resin. Incubate the mixture for 1 h without shaking.
13. Collect the eluate in a microcentrifuge tube by opening the cap of the column and letting the elution buffer (now containing your POI) flow into the tube. Carefully close the cap of the column.
14. Repeat the elution step by adding an additional 1 mL of elution buffer to the resin. Incubate the mixture ON at 4 °C without shaking and collect the eluate the next day.
15. Store the eluates at 4 °C until further use. It is highly recommended to continue with the buffer exchange procedure (*see* **point 17** below, this subheading) directly after collecting the second elute to avoid protein degradation.
16. Purification of GST-Tagged Proteins:

    Follow the same protocol as described above (i.e., purification of MBP-tagged proteins) but using Glutathione Sepharose™ 4B (*see* **Note 2**) instead of the amylose resin as well as the respective buffers listed in Subheading 2.3 (lysis buffer, wash buffer, and elution buffer for the purification of GST-tagged proteins).
17. Buffer exchange using Amicon® Ultra Centrifugal Filters (*see* **Notes 2** and **10**):

    Choose the right size exclusion filter depending on the size of your purified POI (*see* **point 17** in Subheading 2.3). Use the buffer listed under **point 18** in Subheading 2.3 and follow the manufacturer's instructions for an efficient buffer exchange and concentration of your POI.
18. Estimate the concentration and quality of your protein of interest by loading different amounts of your final eluate on a polyacrylamide gel next to a BSA standard curve. Perform

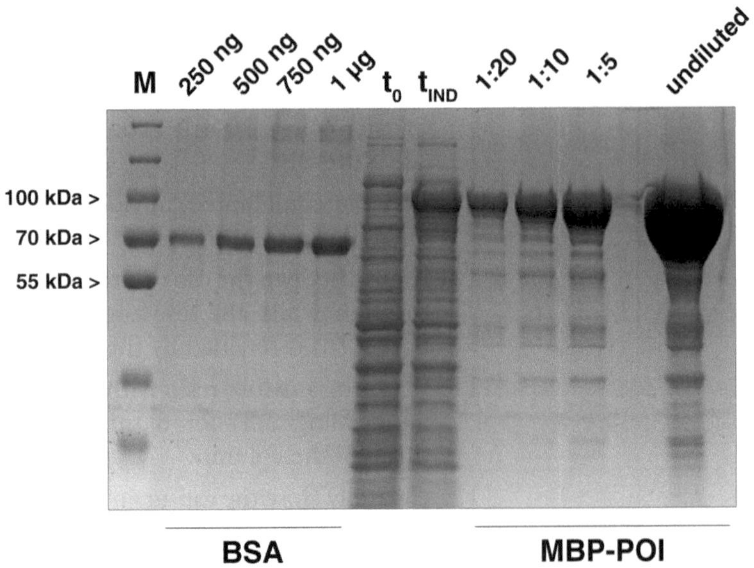

Fig. 3 Protein purification and estimation of protein concentration. To verify successful protein purification of a certain POI, it is recommended to load different dilutions of the final eluate (after buffer exchange as described in the Methods section) on a 12% polyacrylamide gel next to a BSA standard curve of known protein concentrations (lanes 2–5). Here, 1 µL of a 1:20, a 1:10, a 1:5, and of the undiluted elute of a purified MBP-tagged POI was loaded on the gel that was subsequently stained with Coomassie. The induction control samples of the same MBP-tagged POI (t_0 and t_{IND}) are loaded on the same gel to verify that the purified protein corresponds to the POI. In the example shown here, the 1:20 diluted elute sample has an approximate concentration of 1 µg/µL. For the in the *in vitro* ubiquitination assay, the total eluate should therefore be diluted 20 times, and aliquots can be stored at −80 °C. BSA = Bovine Serum Albumin. M = PageRuler™ Plus Prestained Protein Ladder (Thermo Fisher Scientific)

SDS–PAGE and subsequently stain the gel with ready-to-use Coomassie protein stain (e.g., Instant*Blue*™, **Notes 2** and **11**). An example of a successful protein purification and an estimation of protein concentration is given in Fig. 3.

19. Adjust the protein concentrations to approximately 1 mg/mL and store samples at −80 °C until further usage.

3.3 In vitro *Ubiquitination Assay*

The following paragraph provides instructions for performing different types of *in vitro* ubiquitination assays: It explains how to test the enzymatic activity of all components necessary for a successful *in vitro* ubiquitination cascade reaction, how to test for autoubiquitination of a certain studied putative E3 ligase, and how to test whether a putative target substrate is ubiquitinated *in vitro* by a certain E3 ligase.

3.3.1 Enzymatic Activity Assessment

The first step toward performing a successful *in vitro* ubiquitination assay is the assessment of the enzymatic activity of the previously purified components of the ubiquitination cascade (especially E1 and E2; see Subheading 3.2). To do so, use a purified E3 ligase with proven enzymatic activity (*see* **Note 12**) as a control.

1. To test the enzymatic activity of the previously purified E1 and E2 enzymes, as well as the previously purified E3 ligase with known enzymatic activity (*see* **Note 12**), prepare:
2. Prepare the reaction mix containing: 25 mM Tris–HCl pH 7.5, 5 mM MgCl2, 50 mM KCl, 0.6 mM DTT to prepare the ubiquitination buffer, and add 2 mM ATP, 2 μg of ubiquitin, 0.2 μg E1, 1.2 μg E2, and 2 μg E3 (e.g., using stock solutions listed in Subheading 2.3 **point 3**, for a total of 30 μL reaction volume mix 0.8 μL Tris–HCl pH 7.5, 1.5 μL MgCl2, 3 μL KCl, 0.2 μL DTT, 0.6 μL ATP, 0.9 μL ubiquitin and the relative amount of purified E1, E2, and E3 enzymes (For a 30 μL reaction: 0,2 μg E1, 1,2 μg E2 and 2 μg E3); *see* **Note 13**. Add the respective amount of ultrapure water to reach 30 μL of total reaction volume; *see* **Note 14**).
3. Control reactions: Prepare the same reaction mix as described in the previous step, but omitting one of the components in each control reaction: Control reaction 1 contains all components, except ATP. Control reaction 2 contains all components except ubiquitin. Control reaction 3 contains all components except E1. Control reaction 4 contains all components except E2. Control reaction 5 contains all components except E3.
4. After preparing all reactions in individual PCR tubes, mix reaction components by shortly vortexing the tubes. Shortly spin down the tubes in a microcentrifuge.
5. Incubate all six samples (five controls and one sample) for 2 h at 30 °C (*see* **Note 15**).
6. Stop the reaction by adding 10 μL of 4x Laemmli buffer. (This amount varies depending on the initial reaction volume. The Laemmli buffer should be diluted to a final concentration of 1x in the reaction mix; e.g., if the initial reaction volume is 60 μL, add 20 μL of 4x Laemmli to the reaction.)
7. Vortex the samples, shortly spin them down in a microcentrifuge, and denature the samples for 5 min at 95 °C.
8. Proceed with the analysis of the samples via SDS–PAGE and immunoblotting (*see* Subheading 3.4).
9. For the interpretation of obtained results, please refer to Subheading 3.4, **point 10**.

3.3.2 Enzymatic Activity and Autoubiquitination Assessment

Once the enzymatic activity of the components (tested as described in the previous paragraph) has been proven, the *in vitro* ubiquitination assay can be used to test whether a certain POI possesses E3 ligase activity and whether it also has the capacity to attach ubiquitin molecules to itself in a process termed autoubiquitination, which is common for many E3 ligases.

1. To test whether a certain POI possesses E3 ligase activity, prepare the same reaction mix as described in the Subheading 3.3.1, substituting the E3 ligase with known enzymatic function with the respective amount of purified POI.
2. Include the same five control reactions described in Subheading 3.3.1 (*see* **Note 16**).
3. Perform the steps described in the previous Subheading 3.3.1.
4. For the interpretation of obtained results, please refer to Subheading 3.4, **point 10**.

3.3.3 Substrate Ubiquitination Assessment

Experimentally proving an enzymatic E3 ligase activity of a certain POI opens up the possibility to further extend the analysis by using the *in vitro* ubiquitination assay to investigate whether that POI is capable of transferring ubiquitin molecules to a putative target substrate.

1. To test the capacity of a certain POI with E3 ligase activity to *in vitro* ubiquitinate a putative target substrate, proceed as follows:
2. Prepare the reaction mix as described in Subheading 3.3.1 in **step 2** and 0.3 μg of target substrate (e.g., using stock solutions listed in Subheading 2.3, for a total of 30 μL reaction volume, mix 0.8 μL Tris–HCl pH 7.5, 1.5 μL $MgCl_2$, 3 μL KCl, 0.2 μL DTT, 0.6 μL ATP, 0.9 μL ubiquitin, and the relative amount of purified E1, E2, and E3 enzymes and the target substrate; *see* **Note 13**. Add the respective amount of ultrapure water to reach 30 μL of total reaction volume; *see* **Note 14**).
3. Control reactions: Prepare the same reaction mix as in **step 2** of this paragraph, omitting one of the components for each control reaction as described in Subheading 3.3.1, **step 3**.

   Prepare the controls 1-4 as described in the latter-mentioned step. Control reaction 5 contains all components except the target substrate.
4. To prove the specificity of the target substrate ubiquitination by the POI E3 ligase under investigation, an additional reaction mix can be prepared by substituting the POI E3 ligase with an unrelated enzymatically active E3 ligase (used in Subheading 3.3.1). As an additional control, a mutated POI E3 ligase version that is catalytically inactive could be used.
5. Proceed with **steps 4–9** described in Subheading 3.3.1.

3.4 SDS–PAGE and Immunoblotting

The in vitro ubiquitination assay can be analyzed by performing SDS–PAGE of generated samples followed by immunoblotting and detection of suitable chemiluminescent antibodies.

1. Resolve the denatured samples on a polyacrylamide gel (preferentially a 4–20% gradient gel, but also other percentages might be suitable) by SDS–PAGE.
2. In the case of 3.3.1 *Enzymatic Activity Assessment*, load 20 μL per sample on one gel.
3. In the case of 3.3.2 *Enzymatic Activity and Autoubiquitination Assessment*, load 20 μL per sample on one gel and 20 μL of sample on a second gel if the putative E3 is tagged with GST; load 10 μL of sample on the second gel if the putative E3 is tagged with MBP (*see* **Note 17**).
4. In the case of Substrate Ubiquitination Assessment, load 20 μL per sample on one gel, 20 μL of sample on a second gel, and 10 μL of sample on a third gel.
5. Blot the proteins onto a nitrocellulose or a PVDF membrane according to the manufacturer's instructions of the respective blotting system.
6. Block the membrane in 5% skimmed milk in TBS-T, preferentially overnight.
7. Incubate membranes with the antibodies against ubiquitin and the tags fused to the POI with putative E3 ligase activity or the putative target substrate POI (*see* **point 6** under Subheading 2.5). The membranes were 20 μL of sample were blotted onto should be used to incubate the antibodies against ubiquitin and GST, while the membranes were 10 μL of sample were blotted onto should be used for the MBP blot (see above).
8. After 1 to 2 h of antibody incubation, wash the blots with TBS-T buffer, incubate the membrane with an appropriate secondary antibody if needed (for 1 h, then wash again), and detect the chemiluminescent antibody signal using a Western blot imaging apparatus for chemiluminescent image capturing in combination with an appropriate chemiluminescent detection reagent (*see* **Note 18**).
9. Analyze the blots as described below.
10. Interpretation of the results: Enzymatic activity test (anti-Ubiquitin blot): If all necessary components for the ubiquitination cascade (E1, E2, E3, ATP, Ubiquitin, and buffers) are present in a sample and enzymatically active, this results in the detection of a typical ubiquitination smear (see Fig. 4a, lane 1). The addition of ubiquitin molecules increases the molecular weight of the ubiquitinated protein. Thus, a smear represents a successful ubiquitin conjugation reaction yielding high

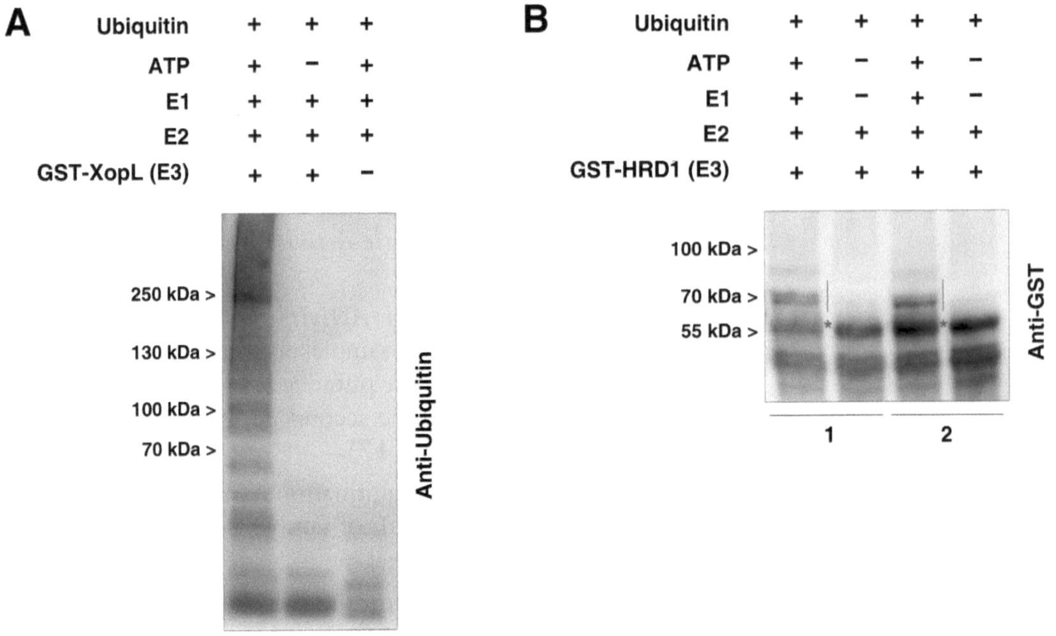

Fig. 4 *In vitro* ubiquitination assay. To test the enzymatic activity of all necessary components required for the ubiquitination cascade, they are combined in a reaction mixture described in the Methods section. (**a**) shows the capacity of the used enzymes to generate ubiquitin conjugates, proving their *in vitro* enzymatic activity and the functionality of the assay. Here, the recombinant GST-XopL purified from *E. coli* was used as E3 ligase. Reaction mixes lacking one of the necessary components (- ATP, lane 3; - E3, lane 4) served as negative controls. After 1 h at 30 °C, 20 μL of each reaction mixture was run on a 4–20% polyacrylamide gradient gel, followed by immunoblot analysis. The activity of the enzymes was analyzed using an anti-ubiquitin antibody and is shown by the typical ubiquitin "smear" in lane 1. In (**b**), the *in vitro* autoubiquitination activity of the E3 ligase HRD1 from Arabidopsis is shown (two replicates, indicated by 1 and 2). The assay was performed as described in Subheading 3.3.2. After 1h at 30 °C, 20 μL of each reaction mixture was run on a 4–20% polyacrylamide gradient gel, followed by immunoblot analysis. In the presence of all necessary components (lanes 1 and 3), a laddering pattern (size distribution indicated by the red line) starting from the size of the unmodified GST-HRD1 (~55 kDa, marked by the red asterisk) appears on membranes that are incubated with an anti-GST antibody against the GST-tagged HRD1 E3 ligase. Reaction mixes lacking one of the necessary components (- ATP and -E1, lanes 2 and 4) served as negative controls. In these negative controls, no laddering pattern reflecting ubiquitinated E3 is detectable, but only unmodified GST-HRD1, and some degradation products can be observed. E1-activating enzyme: His_{6x}-*At*UBA1, E2-conjugating enzyme: His_{6x}-*At*UBC8. Figure 4a was adapted from Ref. [19], while Fig. 4b was adapted from Ref. [20]

molecular weight ubiquitination products (around 100–250 kDa and higher). In case, the tested E3 ligase has the capacity to autoubiquitinate; usually the ubiquitin signal is very intense at and around the size of the E3 ligase. If one component is missing (e.g., ATP or the E3 ligase), this results in the absence of the ubiquitin smear and only free ubiquitin (low molecular weight) is seen in the blot (see Fig. 4a, lanes 2 and 3).

E3 autoubiquitination (anti-GST blot): If the POI with putative E3 ligase activity is fused to a GST tag, the autoubiquitination potential of the protein can be detected by incubating the membrane with an antibody against GST. In the presence of all necessary components of the ubiquitination cascade, autoubiquitination of the E3 is displayed by a typical laddering pattern around the size of the POI on the GST blot, reflecting the unmodified protein as well as the protein attached to different amounts of ubiquitin (Ub_n, see Fig. 4b, lanes 1 and 3). In the absence of one of the necessary components for the reaction, only the band reflecting the unmodified E3 and eventual degradation products (bands below full-length E3), but no laddering pattern should be visible (see Fig. 4b, lanes 2 and 4).

Target ubiquitination (anti-MBP blot): If the putative target substrate POI (fused to MBP) is ubiquitinated by the respective E3 ligase POI under investigation (fused to GST), the substrate ubiquitination can be detected by incubating the membrane with an antibody against MBP. In the presence of all necessary components of the ubiquitination cascade, ubiquitination of the target substrate is displayed by a typical laddering pattern around the size of the target POI on the MBP blot, reflecting the unmodified protein as well as the protein attached to different amounts of ubiquitin (Ub_n, see Fig. 5, lane 1). In the absence of one of the necessary components for the reaction, only the band reflecting the unmodified target substrate, but no laddering pattern should be visible (see Fig. 5, lanes 3 and 4). In these controls, bands of a higher molecular weight indicate unspecific binding of one of the antibodies used in the Western Blot, since they are not a result of the ubiquitination (see Fig. 5, lanes 3 and 4, red asterisks). In the control reaction without added target substrate, no band should be detectable in the anti-MBP blot (see Fig. 5, lane 2).

4 Further Applications

4.1 Identifying Effector Functions Other than E3 Activity or UPS Targeting

A phytoplasma effector protein that is potentially involved in the manipulation of the host immune system by interfering with host protein ubiquitination or by interfering with or exploiting the UPS does not necessarily possess E3 ligase activity to do so. A phytoplasma effector might also have other enzymatic functions, it could act, for example, as an E2-conjugating enzyme together with host E3 ligases to interfere with host target substrate stability, or it could be a DUB removing ubiquitin chains from the host target substrate to also influence substrate protein stability. Another option would be that the phytoplasma effector POI influences host protein degradation by other means, for example, acting as a shuttle factor binding to already ubiquitinated target substrates and leading them

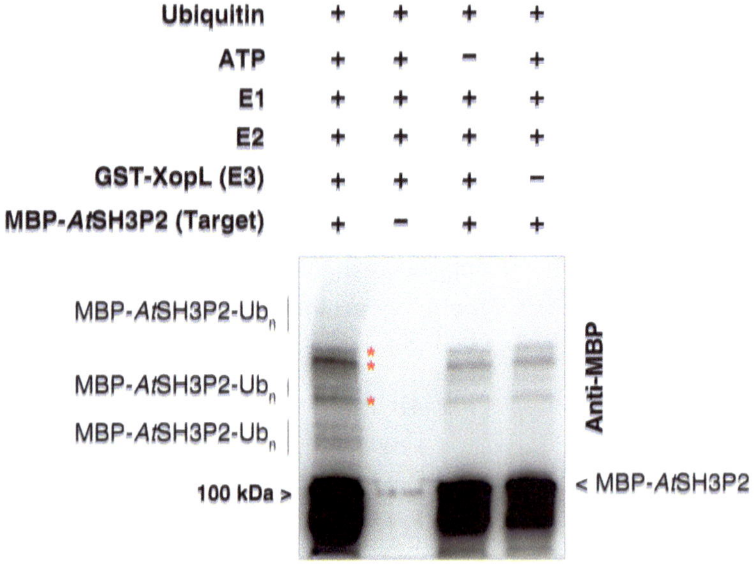

Fig. 5 Identification of target substrate ubiquitination using the *in vitro* ubiquitination assay. The *in vitro* ubiquitination assay described here can be used to verify whether a certain enzymatically active E3 ligase POI is able to ubiquitinate its putative target substrates. As an example, this figure shows that the E3 ligase XopL (fused to a GST-tag, thus not visible in the MBP-blot) ubiquitinates its target SH3P2 from Arabidopsis (*At*SH3P2, fused to an MBP-tag; ~ 100 kDa *in vitro.* E1-activating enzyme: His_{6x}-*At*UBA1, E2-conjugating enzyme: His_{6x}-*At*UBC8. The assay was performed as described in Subheading 3.3.3. After 1h at 30 °C, 10 μL of each reaction mixture was run on a 4–20% polyacrylamide gradient gel, followed by immunoblot analysis. In the presence of all necessary components, a typical laddering pattern starting from an unmodified MBP-*At*SH3P2 protein appears on the membrane that was incubated with an anti-MBP antibody, reflecting *At*SH3P2 protein conjugated to different numbers of ubiquitin moieties (lane 1). Red asterisks indicate unspecific bands that are not ubiquitinated and also occur in the control lanes 3 and 4. In the reaction mix lacking the target MBP-*At*SH3P2, no bands can be detected (lane 2, the very faint band is only spillover from the other lanes). Reaction mixes lacking one of the necessary components (- ATP, lane 3; - E3, lane 4) served as negative controls. In those samples, no laddering pattern reflecting ubiquitin-conjugated MBP-*At*SH3P2 is detectable. Figure 5 was adapted from [19]

to the UPS for degradation. These three options can be tested by adapting the *in vitro* ubiquitination assay described as follows:

4.1.1 E2 Enzymatic Activity of an Effector

Reaction mixes can be prepared as described in Subheading 3.3.3.

In this subsection, it is described how substrate ubiquitination can be assessed by substituting the *At*UBC8 E2 conjugating enzyme with the purified phytoplasma effector POI, which is thus tested for E2 enzymatic activity.

4.1.2 DUB Function of an Effector

Reaction mixes can be prepared as described in Subheading 3.3.3. Prepare an additional reaction containing all components necessary for the ubiquitination and add 2 μg of purified phytoplasma effector POI with putative DUB function. If the POI is an active DUB, the typical ubiquitin smear present when all necessary components of the ubiquitination cascade are present should largely or fully disappear (see Fig. 6, lane 4).

4.1.3 Shuttle Function of an Effector

Another option, i.e., the phytoplasma effector acts as shuttle factor, can be only indirectly tested using the *in vitro* ubiquitination assay described here. If a certain phytoplasma effector was found to interact with E3 ligases and/or potential E3 ligase substrates, purified effector POI can be added to the ubiquitination reaction mix containing those interacting proteins. If the ubiquitination status of the putative E3 target substrate changes in the presence of the phytoplasma effector, this might indicate that it interferes (at least to some extent) with the functional relationship between E3 and its target substrate. However, this only provides hints on a potential effector function that needs to be experimentally proven by other methods.

4.2 Characterizing the Ubiquitination Chemistry

4.2.1 Ubiquitin Chain Linkage

Different ubiquitin chain topologies can be formed on target substrates depending on which lysin residue of a certain ubiquitin moiety is attached to which lysin of another ubiquitin moiety. Revealing the chain topology of ubiquitin chains attached to a certain target substrate is important since the chain topology determines the fate of the ubiquitinated protein (e.g., K48-linked polyubiquitin chains often mark a target substrate for degradation via the 26S proteasome). Ubiquitin chain linkages can be investigated by at least two different methods using the *in vitro* ubiquitination assay described here. However, the obtained results must be interpreted with care and must be confirmed by other methods. Reaction mixes can be prepared as described in Subheading 3.3.3, substituting the ubiquitin with different commercially available ubiquitin mutants (e.g., Ub K48R or Ub K63R, where the respective lysine is mutated to arginine).

To differently analyze the ubiquitin linkage topology, reaction mixes can be prepared as described in Subheading 3.3.3. Subsequently, western blots must be performed using commercially available antibodies binding specific ubiquitin-linkage (e.g., antibodies against K48- or K63-linked ubiquitin chains).

4.2.2 E2 Specificity

The E2 conjugating enzyme *At*UBC8 used here in this protocol is a generic E2 enzyme able to act together with many different E3 ligases to attach ubiquitin molecules to target substrates. However, some E3 ligases require specific E2 conjugating enzymes to autoubiquitinate or actively ubiquitinate their targets. Hints for this might be given by interactome analysis that identifies other E2

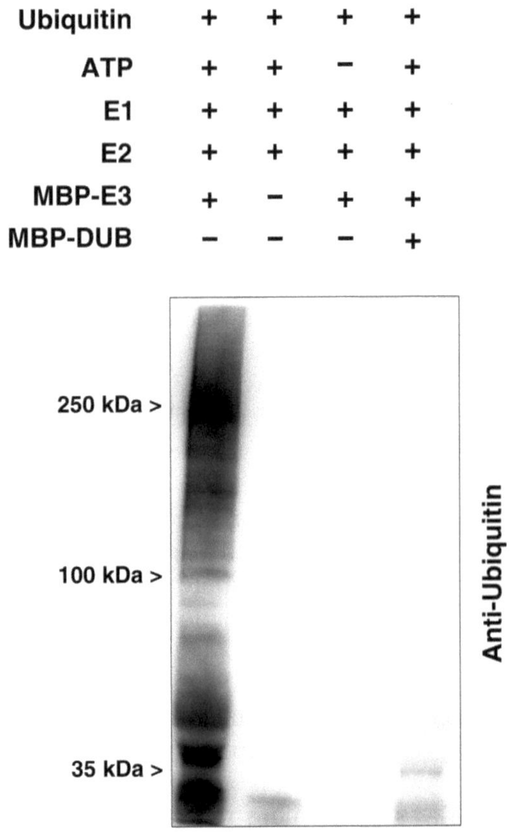

Fig. 6 Analysis of a deubiquitinating activity using the *in vitro* ubiquitination assay. The recombinant proteins required for an "*in vitro* deubiquitination assay" are expressed in *E. coli* and purified via the respective affinity tags as described in the Methods section. E1-activating enzyme: His_{6x}-*At*UBA1, E2-conjugating enzyme: His_{6x}-*At*UBC8, E3 ligase: MBP-tagged E3 ligase with confirmed enzymatic activity *in vitro*, DUB: MBP-tagged POI with putative deubiquitinating (DUB) activity *in vitro*. To verify the enzymatic DUB activity of the POI, this was tested together with the E1, E2, and E3 enzymes and the other necessary components in a reaction mixture as described under **point 1**, Subheading 3.3.3. A reaction mix not containing the putative DUB (lane 1) must be used as a positive control to verify the functionality of the *in vitro* ubiquitination assay. Reaction mixes lacking one of the necessary components (- E3, lane 2; - ATP, lane 3) served as negative controls and do not display the typical ubiquitination "smear." The enzymatic activity of the putative DUB can be analyzed by immunoblotting using an anti-ubiquitin antibody and is shown by the disappearance of the typical ubiquitin "smear," reflecting its potential capacity to trim ubiquitin chains that were previously conjugated by the action of E1, E2, and E3

conjugating enzymes as interaction partners of the studied E3 ligase. To test whether a certain putative E3 ligase POI is active or ubiquitinates its putative target in the presence of another specific E2, this E2 can be cloned, expressed and purified as described for *At*UBC8. The same protocol for *in vitro* ubiquitination assay described here can be used by substituting *At*UBC8 with the more specific E2 to unravel the enzymatic activity of that specific E2-E3 combination (*see* **Note 19**).

4.2.3 Identification of Ubiquitination Sites

One downstream analysis that provides important information when studying the process of ubiquitination is the identification of ubiquitination sites on either the E3 ligase POI or its target substrate. Samples from *in vitro* ubiquitination assays can be analyzed by liquid chromatography tandem mass-spectrometry (LC–MS/MS). Additionally, this method can also be applied to study ubiquitin chain linkage types. Please refer to experts in LC–MS/MS analysis for further information.

5 Notes

1. Storage periods indicate a minimum storage life of respective media/buffers/solutions and can eventually be prolonged if no contamination or precipitates are observed.
2. The product mentioned in this step worked well in the described experimental setting. However, similar products from other companies or suppliers might be equally suitable for the described method.
3. At least three different expression vectors (three different tags) are needed to distinguish the proteins of interest (POIs) when revealing the *in vitro* ubiquitination reaction by immunoblotting later. The expression vectors compatible with the Gateway^{TM} cloning technology used in this protocol were purchased from Thermo Fisher Scientific (former Invitrogen^{TM}) and are the following: pDEST17 (N-term 6x Histidine, His_{6x}), pDEST15 (N-term Glutathione-S-Transferase, GST), pMALc2 (N-term Maltose Binding Protein, MBP). The purification of recombinant proteins described in this protocol is based on affinity chromatography of His-, GST-, and MBP-tagged proteins. For the expression of the POIs, other suitable expression vectors and/or protein tags can be used, adapting the procedure accordingly.
4. Different tags and tag orientations (N-term vs. C-term) do influence the expression and purification efficiency of POIs. Therefore, it is advisable to experimentally test which GOI should be cloned in which expression vector to maximize the purification efficiency of all components used in the reaction.

5. This buffer is used in case His_{6x}-tagged proteins are purified with the ÄKTA™ start protein purification system in combination with the prepacked affinity chromatography columns HisTrap™ FF (catalogue no. GE17-5319-01, Merck). If His_{6x}-tagged proteins are affinity purified with, for example, Nickel NTA agarose beads, respective buffers should be used according to manufacturer's instructions. Imidazole concentrations in both buffers might be adjusted to maximize purification efficiency.
6. The Äkta™ Start system can also be used to purify proteins fused to other affinity tags (e.g., GST or MBP) using the appropriate affinity columns. In this protocol, only His_{6x}-tagged proteins were purified with this system.
7. The enzymatic activity of each component of the ubiquitination cascade (E1, E2, E3) needs to be assessed experimentally. To test the activity of the E1 and E2 enzymes, a known E3 ligase with previously assessed enzymatic activity should be used. Once the enzymatic activity of the E1 and E2 has been proven, the *in vitro* ubiquitination assay can be used to assess the enzymatic activity of POIs with putative E3 ligase activity, as well as the substrate ubiquitination activity of a certain E3 ligase on a putative target substrate.
8. POIs with putative E3 ligase activity and their putative target substrates should be cloned into expression vectors with different tags (e.g., putative E3 ligase tagged with MBP and putative target substrate tagged with GST). It is also recommended to use tags that are different from His_{6x} or HA (tag attached to the E1, E2, and eventually to ubiquitin) to avoid the detection of multiple bands on the western blot that cannot be properly assigned to a specific protein of interest.
9. The right induction condition needs to be assessed individually for each POI. To assess the right induction condition, small culture volumes (e.g., 50 mL) of each expression construct should be induced under the aforementioned conditions to test which condition works best for the respective POI. Once the right condition is assessed, larger culture volumes can be prepared for subsequent protein expression and purification. In case problems with protein induction are encountered using beforementioned conditions, protocols to optimize this step can be found in literature. For His_{6x}-*At*UBA1 and His_{6x}-*At*UBC8, culture growth at 37 °C for 4 h after addition of IPTG at 180 rpm was used to efficiently express those proteins.
10. This step is important for long-term storage of the purified POIs. The buffer used here contains glycerol, which makes the proteins stable enough to be stored at −80 °C for several months to years. Several freeze-thaw cycles do not harm their

enzymatic activity. It is also an important step to remove reagents that have a potential negative impact on the enzymatic activity of certain proteins (e.g., imidazole could eventually inactivate the enzymatic activity of a POI). Additionally, the centrifugal filters used in this step are size exclusion filters and might therefore increase the purity of the POI by removing smaller proteins or degradation products that are co-purified during protein purification using ÄKTA™ Start, amylose resin, or Glutathione Sepharose™ 4B.

11. If the Coomassie protein stain reveals impurities in your purified POI sample (e.g., many degradation products, unspecific bands in the elute sample), it does not mean that there is not enough enzymatically active protein. Compare the band thickness at the size of your estimated full-length protein with the BSA standard curve to assess the concentration of your POI. Enzymatic activity will then be experimentally assessed by performing the *in vitro* ubiquitination assay.
12. Several E3 ligases have been previously shown to have enzymatic activity in *in vitro* ubiquitination assays, being able to either autoubiquitinate themselves or to ubiquitinate target substrates in the presence of all necessary components of a ubiquitination cascade (E1, E2, E3, target substrate, ATP, Ubiquitin, buffers). Using those published active E3 ligases as controls will be helpful for interpreting the data obtained from the performed *in vitro* ubiquitination assay reactions. Some examples of proteins with experimentally proven *in vitro* E3 ligase activity are the bacterial effector protein XopL (from *Xanthomonas campestris* pv. *vesicatoria*) or the Arabidopsis membrane bound E3 ligase HRD1 [19, 20]. In both studies, the *in vitro* ubiquitination assay protocol described here was used to assess ubiquitination activity of respective E3 ligases and their target substrates, demonstrating that this is a powerful tool to unravel the role of a certain protein under investigation in this process.
13. The concentrations of purified E1, E2, and E3 are estimated by comparing the final elutes to a BSA standard curve on polyacrylamide gel stained with ready-to-use Coomassie protein stain. Protein concentrations are ideally adjusted to 1 mg/mL to simplify calculations and pipetting; however, this might not always be possible. First, this protein quantification method does not provide very accurate results, and second, some proteins cannot be easily purified, impacting overall protein yield. If in doubt, it is recommended to add a small excess of E1, E2, and E3 to the reaction mixture.
14. The prepared reaction volume depends on which downstream analysis will be performed. If, for example, more than one

immunoblot will be performed on those samples, or samples are supposed to be subjected to additional analyzes such as mass spectrometry (to identify *in vitro* ubiquitination sites), it is advisable to prepare larger reaction mixes.

15. The optimal incubation time of the reaction mixtures at 30 °C might vary depending on the respective enzymes tested. For example, some E3 enzymes catalyze the ubiquitin conjugation faster than others. The optimal incubation time can be assessed by time-course experiments.
16. Some of the control reactions might not be necessary to be included every time an *in vitro* ubiquitination assay is performed. The control reactions without ATP, without E1, E3, and substrate (in case of *Substrate Ubiquitination Assessment*) are the most important ones. Control reactions without ubiquitin or without E2 can also be skipped to save reaction components.
17. The MBP antibody used in this protocol is more sensitive than the GST antibody; thus, using an individual gel is recommended for optimal interpretation of WB signals. Loading too much of sample might lead to results (*see* **step 10**, Subheading 3.4) that are unable to be interpreted due to very bright chemiluminescent signal intensities. Of note, optimal sample volumes loaded on the respective gels should be determined experimentally (taking into account, for example, which antibodies and which gel pocket sizes are used), the volumes stated here are only for reference.
18. After washing the membranes with TBS-T buffer, an additional washing step with TBS buffer (without Tween-20) could improve imaging results since Tween-20 could interfere with the performance of the chemiluminescent detection reagent.
19. It must be noted that some E2s (e.g., *At*UBC35) require auxiliary factors (ubiquitin conjugating enzyme variants, UEVs) to be enzymatically active [21]. Those factors must be included in the assay to correctly interpret obtained results.

Acknowledgments

This protocol is adapted from Furlan & Trujillo, 2017 [7]. The author would like to thank Prof. Dr. Frederik Börnke for the possibility to experimentally establish this protocol in the lab. The work was supported by funding from the Deutsche Forschungsgemeinschaft (DFG). The author would also like to thank Prof. Dr. Suayb Üstün for supporting the writing of this chapter.

References

1. Raffeiner M, Zhu S, González-Fuente M, Üstün S (2023) Interplay between autophagy and proteasome during protein turnover. Trends Plant Sci 28:698–714. https://doi.org/10.1016/j.tplants.2023.01.013
2. Hatfield PM, Gosink MM, Carpenter TB, Vierstra RD (1997) The ubiquitin-activating enzyme (E1) gene family in Arabidopsis thaliana. Plant J 11:213–226. https://doi.org/10.1046/j.1365-313X.1997.11020213.x
3. Kraft E, Stone SL, Ma L et al (2005) Genome analysis and functional characterization of the E2 and RING-type E3 ligase ubiquitination enzymes of Arabidopsis. Plant Physiol 139:1597–1611. https://doi.org/10.1104/pp.105.067983
4. Ye Y, Rape M (2009) Building ubiquitin chains: E2 enzymes at work. Nat Rev Mol Cell Biol 10:755–764. https://doi.org/10.1038/nrm2780
5. Mazzucotelli E, Belloni S, Marone D et al (2006) The E3 ubiquitin ligase gene family in plants: regulation by degradation. Curr Genomics 7:509–522. https://doi.org/10.2174/138920206779315728
6. Trujillo M, Shirasu K (2010) Ubiquitination in plant immunity. Curr Opin Plant Biol 13:402–408. https://doi.org/10.1016/j.pbi.2010.04.002
7. Furlan G, Trujillo M (2017) In vitro ubiquitination activity assays in plant immune responses. In: Methods in molecular biology. Humana Press Inc., pp 109–121. https://doi.org/10.1007/978-1-4939-6859-6_8
8. Miricescu A, Goslin K, Graciet E (2018) Ubiquitylation in plants: Signaling hub for the integration of environmental signals. J Exp Bot 69:4511–4527. https://doi.org/10.1093/jxb/ery165
9. Kumari S, Nagendran K, Rai AB et al (2019) Global status of phytoplasma diseases in vegetable crops. Front Microbiol 10:441347. https://doi.org/10.3389/fmicb.2019.01349
10. Sugio A, MacLean AM, Kingdom HN et al (2011) Diverse targets of phytoplasma effectors: From plant development to defense against insects. Annu Rev Phytopathol 49:175–195. https://doi.org/10.1146/annurev-phyto-072910-095323
11. Bai X, Correa VR, Toruño TY et al (2009) AY-WB phytoplasma secretes a protein that targets plant cell nuclei. Mol Plant Microbe Interact 22:18–30. https://doi.org/10.1094/MPMI-22-1-0018
12. Janik K, Mithöfer A, Raffeiner M et al (2017) An effector of apple proliferation phytoplasma targets TCP transcription factors – a generalized virulence strategy of phytoplasma? Mol Plant Pathol 18:435–442. https://doi.org/10.1111/mpp.12409
13. Mittelberger C, Stellmach H, Hause B et al (2019) A novel effector protein of apple proliferation phytoplasma disrupts cell integrity of Nicotiana spp. protoplasts. Int J Mol Sci 20:4613. https://doi.org/10.3390/ijms20184613
14. Hoshi A, Oshima K, Kakizawa S et al (2009) A unique virulence factor for proliferation and dwarfism in plants identified from a phytopathogenic bacterium. Proc Natl Acad Sci U S A 106:6416–6421. https://doi.org/10.1073/pnas.0813038106
15. Tomkins M, Kliot A, Marée AF, Hogenhout SA (2018) A multi-layered mechanistic modelling approach to understand how effector genes extend beyond phytoplasma to modulate plant hosts, insect vectors and the environment. Curr Opin Plant Biol 44:39–48. https://doi.org/10.1016/j.pbi.2018.02.002
16. Strohmayer A, Moser M, Si-Ammour A et al (2019) "*Candidatus* Phytoplasma mali" genome encodes a protein that functions as an E3 ubiquitin ligase and could inhibit plant basal defense. Mol Plant Microbe Interact 32:1487–1495. https://doi.org/10.1094/MPMI-04-19-0107-R
17. MacLean AM, Orlovskis Z, Kowitwanich K et al (2014) Phytoplasma effector SAP54 hijacks plant reproduction by degrading MADS-box proteins and promotes insect colonization in a RAD23-dependent manner. PLoS Biol 12:e1001835. https://doi.org/10.1371/journal.pbio.1001835
18. Huang W, MacLean AM, Sugio A et al (2021) Parasitic modulation of host development by ubiquitin-independent protein degradation. Cell 184:5201–5214.e12. https://doi.org/10.1016/j.cell.2021.08.029
19. Leong JX, Raffeiner M, Spinti D et al (2022) A bacterial effector counteracts host autophagy by promoting degradation of an autophagy component. EMBO J 41:e110352. https://doi.org/10.15252/embj.2021110352
20. Langin G, Raffeiner M, Biermann D et al (2023) ER-anchored protein sorting controls

the fate of two proteasome activators for intracellular organelle communication during proteotoxic stress. bioRxiv:12.11.571118. https://doi.org/10.1101/2023.12.11.571118

21. Wen R, Torres-Acosta JA, Pastushok L et al (2008) Arabidopsis UEV1D promotes lysine-63-linked polyubiquitination and is involved in DNA damage response. Plant Cell 20:213–224. https://doi.org/10.1105/tpc.107.051862

Chapter 15

Post-embedding Immunolabeling of Thin Sections of Phytoplasma-Infected Plant Tissues for Transmission Electron Microscopy

Rita Musetti

Abstract

The chapter describes a transmission electron microscopy immunogold-labeling method applied to phytoplasma-infected leaf tissue. The method allows to identify the pathogen *in situ* and characterizes the specific molecular relationships established with the cell of the host plant, contributing to give insights to phytoplasma pathogenesis process.

Key words Antibodies, Immunogold labeling, Phytoplasma, Plant cell, Post-embedding, Transmission electron microscopy, Ultrastructure

1 Introduction

The relationships between plants and pathogenic microorganisms become manifest through the expression of macroscopic symptoms, behind which lie complex interactions regulated at the biochemical, molecular, and ultrastructural levels.

The transmission electron microscope (TEM) can reveal stunning detail of biological and non-biological samples at the atomic scale, by magnifying nanometer structures up to 50 million times. In plant pathology, TEM is a tool that not only contributes to a rapid and accurate detection of new pathogens but is also indispensable for the imaging of cellular structures involved in infectious processes, particularly those related to obligate, non-cultivable, ultramicroscopic pathogens, such as viruses and phytoplasmas. The availability of gold-conjugated antibodies, as tools to localize proteins inside cells and tissues at the ultrastructural level, empowered the use of TEM by providing insights into structure-function associations. In fact, specific antibodies are used to affix small particles of colloidal gold (5–25 nm in diameter),

Katrin Janik and Mattia Tabarelli (eds.), *Phytoplasma: Methods and Protocols*, Methods in Molecular Biology, vol. 3008, https://doi.org/10.1007/978-1-0716-5104-9_15,

which are easily visible under the electron microscope, to the macromolecule of interest.

Phytoplasmas are among the most recently discovered plant pathogens: they are difficult to study because uncultivable *in vitro* and because, in the host plants, they are confined to the sieve elements (SEs) of the phloem, one of the deepest tissues of the plant. Phytoplasmas greatly affect the ultrastructural organization of the host cells, inducing severe functional impairment, with important effects on assimilate translocation, in particular sugar metabolism and transport [1]. Using various methods based on both conventional transmission- and immuno-electron microscopy, it is possible to unravel the identity of the cell structures changed upon infection, thus helping to reconstruct phytoplasma physiopathology.

The aim of this chapter is to describe in detail a post-embedding technique applicable for the immunogold labeling of components of both phytoplasmas and phytoplasma-infected plant cells. According to the post-embedding procedure, tissues are embedded in hydrophilic resins, such as the London Resins, and ultrathin sections are first obtained and then immunolabeled. The method provides a good balance between generating strong immunolabeling and preserving ultrastructure.

2 Materials

2.1 Plant Material

Samples are taken from the main veins, shoots, and petioles of phytoplasma-infected and healthy control plants (*see* **Notes 1** and **2**).

2.2 Buffers and Fixatives

1. Sørensen's phosphate buffer (PB), pH 7.2: prepare a 0.2 M stock solution A (dissolve 2.40 g NaH_2PO_4 or 2.76 g $NaH_2PO_4{\cdot}H_2O$ or 3.12 g $NaH_2PO_4{\cdot}2H_2O$ in 100 mL distilled water) and solution B (dissolve 5.36 g $Na_2HPO_4{\cdot}7H_2O$ or 7.16 g $Na_2HPO_4{\cdot}12H_2O$ in 100 mL distilled water). Mix 28 mL solution A with 72 mL solution B. If necessary, adjust the solution to final pH 7.2 using HCl or NaOH and store at 4 °C. Dilute to 0.1 M with distilled water for washing and the preparation of glutaraldehyde solution.
2. Fixative: 0.2% (w/v) glutaraldehyde. Dilute 2 mL 8% (w/v) glutaraldehyde (stock solution) in 78 mL 0.1 M PB, to make up the final volume to 80 mL (*see* **Note 3**).

2.3 Dehydration Chemicals

50 mL each of 25%, 50%, 75%, and 100% of ethanol or acetone: water solution. Store in hermetically sealed bottles (*see* **Note 4**).

2.4 Embedding Media

1. London Resin White (LRW) Hard Grade (*see* **Note 5**).
2. Benzoyl peroxide (supplied to LRW resin in uncatalyzed form).
3. Bottleneck tip specimen embedding capsules.
4. Oven for resin curing.

2.5 Ultramicrotomy

1. Ultramicrotome (*see* **Note 6**).
2. Razor blades.
3. Diamond knife.
4. Forceps.
5. Carbon formvar-coated nickel grids.
6. Grid case.

2.6 Immunogold Labeling

1. TBS: 10 mM Tris–HCl, pH 7.4, 0.15 M NaCl.
2. Blocking solution: supplement TBS with 0.1% w/v Tween 20, 1% w/v bovine serum albumin (BSA) and 0.05% w/v sodium azide (*see* **Note 6**).
3. Dilution solution for immunogold: supplement TBS with 0.1% w/v Tween 20, 0.1% w/v BSA.
4. Primary antibodies raised against specific phytoplasma or plant proteins (e.g., monoclonal or polyclonal antibodies against phytoplasma membrane proteins [2], monoclonal antibody against plant actin [3]).
5. Primary antibodies diluted to the appropriate concentration with the blocking solution.
6. Washing buffer: 10 mM Tris–HCl, pH 7.4, 0.4 M NaCl.
7. Colloidal gold-labeled (5, 10, or 15 nm) anti-rabbit and anti-mouse secondary antibodies diluted to appropriate concentrations with the blocking solution (*see* **Note 7**).
8. Silicone rubber mats.
9. Moist chamber.

2.7 Stains

1. 3% v/v Glutaraldehyde solution in 0.1 M PB.
2. 1% v/v Osmium tetroxide in 0.1 M PB: mix 2% OsO_4 with an equal volume of 0.2 M PB, pH 7.2.
3. Non-radioactive heavy metal stain for TEM, appropriately diluted (*see* **Note 8**).

3 Methods

3.1 Sampling, Fixation, Washing, and Dehydration Steps

The main steps of the post-embedding immunogold labeling protocol here described are illustrated in Fig. 1.

1. Cut the plant vein collected into small pieces in a size of about 1 mm × 1 mm × 3 mm (the thickness of the leaf is usually lower than 1 mm). This shape facilitates the following sample localization to expose the transversal side of the vascular bundle during sectioning. Sample cutting should be precise and swift by using a very sharp blade without squeezing and dragging. After cutting, promptly process the plant samples with the fixative solution. It is indispensable to include a healthy plant as a control for the differentiation of the phytoplasma-infected from the normal structures.
2. After sampling, immediately place the tissue pieces in 0.2% glutaraldehyde fixative and fix for 1 h at 4 °C. Rinse fixed tissue pieces in 0.1 M PB, pH 7.4, for 30 min, with two or three changes.

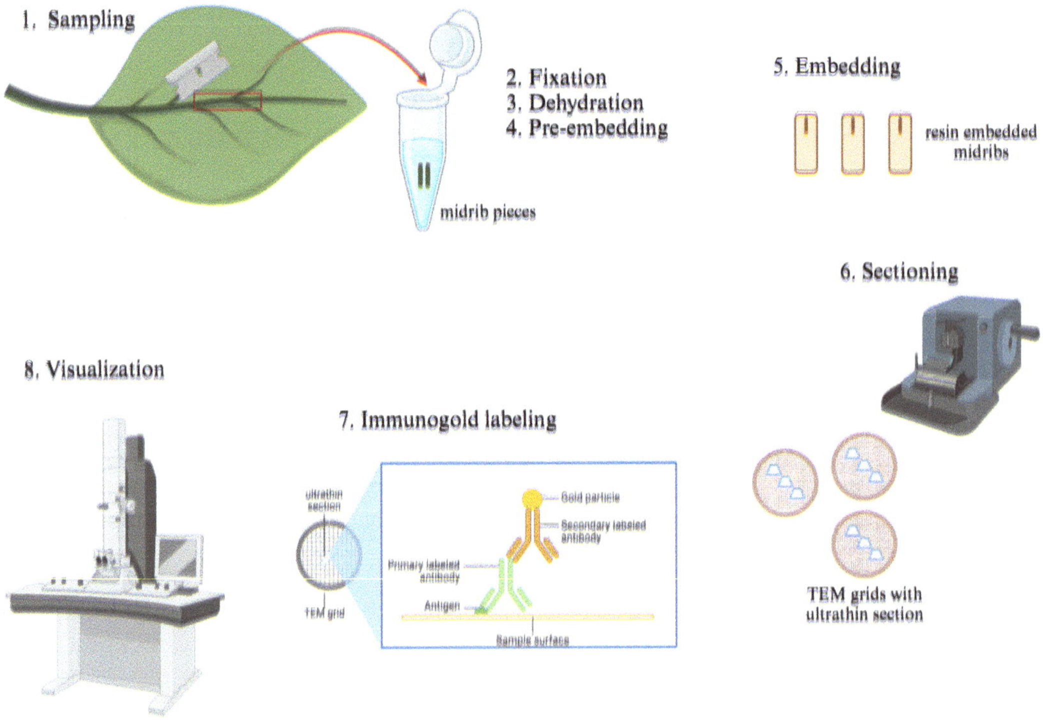

Fig. 1 Schematic illustration of the different steps for the post-embedding immunogold labeling of phytoplasma-infected samples

3. Dehydrate with a graded-ethanol or acetone series at 4 °C using 25%, 50%, and 75% each for 30 min. Leave samples in 100% ethanol or acetone for 1 h, at 4 °C.

3.2 Resin Pre-embedding and Embedding Steps

1. Dip the samples in fresh LRW resin:ethanol (v/v ratio 1:2) for 30 min, LRW:ethanol 2:1 for 30 min, and 100% LRW for 1 h, then in fresh 100% LRW overnight (two changes: the first 1 h, then overnight) at 4 °C.
2. Embed in fresh LRW resin containing 2% benzoyl peroxide (w/w) in bottle neck tip specimen embedding capsules and put samples in oven for 24 h at 60 °C, taking care that embedding capsules remain upright (*see* **Note 9**).

3.3 Immunogold Labeling

1. Cut ultrathin sections (60–70 nm in thickness) and collect them on the carbon formvar-coated nickel grids (*see* **Notes 10** and **11**). Place a 50 μL drop of blocking solution on a silicone rubber mat. Gently place the grid, tissue side down, on the drop and incubate for 2 h, at room temperature (*see* **Note 12**).
2. Place on a silicone rubber mat a 50 μL drop of primary antibody diluted at the appropriate dilution in blocking solution. Gently place the grid, tissue side down, on the drop and incubate for 2 h at room temperature, or overnight at 4 °C. Control grids must be incubated only in blocking solution without primary antibody.
3. Wash off unbound antibody by gently placing the grid, tissue side down, on top of a 50 μL drop of washing solution placed on the silicone rubber mat and wash for 5 min.
4. Repeat the wash step three times.
5. Place on a silicone rubber mat a 50 μL drop of secondary antibody diluted 1:20 in blocking solution and incubate for 1 h at room temperature (*see* **Note 11**).

3.4 Controls

Negative controls are represented by sections incubated with their respective secondary antibody alone (i.e., avoiding the primary antibody incubation step).

3.5 Section Staining and Imaging

1. Place the grids with the tissue side down on a drop of 2% glutaraldehyde for 5 min, then on 1% OsO_4 for 15 min and then on a drop of non-radioactive heavy metal stain for TEM (*see* **Note 8**) for 2 min.
2. Wash the grids by gently placing them for 2 min, tissue side down, on top of a 50 μL drop of distilled water placed on the silicone rubber mat.
3. Examine the sections under a TEM operating at 80–120 kV.

4 Notes

1. Perform the described procedures under fume hood, wearing gloves. Use disposable pipettes to aspirate the solutions in use, ensuring the tissue remains immersed in the liquid at all times. Follow the recommended safety protocols for disposing of all waste.
2. Working conditions for TEM are destructive to living organisms because high vacuum and exposure to high-energy electrons will kill organisms. To generate informative images, special treatments of the biological samples termed sample preparations, such as fixation, dehydration, embedding into media, should be performed to retain the structure for TEM similar to its natural state without distortion. Limited by the transmission capability of the electron beam, the sample for TEM imaging should be thin enough. For biological samples, this is done either by ultrathin sectioning of 50–100 nm in thickness.
3. Dehydration of TEM samples usually requires the use of ethanol or acetone. In order to minimize tissue shrinkage caused by osmotic shock, a graded series of dehydration agents is used.
4. Antigenic sites are easily denatured or masked during chemical fixation. Glutaraldehyde gives good fixation but may mask antigens. To avoid antigen masking, paraformaldehyde is often the better choice but results in poor morphology for electron microscopy. The use of glutaraldehyde at low concentration (e.g., 0.2%) could be a good compromise. Performing sample fixation for immunogold labeling, the use of osmium tetroxide is not recommended, as it is a strong oxidizing reagent and alters epitope structure.
5. One 500 g bottle of LR White Resin requires 9.9 g of catalyst to be added. If bought in pre-measured aliquots, one aliquot should be added to each bottle. The catalyst should be added to resin at room temperature and the resin must be shaken thoroughly immediately after addition of the catalyst. The catalyst will take a full 24 h at room temperature to dissolve completely. During this time, it is helpful if the bottle is shaken from time to time.
6. For post embedding, samples are first fixed, embedded, and sectioned before being subjected to immunolabeling. Post-embedding labeling is performed on sectioned tissue, on grids, allowing access to internal antigens.
7. Depending on the primary antibody type, the secondary antibodies are either goat-anti rabbit (GAR) IgGs or goat-anti

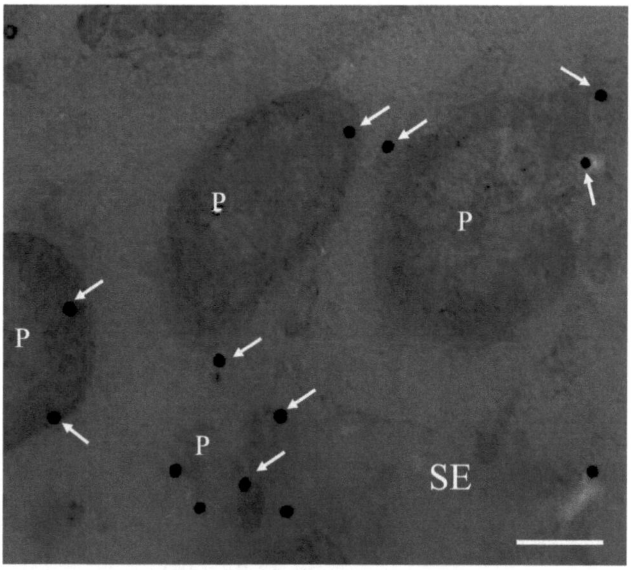

Fig. 2 Immunogold labeling of phytoplasmas associated with Chrysanthemum Yellows in the sieve elements of *Catharanthus roseus*, obtained using 15 nm gold particles. Particles are regularly distributed on the phytoplasma cell membrane. SE: sieve element lumen; P: phytoplasmas; arrows: gold particles. Bar corresponds to 0.1 μm

mouse (GAM) IgGs coupled to 5 nm (or 10 nm or 15 nm) gold particles. Different dilutions of the primary and gold conjugated secondary antibodies are evaluated in order to obtain the best combination. For phytoplasma membrane protein localization the best results are obtained with a dilution of 25 g/mL of the primary monoclonal antibody and 1:20 of the secondary GAM gold conjugated antibody (Fig. 2) [2]. For the actin visualization in phytoplasma-infected tomato sieve-elements, a dilution 1:200 of the primary monoclonal anti-actin antibody and a dilution 1:40 of the secondary GAM gold conjugated antibody are used (Fig. 3), [3].

8. Uranyl acetate replacement stain (UAR–EMS, Electron Microscopy Sciences) diluted 1:3 in ddH_2O, proved highly effective for phytoplasma visualization. Comparable non-radioactive alternatives from other suppliers may also be suitable.

9. Acrylic resins, such as LR White, are used instead of Epoxy resins. LR White is a blend of hydrophilic and acrylic monomers that rapidly penetrate tissue because of their low viscosity. LR White can be cold cured or heat cured. The hydrophilic nature of this resin makes it very appropriate for immunocytochemistry.

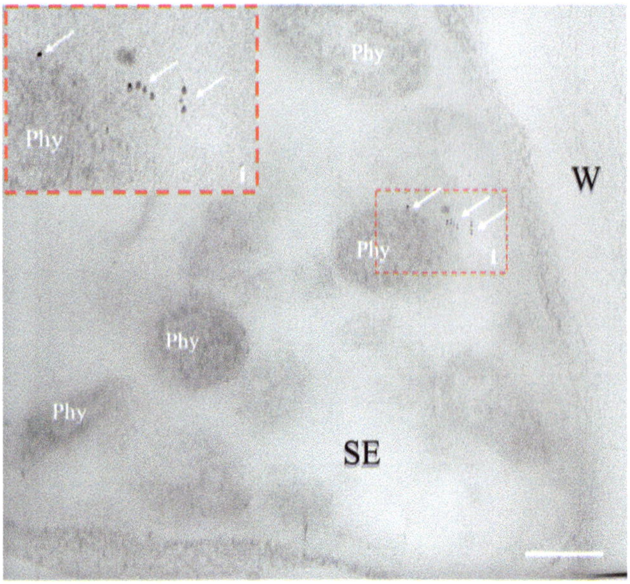

Fig. 3 Sieve element actin in contact with the '*Candidatus* Phytoplasma solani' cells is evidenced by using 5 nm gold particles. The area of interest is magnified in the inset i. SE: sieve element lumen; P: phytoplasmas; W: sieve element wall; arrows: gold particles. Bar corresponds to 200 nm

10. TEM grids are available in different mesh sizes. The smaller the mesh size, the better the sections will be supported by the grids and able to resist tearing when exposed to the electron beam. However, a mesh size too small can result in parts of the section being obscured, and, as we all know, the obscured area is always the one containing the best staining. With larger mesh sizes, such as the commonly used 200, the sections are more likely to tear under the electron beam. To help preventing this short coming, grids coated with a formvar/carbon film can be used. However, the formvar/carbon film will sometimes reduce image contrast. Therefore, decision on what mesh size to use and whether or not to use grids coated with a formvar/carbon film is mostly a matter of personal preference.
11. Do not use copper grids since copper will react with the salt in the antibody solution. Nickel, in contrast, will not react with the salt in the antibody solution.
12. Sections should never dry out during the entire incubation procedure, so it is advisable to carry out the entire reaction in a moist chamber.

References

1. van Bel AJE, Musetti R (2019) Sieve-element biology provides leads for research on phytoplasma lifestyle in plant hosts. J Exp Bot 70: 3737–3755. https://doi.org/10.1093/jxb/erz172
2. Musetti R, Loi N, Carraro L et al (2002) Application of immunoelectron microscopy techniques in the diagnosis of phytoplasma diseases. Microsc Res Tech 56:462–464. https://doi.org/10.1002/jemt.10061
3. Buxa SV, Degola F, Polizzotto R et al (2015) Phytoplasma infection in tomato is associated with re-organization of plasma membrane, ER stacks, and actin filaments in sieve elements. Front Plant Sci 6:650. https://doi.org/10.3389/fpls.2015.00650

Chapter 16

The Use of *Spiroplasma melliferum* as a Model Organism to Study Potential Antimicrobials Against Phytoplasma

Amir Fine, Luca Galtarossa, and Katrin Janik

Abstract

Spiroplasma melliferum is a cultivable mollicute that can be used as a proxy model organism to test the inhibitory effects of filtrates, compounds, or substances against phytoplasma. In this protocol, we describe the use of the resazurin-based alamarBlue™ dye for the monitoring of *Spiroplasma* survival and growth in the presence of presumably inhibitory agents.

Key words Phytoplasma, Biocontrol, *Spiroplasma melliferum*, Resazurin

1 Introduction

Screening for anti-phytoplasma molecules or compounds is a key approach in developing effective control and management strategies against phytoplasma-associated diseases [1]. The non-pathogenic bacterium *Spiroplasma melliferum* (*S. melliferum*) has been shown to serve as a suitable proxy model organism for testing the inhibitory effects of filtrates, compounds, or other substances against phytoplasma [2]. In some studies, a promising potential of naturally occurring microbes such as the fungus *Epicoccum nigrum* and the bacterium *Frateuria defendens* against different phytoplasma has been demonstrated [3–6]. To evaluate the antimicrobial activity of certain metabolites, filtrates, compounds, or other potentially inhibitory substances against phytoplasma, *in vitro* inhibition assays can be performed using a cultivable model microorganism.

Like '*Cand.* Phytoplasma', *Spiroplasma* is a genus within the class Mollicutes. *S. melliferum* is a cultivable member of this genus and was originally isolated from the hemolymph and gut of honeybees [7]. Due to its ease of cultivation [8], this bacterium can be used as an appropriate model organism for screenings to test the potential inhibitory function of an agent against phytoplasma.

Katrin Janik and Mattia Tabarelli (eds.), *Phytoplasma: Methods and Protocols*, Methods in Molecular Biology, vol. 3008, https://doi.org/10.1007/978-1-0716-5104-9_16,

S. melliferum has been previously utilized for this kind of screening purpose [2, 9].

In this chapter, a method for screening the antimicrobial activity of liquid or soluble substances onto the growth and survival of the phytoplasma-model organism *S. melliferum* is described. The survival and growth of *S. melliferum* can be monitored by continuous measuring the redox-potential of the growth medium. Living and proliferating cells create a reductive environment and growth of the microbial model organism is thus indicated by an increase of the reduced form of a redox-sensitive dye associated with an increase of fluorescence. Growth inhibitory effects of a tested substance are indicated by the absence or the interruption of an increase of the fluorescence signal.

In this protocol, a method using the resazurin-based reagent alamarBlue™ as a redox indicator is described. AlamarBlue™ has been widely used for determining cell viability and proliferation. It contains the cell-permeable, non-toxic, blue weakly-fluorescent indicator dye resazurin (IUPAC name: 7-hydroxy-10-oxidophenoxazin-10-ium-3-one). Upon its reduction, resazurin is irreversibly converted into resofurin (IUPAC name: 7-Hydroxy-3H-phenoxazin-3-one), a pink-colored and highly fluorescent dye [10]. This redox-sensitivity makes resazurin an ideal indicator for monitoring the reducing environment surrounding a living cell [11]. The accumulation of resofurin is associated with increasing fluorescence and can be spectrophotometrically measured using an excitation wavelength at 530–560 nm and an emission wavelength at 590 nm [11].

In this protocol, we describe how to use alamarBlue™ to test the antimicrobial activities of solutions of different origin on the growth of the model organism *S. melliferum*.

2 Material

All solutions must be prepared in double-deionized water. All aseptic work, i.e., the preparation of sterile solutions and media, bacterial culture, etc., must be performed in a biological safety cabinet. The viability of the cell was evaluated with alamarBlue™ (Thermo Fisher Scientiic, Waltham, MA, USA), according to the protocol provided by the manufacturer.

2.1 Culturing Spiroplasma melliferum

1. LB broth: Dissolve 5 g NaCl, 10 g tryptone, and 5 g yeast extract in 1 L ddH_2O and autoclave.
2. *Spiroplasma* modified broth (*Spiroplasma* growth medium): Dissolve the following reagents at the given concentrations in 1 L LB-broth, 2.5 g/L Heart infusion broth (powder), 1 g/L Fructose, 1 g/L Glucose, 10 g/L Sucrose, 70 g/L Sorbitol,

4.5 g/L NaCl. Adjust the pH to 7.6, add water to 900 mL and autoclave. After the medium has cooled, add 100 mL of Donor Equine Serum heat inactivated (*see* **Note 1**).

3. *Spiroplasma melliferum*: e.g., the freeze-dried *Spiroplasma melliferum* (Clark et al.) ATCC #33219. Grow the strain follow the culturing and preparation instructions from ATCC. Prepare 20 μL aliquots with a final glycerol concentration of 15–20% (v/v) and store them at −80 °C for long-term preservation (*see* **Note 2**).
4. Ampicillin (100 mg/mL): dissolve 1 g ampicillin sodium salt in 10 mL water and sterilize the solution by passing it through a 0.22 μm filter. Aliquot the solution and store at −20 °C. The aliquots can be stored for several months up to years.
5. 5% (w/v) Thallium acetate: Dissolve 50 mg of thallium acetate in 1 mL water and store at −20 °C (*see* **Note 3**). The aliquots can be stored for several months.

2.2 Inhibition and Viability Assay

1. The solution of interest (analyte) to be tested regarding its activity on *S. melliferum* growth (*see* **Note 4**).
2. Oxytetracycline (50 mg/mL): Dissolve 50 mg of oxytetracycline in 1 mL of dimethyl sulfoxide (DMSO, 100%). Protect from direct light and keep at −20 °C.
3. Deep well plates: polypropylene, 96 deep well with round bottom (*see* **Note 5**).
4. 3M Micropore tape to seal the lid and the plate.
5. Sterile 2 mL Eppendorf tubes.
6. AlamarBlue™ cell viability reagent (Thermo Fisher Scientific).
7. Cell viability assay medium: Dilute the alamarBlue™ cell viability reagent 1:10 in *Spiroplasma* growth medium.
8. Black 96 well plates (polystyrene), flat bottom with lid.
9. 50 mL Conical centrifuge tubes (e.g., from Falcon).
10. 14 mL culture tubes (e.g., Greiner Bio-One Round Bottom Polypropylene culture tube with two-position vent stopper).
11. Microtiter plate reader with incubation and shaking mode and laser and filter sets that allow the excitation and measurement of resazurin/resofurin fluorescence (e.g., the Varioskan LUX multimode microplate reader, Thermo Scientific).
12. Sterile 2 mL reaction tubes.
13. Multichannel pipet for transferring 5 μL/well, 95 μL/well and 180 μL/well sample volumes in a 96 well plate.

3 Methods

3.1 Preparation of the Analyte to Be Tested

Solutions (analytes) of different origins can be tested (*see* **Note 4**).

1. Sterilize the solution with an appropriate method (*see* **Note 6**).
2. Store the solution at appropriate temperature until used in the inhibition assay.

3.2 Cultivation of Spiroplasma melliferum

1. Prepare 40 mL of *Spiroplasma* growth medium as follows: In a 50 mL conical centrifuge tube, mix 40 mL of *Spiroplasma* modified broth, 100 μL 5% thallium acetate and 40 μL 100 mg/mL Ampicillin. Store the solution for further use at 4 °C (*see* **Note 7**).
2. Transfer 5 mL of the medium into a 14 mL culture tube.
3. Thaw on ice an aliquot of *Spiroplasma melliferum*.
4. Inoculate the culture tube with 10 μL of thawed *Spiroplasma melliferum* stock.
5. Grow in a shaker incubator set to 28 °C degrees and shake at 200 rpm (*see* **Note 8**).
6. The growth media in the culture tube should become turbid after ca. 3 days (*see* **Note 9**).

3.3 Assessing Inhibitory Effects on Spiroplasma melliferum Growth

1. To set up a 96 well deep well plate for testing growth inhibition of *Spiroplasma*, refer to the pipetting scheme provided in Fig. 1 (*see* **Note 10**). All samples and controls should be measured in duplicates. As a positive control, use the antibiotic oxytetracycline (*see* **Note 11**).
2. Thaw the analytes that are supposed to be tested for their antimicrobial activity against *Spiroplasma melliferum*.
3. Thaw an aliquot of the oxytetracycline stock at room temperature.
4. Label a 2 mL reaction tube for every analyte (max. 12), label 2 tubes for the oxytetracyline control, and 2 tubes for the negative control (NT).
5. Use the media prepared in **step 1** in Subheading 3.2 and dispense 1 mL of fresh media into each of the 12 reaction tubes prepared for sample analysis, 2 mL into each of the tubes for the oxytetracycline and the NT controls.
6. Vortex the analytes to mix them.
7. Dispense 1 mL of analyte into the corresponding labelled reaction tube, close and mix well.
8. Pipet 2 μL of oxytetracycline into the tube labelled as the oxytetracycline control, close and mix well.

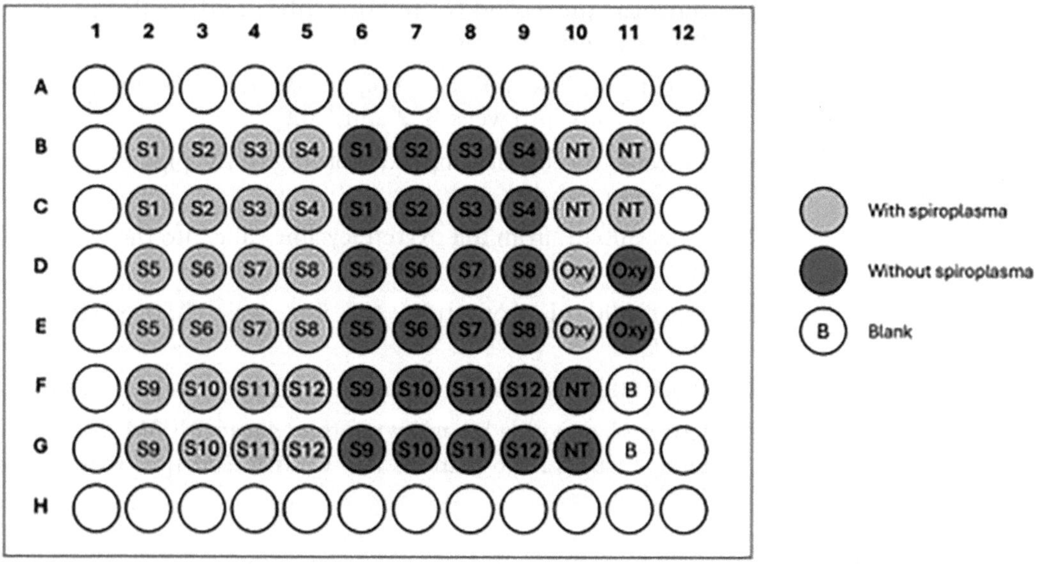

Fig. 1 Example of a plate setup and pipetting scheme for the viability assay. Samples (S1–S12), negative controls with media only (NT), inhibition positive control containing oxytetracycline (Oxy), and blanks (with water instead of sample) are loaded in duplicates. The wells in which *S. melliferum* is added are indicated as light grey wells and the wells serving as controls not containing Spiroplasma are indicated by a dark grey shade. Peripheral wells (white shade, no text) are loaded with water

9. Pipet 1 mL sterile water to all the outer wells of the plate (see Fig. 1).
10. Start with sample 1 and pipet 500 μL of the analyte mix into the wells of the positions B06 and C06. Do not discard the tube.
11. Proceed loading all samples according to the pipetting scheme shown in Fig. 1. Do not discard the tubes.
12. Pipet 500 μL of the oxytetracycline control into the wells at positions D11 and E11. Do not discard the tube.
13. Use one of the NT tubes to pipet 500 μL into the wells in position F10 and G10. Discard the tube.
14. Pipet 500 μL sterile water in the wells F11 and G11.
15. Take the 14 mL culture tube containing the *Spiroplasma* tube growing for three days in the shaking incubator and vortex well. The media should look turbid (*see* **Note 9**).
16. Transfer 500 μL from the 14 mL tube into a fresh 2 mL reaction tube.
17. Inoculate each 2 mL Eppendorf sample tube (with the remaining 1 mL analyte mix) and the oxytetracycline tube with 2 μL from the *Spiroplasma* culture tube and mix well.

18. Inoculate the second NT reaction tube (2 mL) with 4 μL of the *Spiroplasma* culture and mix well.
19. Starting with sample 1 and pipet 500 μL of the analyte/*Spiroplasma* mix into wells B02 and C02.
20. Proceed loading all samples according to the pipetting scheme shown in Fig. 1.
21. Pipet 500 μL from the oxytetracycline tube into the wells D10 and E10.
22. Pipet 500 μL from tube NT into wells B10, C10, B11, and C11.
23. Pipet 500 μL of sterile water in the blank wells F11 and G11.
24. Cover plate with a lid and wrap the edges with 3M Micropore tape. Incubate the plate at 28 °C with 200 rpm shaking for 3–4 days.

3.4 AlamarBlue™ Viability Assay

1. Switch on the plate reader and computer.
2. Set up a fluorescence kinetic protocol on your plate reader (with incubation and shaking function) using an excitation wavelength of 530–560 nm and an emission wavelength of 590 nm to monitor resorufin accumulation.
3. Select the plate type "Nunc 96F well black with lid" in the software.
4. Switch on the incubator of the plate reader for incubation at 28 °C.
5. Set up the microplate reader run according to the applied sample loading scheme (Fig. 1).
6. Prepare a 1:10 dilution working stock of 6.5 mL the of cell viability assay medium by adding 650 μL of alamarBlue™ reagent to 5850 μL of fresh *Spiroplasma* growth medium (from **step 1** in Subheading 2.2). Protect the solution from light.
7. Use a new black 96 well plate and fill the peripheral wells with 100 μL of sterile water according to the pipetting scheme in Fig. 1.
8. Transfer (using a multichannel pipet) 95 μL of the cell viability medium into all wells according to the plate layout but omitting the outer wells containing water only.
9. Remove the 96 deep well plate from the incubator.
10. Resuspend the culture by pipetting the media up and down several times (by using a multichannel pipet set to 180 μL/well). Change tips between every row and work with caution to avoid cross-contamination between wells.

11. Transfer 5 μL of the resuspended culture (using a multichannel pipet) to the corresponding row (i.e., only the wells with *Spiroplasma*) of the black 96 well plate with the cell viability media, mix well by pipetting up down three times. Be careful not to create bubbles.
12. Cover the plate with a lid and wrap it at the edges with 3M Micropore tape.
13. Put the plate in the microplate reader and start the run. The protocol will run for 48 h.
14. After the run has finished, remove the plate and discard it according to the respective guidelines for waste disposal of the used chemicals and biological agents.

3.5 Data Analysis

1. Prior to extracting the run results, it is recommended to run a blank subtraction using the mean values generated from the blank wells (containing water instead of the samples; do not use the peripheral wells containing only water as blanks) as a reference.
2. Extract the run file as a spreadsheet file.
3. It is important to subtract the blank background from the entire data set.
4. For every measurement (at each time point) of the sample *with* added *Spiroplasma* subtract the value of the corresponding control measurement *without Spiroplasma*. For example, subtract the values of sample 1 (wells B02 and C02) from the corresponding control wells (B06 and C06): Calculation for sample 1: Value1 = B02-B06; Value2 = C02-C06.
5. Repeat this for all samples in the run and average the readings of the duplicates (Value1 and Value2 of each sample and control at each timepoint).
6. For the positive inhibition control of the oxytetracycline control (D10/E10), subtract the measurement of wells (D11/E11) for every time point and average the duplicates.
7. For non-treated control wells (B10/B11) and wells (C10/C11), subtract with wells (F10 and G10) and average the wells measurements for every time point.
8. Plot the measurements as fluorescent intensity over time to visualize the *Spiroplasma* growth curve (Fig. 2a).
9. To calculate the percentage of alamarBlue™ reduction, use the following formula (Fluorescence; Fl): [Fl_{590nm} of test agent dilution]/[Fl_{590nm} of untreated control] × 100
10. Plot the percentage of alamarBlue™ reduction for each analyte after 30 h (Fig. 2b) (*see* **Note 12**).

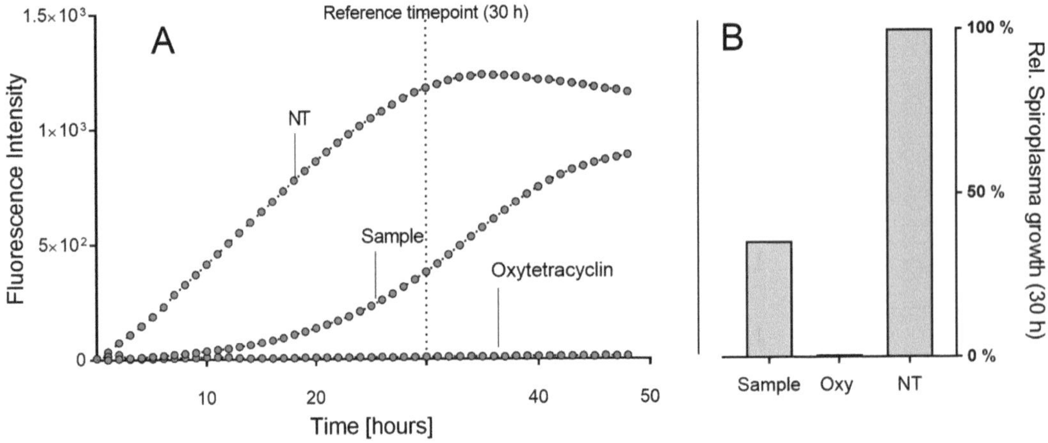

Fig. 2 Inhibitory effects of a sample analyte on *Spiroplasma melliferum* growth. (**a**) Time-course of fluorescence intensity reflecting *S. melliferum* growth in the presence of the analyte (Sample), oxytetracycline (Oxy) as a positive control for growth inhibition, and a non-treated control (NT) containing media only. The analyte sample shows a clear inhibitory effect, reducing fluorescence intensity compared to NT. Oxytetracycline exhibits the strongest inhibition, with fluorescence values remaining near baseline throughout the experiment. The reference time point (30 h) is indicated by a vertical dashed line. (**b**) Relative *S. melliferum* growth in the presence of the sample analyte at the 30-h reference time point, expressed as a percentage of the NT control. The sample analyte leads to a marked *S. melliferum* growth reduction

4 Notes

1. Donor Equine Serum Heat inactivated can be purchased directly from the suppliers. However, it is possible to heat inactivate the Donor Equine serum by placing the thawed bottle of serum into a 56 °C water bath containing enough water to immerse the bottle to just above the level of the serum. Incubate for 30 min swirl the serum every 5 to 10 min to ensure uniform heating. Aliquot and store in −20 °C.
2. Glycerol stocks of *Spiroplasma melliferum* can be stored up to several years. However, it is recommended to perform regular viability tests by streaking the strain on appropriate growth media plates and incubating it for ca. 24 h at 30 °C.
3. Thallium acetate is used as a selective agent against gram-negative bacteria. It is highly toxic, and care must be taken when preparing. Given the extreme toxicity of thallium, our experiments were carried out without its use. Importantly, the outcomes were equally effective and showed no significant differences compared to experiments performed with thallium acetate.
4. When using solutions, filtrates, or other liquid treatments in the assay, it is essential to ensure that these components do not interfere with the alamarBlue™ fluorescence signal.

Appropriate controls must be included to verify that the solvent or matrix itself does not affect cell viability, redox activity, or the assay's readout. This includes potential background fluorescence, quenching, or chemical interaction with the alamarBlue™ reagent.

5. It is recommended to use sterile 96 deep well round bottom (2 mL/well, non-treated) polypropylene plates (e.g., from Nunc). The plates are not supplied with a lid, which needs to be ordered separately.
6. The choice of sterilization method (e.g., filtration through a 0.22 μm membrane, autoclaving, or UV treatment) should be guided by the nature and stability of the solution components. It is crucial to ensure that the sterilization process does not alter the chemical composition or bioactivity of the solution, particularly in ways that might affect cell viability or assay readouts.
7. The modified culture medium solution can be stored at 4 °C for up to one month, provided it is kept sterile and properly sealed.
8. Although 30 °C is reported as the optimal growth temperature for *S. melliferum*, all cultures were incubated at 28 °C. No significant differences in growth performance were observed under these conditions.
9. Turbidity of the culture in the tube is not a consistent parameter. In some cases, the culture may not appear visibly turbid because the bacteria precipitated, in that case, vortexing is recommended to resuspend the pellet; however, this does not affect the outcome of the experiment.
10. Figure 1 is an example of a plate layout. It can be freely modified or adapted according to the specific requirements of the experiment.
11. Considering all necessary controls and measurement duplications, up to twelve samples can be measured on one 96 well plate. It is crucial to measure controls without added *Spiroplasma* for every sample to subtract any fluorescent background signal caused by the sample analyte.
12. The 30-h time point was selected as it corresponds to the onset of the plateau phase in the growth curve. At this stage, cell proliferation stabilizes while metabolic activity remains substantial. This allows for more reliable comparison between samples, minimizing artifacts associated with both exponential growth and the onset of cellular decline.

Acknowledgments

The work was supported by the Autonomous Province of Bozen/Bolzano, Italy, and the South Tyrolean Apple Consortium (project APPLiv).

References

1. Bertaccini A (2021) Containment of phytoplasma-associated plant diseases by antibiotics and other antimicrobial molecules. Antibiotics (Basel) 10. https://doi.org/10.3390/antibiotics10111398
2. Naor V, Ezra D, Zahavi T (2011) The use of *Spiroplasma melliferum* as a model organism to study the antagonistic activity of grapevine endophytes against phytoplasma. Bull Insect 64:S265–S266
3. Musetti R, Grisan S, Polizzotto R et al (2011) Interactions between '*Candidatus* Phytoplasma mali' and the apple endophyte *Epicoccum nigrum* in *Catharanthus roseus* plants. J Appl Microbiol 110:746–756. https://doi.org/10.1111/j.1365-2672.2011.04937.x
4. Musetti R, Farhan K, de Marco F et al (2013) Differentially-regulated defence genes in Malus domestica during phytoplasma infection and recovery. Eur J Plant Pathol 136:13–19. https://doi.org/10.1007/s10658-012-0147-6
5. Naor V, Iasur-Kruh L, Barkai R et al (2019) Introducing the potential biological control agent *Frateuria defendens* into pot- and field-grown grapevines. Phytopath Mediterr:341–346. https://doi.org/10.14601/PhytopatholMediter-10620
6. Naama-Amar A, Gitman S, Shoshana N et al (2020) Antimicrobial activity of metabolites secreted by the endophytic bacterium *Frateuria defendens*. Plants (Basel) 72:1–13. https://doi.org/10.3390/plants9010072
7. Clark TB, Whitcomb RF, Tully JG et al (1985) *Spiroplasma melliferum*, a new species from the honeybee (*Apis mellifera*). Int J Syst Bacteriol 35:296–308. https://doi.org/10.1099/00207713-35-3-296
8. Gasparich GE (2010) Spiroplasmas and phytoplasmas: microbes associated with plant hosts. Biologicals 38:193–203. https://doi.org/10.1016/j.biologicals.2009.11.007
9. Iasur-Kruh L, Naor V, Zahavi T et al (2016) Bacterial associates of *Hyalesthes obsoletus* (Hemiptera: Cixiidae), the insect vector of bois noir disease, with a focus on cultivable bacteria. Res Microbiol 168:94–101. https://doi.org/10.1016/j.resmic.2016.08.005
10. Bueno C, Villegas ML, Bertolotti G et al (2002) The excited-state interaction of resazurin and resorufin with aminesin aqueous solutions. Photophysics and photochemical reaction. Photochem Photobiol 76:385–390. https://doi.org/10.1562/0031-8655(2002)0760385TESIOR2.0.CO2
11. Rampersad SN (2012) Multiple applications of Alamar Blue as an indicator of metabolic function and cellular health in cell viability bioassays. Sensors (Basel) 12:12347–12360. https://doi.org/10.3390/s120912347

Part IV

Genomics and Sequencing: Advanced Sequencing and Genome Reconstruction Techniques

Chapter 17

High-Resolution Identification of Phytoplasmas with Multiplexed Multilocus Sequence Typing Using Hybridization Probe-Based Enrichment

Daniel Sanderson, Harvinder Bennypaul, Christine Hammond, Kevin Muirhead, and Tim J. Dumonceaux

Abstract

We recently described a method for PCR-independent sequence determination of a set of seven taxonomic markers for phytoplasmas and determined its utility for high-resolution strain differentiation of phytoplasma groups, including 16SrI, 16SrIII, 16SrX, and 16SrXII. Here we describe a protocol for high-throughput strain characterization with multiplexed hybridization reactions and a data analysis method that assembles five protein-coding gene sequences and concatenates them into a single phylogenetic marker of approximately 7.7 kb. In combination with the RFLP typing of the determined 16S and *cpn60* sequences, this method provides detailed sequence data on phytoplasmas of known or unknown taxonomic affiliations and can differentiate closely related but distinct phytoplasmas within the same group, for example, readily differentiating the ribosomal subgroup 16SrI phytoplasmas '*Candidatus* Phytoplasma asteris' and '*Candidatus* Phytoplasma triticii'.

Key words Taxonomy, MLST, Plant disease, Insect vector

1 Introduction

Plant pathogenic phytoplasmas are unculturable and therefore classified solely based on their DNA sequences [1]. The molecular identification, taxonomic characterization, biochemical properties, and virulence factors are important aspects of disease management, facilitating the study of disease transmission, vector identification, and the impact of these bacteria on ecological health and agricultural productivity. Approximately 50 '*Candidatus* Phytoplasma' species have been identified using molecular characteristics determined from DNA sequences, primarily (but not exclusively) those of the 16S rRNA-encoding gene locus [1]. Bertaccini et al. [2] have established that a sequence identity of >98.65% across the full-length 16S rRNA-encoding gene is suitable for establishing species

Katrin Janik and Mattia Tabarelli (eds.), *Phytoplasma: Methods and Protocols*, Methods in Molecular Biology, vol. 3008, https://doi.org/10.1007/978-1-0716-5104-9_17,

identity. A ~1200 bp fragment of the 16S rRNA gene, defined by the sequences between the universal primers R16F2n/R16R2 (termed the F2nR2 fragment) [3], has been used to establish a complementary classification scheme based on *in vitro* or *in silico* restriction fragment length polymorphism (RFLP) analysis [4]. The determination of such "ribosomal groups" in unknown phytoplasmas is facilitated by the availability of the online classification tool iPhyClassifier [5], which automatically performs *in silico* RFLP analysis, determines a similarity match to known phytoplasma sequences, and calculates a similarity coefficient. These two distinct classification schemes, based on taxonomy and ribosomal groups, have formed the basis for phytoplasma classification as it is currently understood.

The classification of phytoplasmas has therefore relied predominantly on the sequence of the 16S rRNA-encoding gene. The utility of this sequence stems from its ubiquity within phytoplasmas (and other bacteria), the sequence differences that define the various groups, and the accessibility of the sequence by PCR using well-described and commonly used universal primers. However, there are limitations to the use of this sequence for differentiating phytoplasmas, especially since its sequence divergence is not as high between distinct but related phytoplasmas as it is at other, protein-coding loci. In addition, phytoplasma genomes typically contain two 16S rRNA-encoding genes, and they are not always identical in sequence [6–8]. This 16S rRNA gene heterogeneity implies that two separate, and sometimes divergent, sequences are amplified from a single phytoplasma when the universal primers are applied. These limitations have led to the search for other, supplementary markers to use for the classification of phytoplasmas. Several suitable taxonomic markers have been described that are present in most or all phytoplasmas, including *tuf* [9], *secY* [10], *rp* [11], *cpn60/groEL* [12, 13], and others. The availability of such classification schemes has facilitated the implementation of multiple loci for even higher resolution differentiation of phytoplasmas. Due to the lack of a standardized set of taxonomic markers, multilocus sequence typing (MLST) of phytoplasmas has been employed using a variety of gene combinations [14–17]. Bertaccini et al. recently suggested that a suitable set of markers for providing complementary taxonomic information to 16S rRNA gene sequences is *tufB*, *secY*, *secA*, *rplV-rpsC* (*rp*), and *groEL/cpn60* [2]. Although whole-genome sequencing is increasingly used for taxonomic delineation of phytoplasmas [18], the challenges associated with generating such sequences [19] mean that large-scale application of this method has remained limited, with comparatively few phytoplasma genomes (fewer than 50) reported between 2003 and 2022 [1].

Therefore, the generation of sequences for MLST has relied on PCR and sequencing. This requires the design and application of universal primers that can amplify the target gene from all of the phytoplasmas of interest, as well as the generation of comparatively short (normally under 1kb) sequences that are then used for phylogenetic analysis. This restriction limits the achievable phylogenetic resolution and renders the generation of the sequences iterative and error-prone. In addition, the inference of phylogeny from multiple genes presents a choice between two modalities: the concatenation of gene sequences into a single, long gene consisting of all sequences put together in a specific, defined order (concatenated sequence tree), or the inference of separate phylogenies for each individual gene, which is then represented as a consensus gene tree [20]. The concatenation method has been shown to provide a high accuracy with a low number of gene targets [20], so it is a preferred method for constructing phylogenies from multiple genes.

We recently reported a method for the simultaneous, PCR-free determination of DNA sequences for a set of taxonomic markers for different phytoplasma groups [21]. This hybridization-based approach uses a set of probes to capture phytoplasma taxonomic markers from infected plants or insects, then assemble the individual gene sequences, including the 16S rRNA gene, *cpn60, tuf, secY, secA, nusA*, and *rp*. The probes are designed to detect these markers from any of the ribosomal groups 16SrI, 16SrIII, 16SrX, and 16SrXII, but the method is easily scalable to include other ribosomal groups as required. While the method is simple to perform and can profile many different samples at the same time, manual data analysis proved to be time-intensive and error-prone. Here, we describe this method along with an updated analytical code that automatically trims, maps, and assembles the sequences from each sample. In addition, where possible, the entire open reading frame (ORF) of each gene is selected and used to provide a concatenated sequence of a defined order of genes, which facilitates the phylogenetic analysis of the phytoplasma samples using the concatenated sequence tree approach. For this reason, the current method focuses on the main protein-coding genes of this panel (*cpn60, tuf, secY, secA, and nusA*). Sequences corresponding to 16S rRNA genes and the *rp* locus (which consists of four small ORFs) are handled separately due to their unique properties. Examples of typical results are provided, along with application notes to facilitate the reproducibility of the method.

2 Materials

2.1 DNA Extraction

2.1.1 Reagents

1. Kit or reagents for Plant DNA extraction (*see* **Note 1**).
2. Liquid nitrogen (optional). This can be used to help with tissue homogenization prior to DNA extraction. Alternatively,

homogenization can be performed using a bead beater, plastic pestle, or other suitable method.

3. Ten mM Tris–HCl, pH 8.0.
4. Reagents for fluorometric determination of DNA concentration.
5. Amplification mastermix for quantitative PCR.
6. Primers and FAM-labeled probe for quantitative PCR (Table 1).

Table 1
Primers, probes, and amplification conditions used for the detection of phytoplasma sequences in extracted and hybridization-enriched samples

Primers/probe	Amplification conditions	Phytoplasma	Target gene	Reference
TGGAGTTATTAATG TTGATG GGAGAAGCATATCC TTTA 5′FAM-ATCC TTCAACAACTTC TAATTCTG-3′BHQ1	95 °C, 3 min (1x); 95 °C, 10 s, 59 °C, 10 s, 72 °C, 30 s (40x)	'*Ca.* P. asteris' (16SrI)	*cpn60*	[30]
TGGTGTTATTAACG TTGA TGGTCTGTTAC TAAAACTA 5′FAM-CGACTGACA TAGTTTCCCTATTG T-3′BHQ1	95 °C, 3 min (1x); 95 °C, 10 s, 57 °C, 10 s, 72 °C, 30 s (40x)	'*Ca.* P. mali'; '*Ca.* P. prunorum'; '*Ca.* P. pyri' (16SrX)	*cpn60*	This work
TCCGACAGAGAAAG TATG TCAGCGATGATCAA TAAAG 5′FAM-TTTGATTC TTTGACCACTCC TTCTAAA-3′BHQ1	95 °C, 3 min (1x); 95 °C, 10 s, 59 °C, 10 s, 72 °C, 30 s (40x)	'*Ca. P. solani*' (16SrXII)	*cpn60*	This work
CGTACGCAAGTA TGAAACTTAAAGGA TCTTCGAA TTAAACAACATGA TCCA 5′FAM/TGA+CGG+GA +CTCCG+CA /3′IABKFQ	95 °C, 5 min (1x); 95 °C, 15 s, 64 °C, 1 min (40x)	'*Ca. P. pruni*' (16SrIII) and others	16S rRNA gene	[22]

FAM 6-carboxyfluorescein, *BHQ1* black hole quencher 1, *IABKFQ* Iowa black fluorescence quencher
"+" in front of a nucleotide indicates a locked nucleic acid modification

2.2 DNA Preparation for Hybridization and Sequencing

2.2.1 Reagents and Consumables

1. Illumina-compatible DNA library preparation reagent kit.
2. SPRI beads for size selection and DNA purification.
3. Centrifugal filter units with 30 kDa cutoff.
4. Dual-index sequencing oligos for multiplexed DNA analysis.
5. Nucleic acid hybridization and washing reagent kit (*see* ***Note 2***).
6. Magnetic streptavidin-coated beads for nucleic acid capture.
7. Primers for DNA library amplification.
8. Oligonucleotide blockers for PCR and hybridization applications (*see* **Note 2**).
9. High-fidelity DNA polymerase for PCR amplification.
10. NGS reagent kit for MiSeq Nano sequencing.
11. DNA low-binding 1.5 mL microcentrifuge tubes.

2.3 Equipment

1. Ceramic mortar and pestle for tissue homogenization in liquid nitrogen or tissue homogenizer, e.g., a ball mill homogenizer.
2. Microcentrifuge that can reach 14,000× *g* with a rotor that is suitable for 1.5 mL tubes.
3. Fluorometer for measuring DNA concentration.
4. Ultrasonic tissue homogenizer or ultrasonic cell disruptor.
5. Vacuum concentrator.
6. Automated electrophoresis system for high-sensitivity detection of nucleic acids.
7. MiSeq instrument (Illumina).
8. Thermocycler for quantitative real-time PCR.

3 Methods

3.1 DNA Extraction

A wide variety of methods can be used for DNA extraction from plants and insects. The method we commonly use is described below (*see* **Note 1**).

1. Select the appropriate tissue for DNA extraction (*see* **Note 1**).
2. Homogenize the tissue with physical disruption using a mortar/pestle and liquid nitrogen, or using beads and a vortex with adaptor, or a sample homogenizer at 30 Hz for 1 min.
3. Proceed with the kit-based DNA extraction using up to 100 mg (wet weight) of plant tissue and eluting in 50 μL of elution buffer or 10 mM Tris–HCl, pH 8.0 (*see* **Note 1**).
4. Measure DNA concentration using a fluorometric method.
5. Determine the phytoplasma levels using qPCR. A convenient method based on 16S rRNA genes to detect a wide range of

phytoplasmas has been described [22]. If the taxonomy of the phytoplasma under analysis is known, a taxon-specific qPCR can be used. The primers and probes we have used are provided in Table 1 (*see* **Note 1**).

3.2 Preparation of DNA for Hybridization

Several options are available for shearing and index addition. A fragment size of ~100–1000 bp is ideal, with a peak around 200–400 bp. We have typically prepared the samples as described below. The following description is based on a method used to prepare DNA samples for hybridization with probes corresponding to taxonomic markers for all bacteria [23, 24] (*see* **Note 3**).

1. Bring the DNA to a total volume of 100 μL in 10 mM Tris–HCl, pH 8.0, at a concentration of 2.5 ng/μL. If the concentration is too low, simply add buffer to a final volume of 100 μL and take note of the total amount of DNA going into the shearing procedure.
2. Sonicate the sample using the "high" power setting with 30 cycles of 30 s on and 45 s off. Keep the sonicator's water bath set to 4 °C or, alternatively, keep the sample in an ice-water bath to avoid heating of the sample.
3. Determine the results of the shearing by analyzing 1 μL of the sheared DNA using an automated electrophoresis system for nucleic acids and an appropriate reagent kit. Successful results are indicated by a smear of DNA fragments below 1000 bp, as shown in lanes 3–6 of Fig. 1. Once the conditions are established and verified, this does not need to be repeated for each new sample that is analyzed.
4. Concentrate the DNA to a volume less than 50 μL using a 30 kDa molecular weight cutoff filter unit or a vacuum concentrator. Adjust the volume to 55 μL with 10 mM Tris–HCl, pH 8.0.
5. Determine the DNA concentration using a fluorometric method. Perform a single measurement (2 μL) to preserve sample volume. Typically, the concentration of sheared DNA is 2–4 ng/μL at this stage, so that 100–200 ng of DNA is used as input for adding Illumina adaptors (*see* **Note 4**).
6. Use the sheared DNA (~50 μL) as input for an Illumina-compatible DNA library preparation reagent kit (*see* **Note 5**).
7. The adaptor-ligated DNA is size-selected for a fragment size distribution of approximately 300–500 bp. To achieve this, use 35 μL of SPRI beads for the first bead addition and 25 μL for the second bead addition. We have had very good success with peak sizes in the range of 300–400 bp using these SPRI bead volumes (Fig. 2).

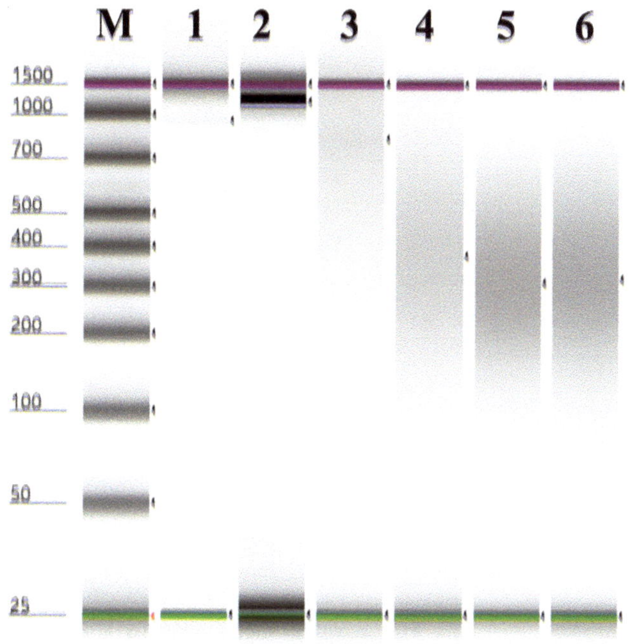

Fig. 1 Virtual electrophoresis image of typical shearing results. M, molecular weight marker (bp); Lane 1, unsheared infected canola; lane 2, unsheared infected caraway; lanes 3 and 4, sheared infected canola; lanes 5 and 6, sheared infected milkweed

8. Add the indices to the adaptor-ligated DNA fragments by following the NEBNext Ultra II DNA Library Prep Kit protocol, with eight cycles of PCR. Purify the indexed PCR products using SPRI beads as recommended in the protocol. The final volume after elution is 30 μL. Typical post-indexing results are shown in Fig. 2. The indexed libraries can be stored at −20 °C at this point.
9. Determine the DNA concentration using a fluorometric method (perform a single measurement with 2 μL of sample, to preserve sample volume) (*see* **Note 6**).

3.3 Hybridization and Sequencing

To examine the reproducibility of the method, we analyzed the same set of three infected samples twice. Samples from subgroups 16SrIII, 16SrI, and 16SrXII yielded assemblies of similar quality in two independent replicates (Table 2) (*see* **Note 7**).

1. Add 5 μL human Cot DNA (from xGen kit) and 2 μL of the xGen blocking oligonucleotides (selected based on the adaptors used) to a 1.5 mL tube that features low DNA binding. This is the TS mix for the NEBNext Ultra II protocol described above.

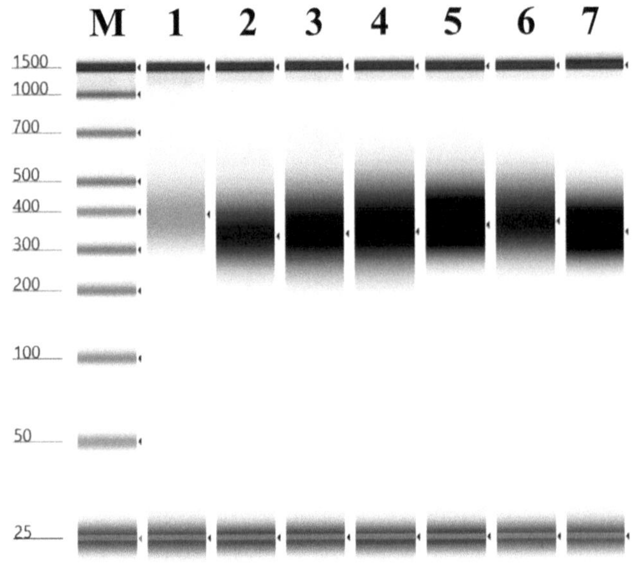

Fig. 2 Virtual electrophoresis image of typical results observed after index addition. M, molecular weight marker (bp); lane 1, infected lilac ('*Ca.* P. pruni,' 16SrIII-A); lanes 2–6, infected insects (*Macrosteles quadrilineatus*) from various geographic locations; lane 7, infected strawberry ('*Ca.* P. triticii,' 16SrI-(R/S)-C)

2. Add 500 ng of the adaptor-ligated, indexed DNA fragments. Up to five uniquely indexed samples can be added to a single hybridization; in the case of such multiplexing, add 500 ng of each library. If the concentration of indexed DNA does not permit the addition of 500 ng, add the entire amount and note the amount added to the hybridization reaction.
3. Dry the solution using a vacuum concentrator, with heating set to "high." The tube should be completely dry after approximately 10–15 min.
4. Follow the xGen hybridization and wash protocol precisely from this point forward (*see* **Note 8**). The hybridization probes (4 μL) are added in the first step, when rehydrating the reagents dried in the previous step.
5. After the final elution from the SPRI beads, determine the concentration of DNA in the captured library using a fluorometric method capable of detecting low concentrations of DNA (*see* **Note 9**).
6. Measure the amount of phytoplasma DNA in the sample using qPCR. See Table 3 and Fig. 3 for pre- and post-hybridization results on samples with various levels of phytoplasma.
7. Sequence the captured libraries using a NGS reagent kit for MiSeq (e.g., Illumina MiSeq Nano kit v2, 500 cycles,

Table 2
Reproducibility of the hybridization, sequencing, and assembly methods

	16S[a]		*cpn60*		*secY*		*secA*		*tuf*		*nusA*		*rp* (#ORF)	
Sample	rep1	rep2	rep1	rep2	rep1	rep2	rep1	rep2	rep1	rep2	rep1	rep2	rep1	rep2
2A1-16SrIII	2413 (III-A)	2288 (III-A)	–	–	**1468**	**1387**	**2902**	**2676**	**1667**	**1631**	1529	1276	**1748 (3)**	**1671 (3)**
BnAYTW1-16SrI	2596 (I-B)	2544 (I-B)	**2069 (I-IB)**	**2006 (I-IB)**	**1917**	**1732**	**3082**	**2846**	**1903**	**1846**	**1655**	**1466**	**1851 (4)**	**1798 (4)**
NA18199-16SrXII	2525 (XII-A)	1702 (XII-A)	**1992 (XII-IA)**	**1870 (XII-IA)**	977	1032	2099	1335	**2286**	**2158**	**1676**	1772	**757 (2)**	**1853 (3)**

Target assembly lengths are provided in bp. Assemblies containing a complete ORF are shown in bold

[a]16S, 16S ribosomal RNA-encoding gene; *cpn60*, chaperonin-60; *secY*, protein translocase SecY; *secA*, protein translocase SecA; *tuf*, elongation factor Tu; *nusA*, transcription factor NusA; *rp*, ribosomal protein complex

Table 3
C_q values before and after hybridization in samples of *Brassica napus* spiked with various levels of AY phytoplasma

Sample	C_q (prehyb)[a]	C_q (posthyb)	Reads/mapping proportion
B. napus AY (high)	18.38	5.57	32965/70.41%
B. napus AY (medium)	21.04	10.10	43456/45.59%
B. napus AY (low)	24.85	13.69	2723/3.04%
B. napus AY (very low)	28.23	16.53	640/0.6%

Also shown are the number of reads that mapped to the target (sum of all genes) at each spike-in level, and the proportion of the reads obtained in each spike level that mapped to all targets. All of these samples yielded suitable assemblies [21]
[a]*Prehyb* prehybridization, *posthyb* posthybridization

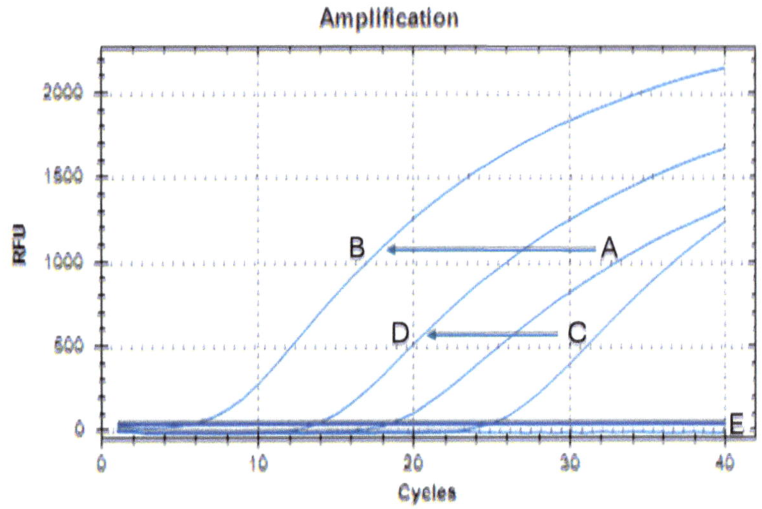

Fig. 3 *cpn60*-targeted amplification curves of a qPCR detecting 16SrI phytoplasma (Aster Yellows) before and after hybridization. See Table 2 for C_q and sample identification. Amplification curves generated from two of the four AY spiked samples described in Table 3 are shown before and after hybridization. Traces are identified as follows: A, AY-high prehybridization; B, AY-high posthybridization; C, AY-low prehybridization; D, AY-low posthybridization; E, unspiked control. The fluorescence threshold is indicated by a thick line above trace E

Illumina), with 25% PhiX DNA provided in the kit. First, determine the concentration and fragment length of hybridized samples using an automated electrophoresis and fragment analysis system.

8. Dilute each sample to the appropriate concentration such that the equimolar pool final total is 4 nM.
9. Confirm the concentration of the 4 nM pool using a fluorometric method capable of detecting low levels of DNA.

10. Prepare samples for MiSeq sequencing by following the manufacturer's instructions (*see* **Note 10**).

3.4 Data Analysis—Download and Install Analysis Code

The data can be analyzed in a variety of ways. Here we provide our standard analysis procedure. We have run the analysis on a server running Red Hat Enterprise Linux release 8.10 (Ootpa).

1. Download the scripts, files, and instructions using git clone https://github.com/kevmu/phytoplasma_hybrid_genes. This will generate a folder called phytoplasma_hybrid_genes that contains the code.
2. Follow the instructions in the README file to install the appropriate conda environments for the shell and Python scripts. This includes EMBOSS and biopython.

3.5 Data Analysis—Execute Shell Script and Retrieve Results

1. Place the fastq files from the Illumina sequencing run into a known, accessible folder (i.e., path/to/fastq/files, which will depend on the storage system).
2. Generate the file *_fastq_files.txt, which will contain a list of the sequences to be analyzed. An example of how this file can be generated is:

   find /path/to/fastq/files -type f -name "*.fastq.gz" | rev | cut -d '/' -f1 | rev | sed 's/_L001_R[1-2]_001.fastq.gz//g' | sort -V | uniq | grep "BbSP_S17\|flaxBR1_S18\|Lilac2A1_S9\|Mq13_S19" > $HOME/phytoplasma_hybrid_genes/fastq_files_list.txt

   where "BbSP_S17," "flaxBR1_S18," "Lilac2A1_S9," and "Mq13_S19" correspond to the sample name prefixes from the fastq files for each of the samples to be analyzed from that run.
3. Use a text editor to edit the file phytoplasma_hybrid_genes.sh, indicating the location of the *.fastq_files.txt generated in step two and of the input directory. The input directory is the location of the fastq files generated for the hybridized samples (path/to/fastq/files). In addition, change the fastq suffixes for reads one and two and the file extension suffix to correspond to those in the fastq files to be analyzed.
4. A base output directory is specified as "output," and this should be changed with each analysis to identify the analysis and prevent overwriting of results.
5. Execute the code using sh phytoplasma_hybrid_genes.sh.
6. Examine the file read_count_summary.tsv. This will have the number of reads that mapped to each of the corresponding genes in each sample and gives an indication of the robustness of the assemblies (*see* **Note 11**).

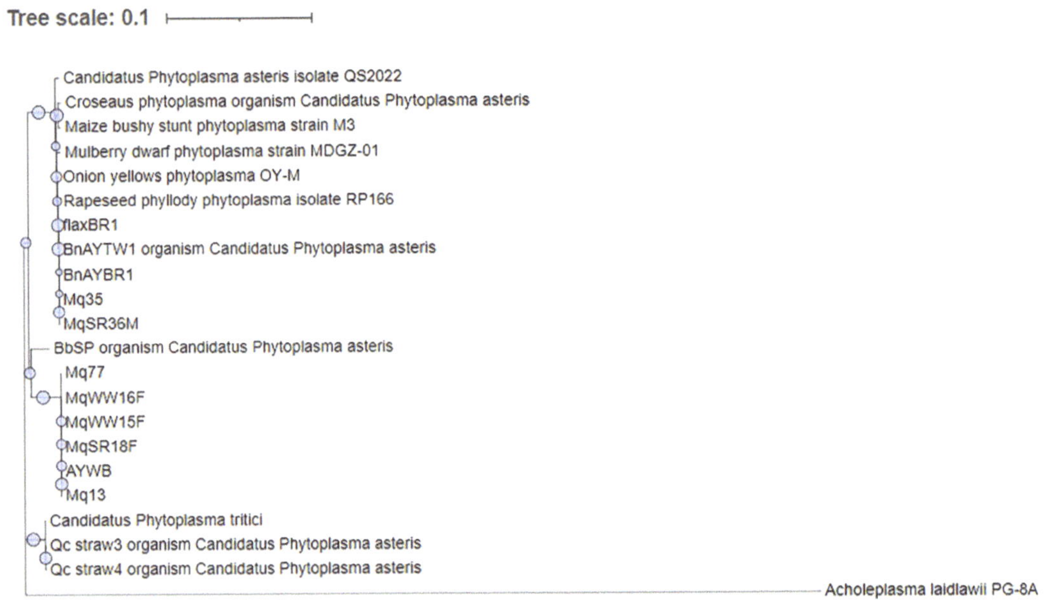

Fig. 4 Concatenated sequence tree of the ~7.7 kb taxonomic marker generated from hybridization and sequencing of various samples from ribosomal group 16SrI. The concatenated sequence consists of the entire ORF of five taxonomic markers, in the order *cpn60/secY/secA/nusA/tuf*. Samples from plants included blueberry (BbSP [28]); flax (BR1 [31]); canola (BnAY [31, 32]); and strawberry (Qc straw [33]). Sequences labeled Mq were from infected *M. quadrilineatus* collected in Saskatchewan (Mq77, Mq13, Mq35) or Manitoba (MqSR36M, MqWW15F, MqWW16F, MqSR18F), Canada. Reference sequences were obtained from the GenBank database [34]. Strong differentiation is observed between ribosomal groups 16SrI-A (reference strain AYWB) and 16SrI-B (reference strain OY-M). The 16S heterogeneous strain BbSP (16SrI(E/AI)AI) is also easily differentiated, and the recently described group 16SrI species '*Ca.* P. tritici' [35] forms a distinct group with the strawberry samples taken in Quebec, Canada [33]

7. Examine the output (*see* **Note 12**).
8. Examine the assemblies for the presence of complete open reading frames (ORF). If these are not present, trim the sequences to the length of the shortest assembly (*see* **Notes 13**, **14**, **15**, and **16**). An example of the results that are observed is shown in Fig. 4.
9. Examine the *cpn60* gene sequence by phylogenetic and RFLP analysis (excluding group 16SrIII) (*see* **Note 17**). Other genes may be analyzed by BLAST analysis to determine the approximate taxonomic placement of the strain under analysis.

3.6 Analysis of the 16S rRNA Gene Sequences

The high conservation of 16S rRNA-encoding gene sequences is challenging for the hybridization and assembly steps (*see* **Notes 18** and **19**).

1. Using bowtie2-build [25], create reference files from the 16S genes 16S_ref.fasta.

2. Map the merged Illumina reads (*.merged.extendedFrags.fastq) against the 16S reference genes.
3. Extract the mapped 16S reads for all samples using samtools [26] (*see* **Note 20**).
4. Assemble the mapped, extracted reads using an appropriate assembly tool. We have used TransAbyss [27].
5. Trim the assemblies to retain only those that are longer than 500 bp.

4 Notes

1. The tissue selected for DNA extraction could be an infected leaf midrib or petiole, or root tissue. While stem tissue can contain phytoplasma DNA [28], it is typically more challenging for extraction, and the concentration of phytoplasma may be lower. The DNA yield can vary considerably, but in most cases, we have worked with starting concentrations above 50 ng/μL. Indexing kits are typically flexible regarding the amount of input DNA. We compared the performance of various DNA extraction methods with respect to the yield of reads obtained, and the proportion of the reads that mapped to the target, and all of the evaluated methods provided a sufficient number of reads for target assembly (Sanderson et al., in preparation). The most important parameters for successful target assembly include the match of the phytoplasma in the sample to the probe and the phytoplasma levels in the infected sample. Generally, lower Cq values (reflecting higher phytoplasma levels) will provide better results, but we have produced successful assemblies with samples featuring Cq as high as ~27 [21]. However, samples with lower concentrations of phytoplasma may not provide assemblies of sufficient length for ORF extraction and analysis. In those cases, a smaller fragment of the gene may be used for phylogenetic analysis. In our setting, the Qiagen Plant Pro Kit performed best.
2. We have used the xGEN hybridization and wash kit (IDT). This kit can be purchased with or without the streptavidin-coated magnetic beads, which can be purchased separately from other suppliers. Regardless of the kit used, it is crucial to select a compatible set of oligonucleotide blockers based on the indices added. For example, we typically use the TS mix (IDT), which is compatible with ligation-based library preparation methods. In other cases, the NXT mix might be used for Illumina Nextera adapter sequences (8 and 10 bp indices). It is crucial to match the blocking oligonucleotide to the sequencing index

used—checking with technical support of the supplier of the hybridization and wash kit is strongly recommended.

3. Chemical or enzymatic DNA fragmentation can also be used (*see* **Note 1**), and a wide variety of kits and protocols are available for this. For tagmentation-based kits, we very strongly recommend the use of library prep kits that are specifically designed for downstream enrichment steps, such as the Illumina DNA Prep with Enrichment kit. That said, with work-around steps, we found that the more general Illumina DNA Prep (M) Tagmentation kits will produce libraries that can be made compatible with this protocol by pre-incubating the fragmented and indexed samples with soluble streptavidin. To do this, reconstitute the streptavidin to 1 mg/mL and add an equal volume of this solution to the prepared sample (e.g., 10 μL sample with 10 μL streptavidin). Incubate at room temperature for 20 min, then purify the DNA using SPRI beads (e.g., from AmPure, Nucleomag, Cytiva, etc.) for full DNA recovery without size selection. Wash the beads three times with 80% ethanol, and elute the DNA in 20 μL of 10 mM Tris–HCl, pH 8.0. Determine the DNA concentration using fluorometry, then use this pre-incubated sample for hybridization.

4. At this point, it is critical to record which index is to be used for each sample so that the data can be demultiplexed later. The hybridizations can be planned such that up to five samples can be pooled into a single reaction. The total number of samples to be sequenced on a single run will depend on the Illumina sequencing chemistry that is used; higher-yield sequencing kits will, of course, permit more samples to be pooled into a single sequencing reaction. We typically obtain 1 M reads per run when using the MiSeq Nano kit to sequence up to 30 samples, consisting of six hybridizations of five pooled samples each. This is a cost-effective strategy for profiling samples at moderate throughput, but we have not investigated higher throughput levels using other Illumina sequencing chemistries.

5. Here, we provide a detailed protocol for library preparation using the NEBNext Ultra II DNA Library Prep kit (NEB). However, appropriate kits from other suppliers might be equally suitable for this procedure. We have followed the NEBNext Ultra II DNA Library Prep kit protocol (revision 7.0, September 2022), with the options described in the steps below.

6. Typical concentrations observed are 20–40 ng/μL. Due to the volume available (~30 μL), concentrations below ~17 ng/μL will not provide sufficient indexed DNA for 500 ng in the hybridization reaction. However, lower amounts can still give

successful sequencing results; we have had successful results with as little as ~150 ng in the hybridization reaction.

7. In this subsection, we provide a detailed protocol for hybridization using the xGEN hybridization and wash kit (IDT) in combination with reagents from the library preparation kit NEBNext Ultra II kit (NEB), which is what we typically use. However, appropriate kits from other suppliers might be equally suitable for this procedure.
8. We have used hybridization times of both 4 h and 16 h with success; users may further optimize the hybridization time or temperature. Since the panel size is under 500 probes, use 14 cycles of PCR in the final amplification step. Probe design details and the sequences of the designed probes have been published [21] and are publicly available.
9. The post-hybridization yield can vary considerably, from ~1 ng/μL to ~80 ng/μL. If the yields are quite low (under 10 ng/μL), we have had success in preparing a dilution of the post-hyb fragments (1:5 in 20 μL), then repeating the amplification using the xGen library amplification primer. This increases the yield considerably, and we have not noticed any deleterious effects on the sequences generated [21]. When the hybridization has worked as expected, the qPCR in step five of the procedure will provide very low Cq values—often around five to eight, depending on the levels of phytoplasma in the original sample.
10. We have used Illumina protocol 15039740 rev. D, using a final concentration of 8 pM of denatured library, including 25% 12.5 pM PhiX (if using V2 chemistry).
11. Reads corresponding to the 16S rRNA gene are specifically excluded from the script due to their special features requiring additional processing steps. The number of reads obtained for each gene will depend on many factors, including the phytoplasma levels in each sample, the yield of reads from the sequencer, and the multiplexing strategy. Typically, thousands of reads are obtained per gene, resulting in long assemblies containing the entire open reading frame (ORF) of the target gene. Generally, if a gene contains fewer than 100 reads, the assembly will not be successful.
12. The folder called gene_sequences will contain sub-folders with the sequences of each gene from each sample. It will also contain a file (*_gene_stats.tsv) with the assembly lengths obtained for each gene from each sample. Finally, this folder will contain a file (*_concat_hybrid_genes.fasta) with the concatenated sequences obtained. This sequence will correspond to the extracted ORF from each gene, in the order *cpn60/secY/*

secA/nusA/tuf, and will be approximately 7.7 kb in length for most samples.

13. The results of the analysis to this point include information on the read numbers, assembly lengths, the individual assembled gene sequences, and the concatenated gene sequence that can be used for phylogenetic analysis. Individual genes can also be used for phylogenetic analysis if preferred, but the ends of the assemblies are not defined by PCR primer-binding sites as would be typical for PCR-based MLST. Therefore, it will be important to trim the assembled sequences to the same size as the reference sequences for the phylogenetic tree. Alternatively, if the ORF is retrieved, it can be extracted and used for phylogenetic analysis.
14. Samples that are '*Ca.* P. pruni' (16SrIII) will not contain a *cpn60* gene, but a few reads may map anyway. These will be artefactual reads that will not assemble into anything meaningful. For this reason, assemblies from 16SrIII samples will be shorter (approximately 6.1 kb), since they lack the *cpn60* gene. Therefore, the concatenated sequence will correspond to *secY/secA/nusA/tuf*.
15. In some cases, insufficient reads are produced to enable ORF extraction from all genes. In these cases, the automated concatenation approach will not work, and the user will have to decide how best to trim individual (not concatenated) sequences for phylogenetic analysis.
16. Assemblies corresponding to the *rp* locus are specifically excluded from the concatenation process, because this locus contains up to four small ORFs (*rpsS*, ~280 bp; *rplV*, ~420 bp; *rpsC*, ~80 bp; and *rplP*, ~330 bp). All four ORFs are not always obtained. The *RP* sequence can be phylogenetically analyzed separately, as described [29].
17. For samples that contain a *cpn60* gene (all but 16SrIII), the assembled *cpn60* sequence can be analyzed using *in silico* RFLP using the *cpn60* classifier (http://cpnclassiphyr.ca) [26]. This will provide information on the *cpn60*-based RFLP group of each sample and also features automated phylogenetic analysis and the ability to download the trimmed universal target sequence.
18. We have found that the probes commonly bind to sequences corresponding to host chloroplast and mitochondrial genes, as well as to 16S rRNA genes of non-phytoplasma bacteria. This results in many more reads mapping in the 16S dataset, most of which are not from the target microorganism, and often provides truncated, poor assemblies that are neither phylogenetically useful nor of sufficient length for RFLP typing using the iPhyClassifier [5]. In addition, as in the case of PCR

amplification and direct sequencing of 16S rRNA-encoding genes, the hybridization procedure cannot differentiate heterogeneous sequences deriving from the two 16S rRNA loci found in many phytoplasma genomes [27], including those represented on the hybridization panel [6, 28, 29]. As a result, the sequence that is determined should be used for information only and should not be deposited in public databases. For the analysis of 16S rRNA genes from samples featuring heterogeneity, it is best to use PCR amplification and cloning to determine the sequences of the individual loci.

19. In certain instances, the lack of specificity of the 16S rRNA gene probes can be a benefit. For example, we have tested the ability of the hybridization probes to detect the presence of "off-target" phytoplasmas, for which no probes are included in the hybridization array. We applied this protocol to samples infected with '*Ca.* P. rubi' (16SrV-C) and observed that no reads mapped to any of the protein-coding genes, but a sufficient number of reads mapped to 16S to enable the assembly of a gene fragment that was typable using the iPhyClassifier—see Table 4. Therefore, this method can provide information on unknown samples and can also identify cases of mixed phytoplasmas that contain taxa that are not represented in the array [21].
20. In some samples, this mapping and read extraction need to be done twice. The first round will assemble the target 16S gene plus host mitochondria/chloroplast and other bacterial 16S genes. If this is a problem, the nontarget sequences can be

Table 4
Reads mapping using 16SrV samples

Plant	Group	Total reads generated	% of reads mapping (all markers)	Number of reads mapping to 16S	Assembly length [bp]	Type (iPhyClassifier)
RubusA	16SrV	2019	2.67	54	1310	Not typed[a]
RubusB	16SrV	37,497	3.03	1138	1930	16SrV-C F = 1.00
RubusC	16SrV	10,9931	2.14	2357	2028	16Sr V-C F = 1.00
RubusD	16SrV	91,637	1.46	1340	1981	16Sr V-C F = 1.00
RubusE	16SrV	63,402	0.88	557	2124	Not typed[a]

RubusA-E correspond to five different leaf samples extracted from 16SrV-infected raspberry plants

[a]iPhyclassifier was unable to assign a group/subgroup to this sequence due to the fact that the recovered fragment, despite its length, did not cover the entire F2nR2 fragment. However, in each case, BLAST analysis confirmed that the recovered gene fragment derived from a 16SrV phytoplasma

used to make a new mapping file (e.g., 16S_junk.fasta) which includes the target 16S gene and the assembled nonspecific sequences (start at bowtie2-build). Map again against this file (16S_junk.fasta) and extract the mapped reads. This should "clean up" the mapped files when the target sequences are selected in the samtools view step, and result in improved target assemblies.

References

1. Wei W, Zhao Y (2022) Phytoplasma taxonomy: nomenclature, classification, and identification. Biology (Basel) 11(8). https://doi.org/10.3390/biology11081119
2. Bertaccini A, Arocha-Rosete Y, Contaldo N et al (2022) Revision of the '*Candidatus* Phytoplasma' species description guidelines. Int J Syst Evol Microbiol 72(4). https://doi.org/10.1099/ijsem.0.005353
3. Gundersen DE, Lee IM (1996) Ultrasensitive detection of phytoplasmas by nested-PCR assays using two universal primer pairs. Phytopathol Mediterr 35(3):144–151. https://doi.org/10.2307/42685262
4. Bertaccini A (2022) Plants and phytoplasmas: when bacteria modify plants. 11(11):1425. https://doi.org/10.3390/plants11111425
5. Zhao Y, Wei W, Lee IM et al (2013) The iPhyClassifier, an interactive online tool for phytoplasma classification and taxonomic assignment. Methods Mol Biol 938:329–338. https://doi.org/10.1007/978-1-62703-089-2_28
6. Zwolińska A, Krawczyk K, Borodynko-Filas N et al (2019) Non-crop sources of rapeseed Phyllody phytoplasma ('*Candidatus* Phytoplasma asteris': 16SrI-B and 16SrI-(B/L)L), and closely related strains. Crop Protection 119:59–68. https://doi.org/10.1016/j.cropro.2018.11.015
7. Pérez-López E, Dumonceaux TJ (2016) Detection and identification of the heterogeneous novel subgroup 16SrXIII-(A/I)I phytoplasma associated with strawberry green petal disease and Mexican periwinkle virescence. Int J Syst Evol Microbiol 66(11):4406–4415. https://doi.org/10.1099/ijsem.0.001365
8. Liefting LW, Andersen MT, Beever RE et al (1996) Sequence heterogeneity in the two 16S rRNA genes of Phormium yellow leaf phytoplasma. Appl Environ Microbiol 62(9):3133–3139. https://doi.org/10.1128/aem.62.9.3133-3139.1996
9. Torres L, Galdeano E, Fernandez F et al (2011) Establishment of the new subgroup 16SrI-S (rr-rp) tuf-H belonging to '*Ca.* phytoplasma asteris' in wild and cultivated plants in Argentina. J Plant Pathol 93(2):311–320
10. Lee IM, Bottner-Parker KD, Zhao Y et al (2010) Phylogenetic analysis and delineation of phytoplasmas based on *secY* gene sequences. Int J Syst Evol Microbiol 60(Pt 12):2887–2897. https://doi.org/10.1099/ijs.0.019695-0
11. Martini M, Bottner-Parker KD, Lee IM (2019) PCR-based sequence analysis on multiple genes other than 16s rRNA gene for differentiation of phytoplasmas. Methods Mol Biol 1875. https://doi.org/10.1007/978-1-4939-8837-2_8
12. Pérez-López E, Olivier CY, Luna-Rodríguez M et al (2016) Phytoplasma classification and phylogeny based on *in silico* and *in vitro* RFLP analysis of *cpn60* universal target sequences. Int J Syst Evol Microbiol 66(12):5600–5613. https://doi.org/10.1099/ijsem.0.001501
13. Mitrović J, Kakizawa S, Duduk B et al (2011) The *groEL* gene as an additional marker for finer differentiation of '*Candidatus* Phytoplasma asteris'-related strains. Ann Appl Biol 159(1):41–48. https://doi.org/10.1111/j.1744-7348.2011.00472.x
14. Contaldo N, Sciovolone A, Barbieri C et al (2021) Multilocus typing of aster yellows phytoplasmas infecting lettuce with yellowing in Calabria, Italy. Phytopathogen Mollicutes 11(2):105–111. https://doi.org/10.5958/2249-4677.2021.00017.7
15. Rihne T, Namita SKP et al (2021) Multilocus gene typing, mixed infection of phytoplasma strains associated with rose genotypes and confirmation of their natural reservoir sources. Trop Plant Pathol 46(6):596–607. https://doi.org/10.1007/s40858-021-00457-x
16. Ćurčić Ž, Kosovac A, Stepanović J et al (2021) Multilocus genotyping of '*Candidatus* Phytoplasma solani' associated with rubbery taproot disease of sugar beet in the pannonian plain. Microorganisms 9(9). https://doi.org/10.3390/microorganisms9091950

17. Noorizadeh S, Khakvar R, Golmohammadi M et al (2021) Multilocus genotyping of '*Candidatus* Phytoplasma aurantifolia' associated with witches' broom disease in Citrus spp. Trop Plant Pathol 46(2):218–231. https://doi.org/10.1007/s40858-020-00404-2
18. Cho S-T, Kung H-J, Huang W et al (2020) Species boundaries and molecular markers for the classification of 16SrI phytoplasmas inferred by genome analysis. Front Microbiol 11. https://doi.org/10.3389/fmicb.2020.01531
19. Nijo T, Iwabuchi N, Tokuda R et al (2021) Enrichment of phytoplasma genome DNA through a methyl-CpG binding domain-mediated method for efficient genome sequencing. J Gen Plant Pathol 87(3):154–163. https://doi.org/10.1007/s10327-021-00993-z
20. Gadagkar SR, Rosenberg MS, Kumar S (2005) Inferring species phylogenies from multiple genes: concatenated sequence tree versus consensus gene tree. J Exp Zool B Mol Dev Evol 304(1):64–74. https://doi.org/10.1002/jez.b.21026
21. Pusz-Bochenska K, Perez-Lopez E, Wist TJ et al (2022) Multilocus sequence typing of diverse phytoplasmas using hybridization probe-based sequence capture provides high resolution strain differentiation. Front Microbiol 13. https://doi.org/10.3389/fmicb.2022.959562
22. Bennypaul H, Sanderson D, Donaghy P et al (2023) Development of a real-time PCR assay for the detection and identification of rubus stunt phytoplasma in Rubus spp. Plant Dis 107(8):2296–2306. https://doi.org/10.1094/pdis-09-22-2193-re
23. Dumonceaux TJ, Links MG, Town JR et al (2017) Targeted capture of *cpn60* gene fragments for PCR-independent microbial community profiling. Protoc Exch. https://doi.org/10.1038/protex.2017.100
24. Links MG, Dumonceaux TJ, McCarthy EL et al (2021) CaptureSeq: hybridization-based enrichment of cpn60 gene fragments reveals the community structures of synthetic and natural microbial ecosystems. Microorganisms 9(4):816. https://doi.org/10.3390/microorganisms9040816
25. Langmead B, Salzberg SL (2012) Fast gapped-read alignment with Bowtie 2. Nat Meth 9(4):357–359. https://doi.org/10.1038/nmeth.1923
26. Li H, Handsaker B, Wysoker A et al (2009) The sequence alignment/map format and SAMtools. Bioinformatics 25(16):2078–2079. https://doi.org/10.1093/bioinformatics/btp352
27. Robertson G, Schein J, Chiu R et al (2010) De novo assembly and analysis of RNA-seq data. Nature Meth 7(11):909–912. https://doi.org/10.1038/nmeth.1517
28. Hammond C, Pérez-López E, Town J et al (2021) Detection of blueberry stunt phytoplasma in Eastern Canada using *cpn60*-based molecular diagnostic assays. Sci Rep 11(1):22118. https://doi.org/10.1038/s41598-021-01439-4
29. Martini M, Lee IM, Bottner KD et al (2007) Ribosomal protein gene-based phylogeny for finer differentiation and classification of phytoplasmas. Int J Syst Evol Microbiol 57(9):2037–2051. https://doi.org/10.1099/ijs.0.65013-0
30. Olivier C, Dumonceaux T, Perez-Lopez E et al (2017) Detection, symptomatology and management of Aster yellows disease in canola. In: Reddy GVP (ed) Integrated management of insect pests on canola and other brassica oilseed crops. CABI
31. Romero B, Dumonceaux T, Olivier C et al (2023) Development of aster yellows on crop and noncrop species from the Canadian prairies. PhytoFrontiers™ 3(3):679–687. https://doi.org/10.1094/phytofr-01-23-0008-r
32. Town JR, Wist T, Perez-Lopez E et al (2018) Genome sequence of a plant-pathogenic bacterium, "*Candidatus* Phytoplasma asteris" strain TW1. Microbiol Resour Announc 7(12). https://doi.org/10.1128/mra.01109-18
33. Plante N, Durivage J, Brochu A-S et al (2024) Leafhoppers as markers of the impact of climate change on agriculture. Cell Rep Sustain 1(2):100029. https://doi.org/10.1016/j.crsus.2024.100029
34. Benson DA, Cavanaugh M, Clark K et al (2012) GenBank. Nucleic Acids Res 41(D1):D36–D42. https://doi.org/10.1093/nar/gks1195
35. Zhao Y, Wei W, Davis RE et al (2021) The agent associated with blue dwarf disease in wheat represents a new phytoplasma taxon, '*Candidatus* Phytoplasma tritici'. Int J Syst Evol Microbiol 71(1). https://doi.org/10.1099/ijsem.0.004604

Chapter 18

Phytoplasma Genome Reconstruction: From DNA Enrichment to Sequencing and Genome Assembly

Zahra Golestani Hotkani, Matteo Brilli, and Mirko Moser

Abstract

Phytoplasmas are plant pathogens transmitted by insects that pose significant challenges for the most common experimental tasks. This is especially true concerning analyses that aim at genome sequencing, due to the phytoplasma's low abundance and its association with the host cellular matrix. This protocol provides a roadmap for researchers in the field of phytoplasma genomics, focusing on three crucial aspects: (1) DNA extraction from plant hosts, including a protocol for low-concentration samples, (2) Sequencing strategies, emphasizing accurate base calling (particularly for Oxford Nanopore) and addressing mixed plant/phytoplasma samples, (3) Phytoplasma genome assembly and analysis, detailing methodologies and exploring computational/comparative techniques to decode phytoplasma genomes. With a focus on tackling pivotal challenges and harnessing possibilities proposed by these advancements, this approach equips researchers to unlock the secrets of these enigmatic plant pathogens.

Key words Phytoplasma genomics, DNA enrichment protocols, Oxford Nanopore sequencing, Genome assembly techniques, Low-concentration DNA samples

1 Introduction

Sequencing the genome of phytoplasmas can provide valuable insights into phylogenetics, pathogenicity, and host–phytoplasma interactions [1, 2]. However, the isolation of the bacterium and subsequent axenic culture have not been achieved yet. This translates into a relatively small number of available, high-quality genomes from members of the genus [3]. The main problems for the genomic analysis of phytoplasma are: (i) phytoplasma must be sequenced from infected plants or insect vectors, (ii) the bacterial DNA concentration from infected samples is usually much lower than that of the host, and (iii) the genome is very small and has a highly biased nucleotide composition. The first problem to deal with in genome sequencing is the overwhelming quantity of host DNA (comprising nucleic, mitochondrial, and/or plastid DNA), which makes direct sequencing of phytoplasma genomes inefficient

Katrin Janik and Mattia Tabarelli (eds.), *Phytoplasma: Methods and Protocols*, Methods in Molecular Biology, vol. 3008, https://doi.org/10.1007/978-1-0716-5104-9_18,

and requires careful optimization of the experimental protocol [4, 5]. Despite years of effort, recovering complete phytoplasma genomes from naturally infected plants and insects remains a technical challenge, even with third-generation sequencing approaches and recent innovations like adaptive sampling (Oxford Nanopore Technology) [6]. Success in this task strictly depends on phytoplasma concentrations in the respective DNA sample. Seasonal variations can significantly impact both the location of the phytoplasma and its overall concentration within the plant; therefore, the timing of sample collection is a crucial aspect to account for getting the highest possible phytoplasma to host DNA abundance ratio [7]. Targeting the tissues where the bacterium is more abundant can also improve the outcome, and although leaves of symptomatic plants are generally the most reliable and easy-to-process diagnostic source, the ideal tissue to collect and extract is the phloem sieve elements; consequently, researchers often target the leaf petioles, mid veins, stems, or inner bark to increase the phytoplasma to host DNA abundance ratio [8]. Many draft genomes have been obtained by sequencing unenriched DNA from natural host plants or from artificially infected model plants such as periwinkle (*Catharanthus roseus*) and *Nicotiana* spp., in which phytoplasma can reach much higher concentrations [9]. The proportion of phytoplasma DNA compared to the surrounding host DNA is minimal. Therefore, protocols for efficiently depleting host DNA and/or enriching phytoplasma DNA are still crucial for successful genome sequencing of these elusive pathogens [10].

If the phytoplasma DNA concentration is too low, it may become impossible to obtain complete genome assemblies; in this case, one possible approach is to enrich bacterial DNA or deplete the host DNA in the sample. Established phytoplasma genome enrichment methods, like density gradient centrifugation and pulse field gel electrophoresis (PFGE), require several days of wet lab processing, specialized lab equipment, harsh chemicals, and are only effective when the phytoplasma concentration in the host plant is sufficiently high [6]. Recent advances aim to overcome these limitations by the implementation of new enrichment methods such as non-ionic iodixanol gradients [6], filter-based approaches [10], methyl-CpG binding (MBD) [9], and immunoprecipitation [11] for depleting host DNA before sequencing. Currently, there are four commercially available microbiome DNA enrichment methods on the market [12]. The MBD method is relatively easy to apply and useful for the enrichment of bacterial DNA by efficient separation of prokaryotic and eukaryotic DNA [9]. CpG sites in the genome are naturally methylated in eukaryotes [13] but not in prokaryotes; moreover, DNA methylation is likely absent in phytoplasma, and its genome composition is such that CpGs are likely absent. The MBD enrichment method leverages methyl-CpG binding domain proteins to selectively capture and remove CpG-methylated DNA of eukaryotic origin: evidence indeed shows that phytoplasma DNA is 10- to 30-fold more abundant after MBD enrichment [9].

Here we present two methods to obtain high-quality phytoplasma DNA for Oxford Nanopore Technology (ONT) sequencing: (1) Direct extraction of total DNA (i.e., of the host and the phytoplasma) using a commercial kit that provides high-quality, high molecular weight DNA with a modification to improve sequencing yield and (2) a method for depleting host-DNA to increase the proportion of phytoplasma DNA in the sample.

Enriching phytoplasma DNA in the sample is, however, only one side of the coin. The selection of the most appropriate sequencing technology after DNA extraction and enrichment is also crucial for the quality of draft genome assemblies. Technological advances have revolutionized this task. Short-read sequencing, like Illumina, has been instrumental in the de novo assembly of bacterial species. However, its limitations become apparent for phytoplasma genome sequencing due to the high AT content and the presence of low complexity regions characterized by homopolymeric sequences, which prevent the generation of unambiguous contigs and disrupt assembly continuity when using short reads [2–14].

Whereas early long-read sequencers had significantly higher error rates compared to short-read technologies, the recent advancements in nanopore chemistry have substantially mitigated this limitation. Thus, nowadays, long-read sequencing technologies, such as PacBio and ONT, offer an appealing alternative to short-read sequencing. By generating reads exceeding 10 kb and reaching up to 2.3 Mb, they can span problematic AT-rich regions with an accuracy up to 99.9% [15]. This advantage translates into more contiguous and accurate genome assemblies [15]. ONT's MinION, in particular, has attracted considerable attention due to its portability and low cost, making it a suitable option for various genomic applications [16].

While advancements have focused on nanopore engineering and on improving the default basecalling algorithms, species-specific basecalling models remain largely unexplored. This is not necessarily a problem, and the standard models work well with most bacterial species. However, since most Mollicutes—and therefore phytoplasmas—are characterized by very low GC content, pre-trained models might fail to translate electrical signals into the correct bases principally because of two possible drawbacks: the model was primarily trained on genomes with more balanced GC content and the intrinsic problem of the nanopore in sequencing homopolymeric or low complexity genomic regions. To avoid these pitfalls, a taxon-specific custom-trained Bonito model (custom-pg45) [17] was implemented in various whole-genome sequencing (WGS) assembly pipelines involving organisms of the mollicutes group. The use of this specifically trained model showed improved consensus accuracies for *Mycoplasma bovis* genomes, which have the same biases as phytoplasmas in terms of nucleotide composition. This highlights the potential benefits of custom models tailored to specific taxa for improving basecalling outcomes, especially when the genome composition is extremely biased [17, 18].

The step after the basecalling is genome assembly. Currently, two primary strategies dominate the field: reassembly and de novo assembly. Reassembly is based on aligning the reads to a reference genome, analogous to aligning fragments of a historical document to a known manuscript. This approach offers efficiency but can result in missed information if the target genome has additional genes with respect to the reference genome. Consequently, this strategy usually has an additional step to assemble reads that were not included in the reference-based assembly. De novo assembly, in contrast, resembles the meticulous reconstruction of a massive jigsaw puzzle, and while it offers a clear advantage over reference-based assembly methods, it presents significant challenges [19].

Two main approaches for genome assembly with long reads can be defined: (1) based on long reads only, or (2) by combining short and/or long reads to enhance the assembly process, also known as hybrid assembly. Hybrid approaches are no longer required, as the fidelity of the long reads has significantly improved, with error rates now comparable to those of short-read sequencers [15]. Software for both kinds of approaches is available and can be used both as stand-alone solutions and in combination. This chapter will focus on the long-read assembly algorithm Flye [20], which was recently compared to other assemblers and identified to provide the most accurate and contiguous assemblies [21].

After the genome assembly process, polishing is a crucial step that involves refining an initial draft sequence to improve its accuracy. The draft sequence often contains errors and artifacts such as wrong bases, insertions, deletions, and structural errors introduced during the assembly process. Assembly polishing aims to correct these errors to enhance both local accuracy and overall consensus accuracy, eventually leading to a better gene prediction/annotation. Some of the assembly polishing tools are Pilon [22], Racon [23], and Medaka [24] (widely adopted standards for nanopore-based polishing) [19].

This protocol aims to provide a comprehensive workflow for improving the sequencing and analysis of phytoplasma genomes. Thus, the primary objectives of this book chapter are to: (1) present a suitable and efficient DNA extraction protocol for phytoplasma DNA recovery from low-concentration samples, (2) provide a sequencing and basecalling strategy to improve the assembly accuracy and effectively sequence mixed plant and phytoplasma DNA, and (3) provide a pipeline for genome assembly and analysis.

2 Materials

The materials from the suppliers mentioned have been successfully used in the described procedure. However, other suppliers can be equally considered if the respective quality checks and controls are performed to guarantee efficient performance.

2.1 General Equipment

1. Microcentrifuge.
2. Vortex mixer.
3. P1000, P100, P20, and P10 pipettes and tips.
4. Spectrophotometer or Fluorometer: for DNA quantification.
5. 0.2 mL thin-walled PCR tubes.
6. 1.5 mL DNA low-binding tubes.
7. Thermal cycler.
8. Rotating mixer for mixing beads and reagents.
9. Ice bucket with ice.
10. Magnetic rack: to separate magnetic beads from solutions.

2.2 Genomic DNA Extraction

1. Genomic DNA isolation kit (*see* **Note 1**).
2. Polyvinylpyrrolidone (PVP40): It should be added to the first buffer of the extraction kit (e.g., lysis buffer) at 1% w/v.

2.3 Enrichment of Phytoplasma DNA Using MBD-Based Depletion of Methylated DNA

1. TE Buffer (10 mM Tris–HCl, 1 mM EDTA), DNase-free, pH 7.5 or 8.0, both can be used.
2. Solid-phase reversible immobilization of paramagnetic beads (*see* **Note 2**).
3. Protein A-conjugated magnetic beads (such as NEBNext Protein A Magnetic Beads): Resuspend using gentle pipetting or a rotating mixer, avoid vortexing.
4. Binding and wash buffer (such as 5X NEBNext Bind/wash Buffer): prepare and use according to the manufacturer's instructions.
5. MBD2-Fc Protein: Prepare the protein and add the required amount according to the manufacturer's instructions.
6. 100% Ethanol: for DNA precipitation.
7. 80% Ethanol: water nuclease-free solution (v/v): Prepare fresh. To be used for bead washing.
8. Microcentrifuge tubes: for sample handling and storage.
9. Agarose Gel Electrophoresis Materials: agarose gel, DNA marker suitable for the detection of fragments from 200 bp to 20 kb, and Gel Loading Dye (6X).

10. qPCR Reagents: Primers (e.g., 16S rRNA for the phytoplasma control, a plant-specific gene for host DNA control) and qPCR kit components.
11. DNA size and quality assessment: Spectrophotometer and an automated nucleic acid fragment analyzer with an adequate genomic DNA analysis kit.

2.4 Library Preparation: Initial Preparation of the DNA, Adapter Ligation, and Clean-Up

All reagents should be prepared using nuclease-free water to ensure the integrity of the DNA throughout the process.

1. Genomic DNA (gDNA).
2. Paramagnetic beads: for purification.
3. DNA repair enzyme mix (such as NEBNext FFPE DNA Repair Mix).
4. End repair and A-tailing enzyme mix (such as NEBNext Ultra II End Repair/dA-Tailing Module).
5. Ligation sequencing kit (such as Ligation Sequencing Kit V14 by Oxford Nanopore).
6. DNA ligation enzyme mix (such as NEBNext Quick Ligation Module).
7. High-sensitivity DNA quantification kit appropriate for the instrument.
8. Recommended: fluorescence-compatible assay tubes.
9. 80% Ethanol: nuclease-free water solution: prepare freshly before use.

2.5 Flow Cell Priming and Loading

Nuclease-free water and DNA low-binding tubes should be used throughout the procedure.

1. Flow cell priming reagent (included in the ligation sequencing kit or a dedicated flow cell priming kit).
2. Nanopore sequencing flow cell (such as MinION™ or P2Solo™ flow cell).
3. Compatible nanopore sequencing device (such as MinION™ or P2Solo™).

2.6 Basecalling Software and Dependencies

1. Fast-Bonito: Basecalling software.
2. Python: The core programming language for running fast-bonito.
3. Python packages: ont_fast5_api==3.0.1, fast-ctc-decode==0.2.5, torch==1.4.0, tqdm==4.31.1, numpy==1.19.2, scipy==1.5.2, tensorflow-gpu==1.14.0, mappy==2.17. These packages handle various tasks such as data manipulation, machine learning, and file I/O.

2.7 Genome Assembly (Long Read)

1. Host Genome Reference Database (in FASTA format).
2. Target Organism Reference Genome (in FASTA format).
3. Oxford Nanopore MinION device (MinION™, P2-Solo™, etc.) able to provide sufficient long-read sequencing output.
4. BLASTN: for sequence alignment.
5. Flye: Software for genome assembly.
6. Custom scripts for data filtering and analysis, in R or Python environments. Long reads in FASTQ format must be used as input.

2.8 Genome Assembly (Hybrid)

1. Linux or macOS as a computational environment.
2. Python 3.4 (or later).
3. A C++ compiler with C++14 support. Suitable compilers include GCC (version 4.9.1 or later), Clang (version 3.5 or later), or Intel's ICC (specific version requirements may vary).
4. Setup tools (only required for installation of Unicycler).
5. SPAdes (version 3.14.0 or later; accessed by spades.py): for short-read or hybrid assembly.
6. Racon (racon): for long-read or hybrid assembly polishing.
7. BLAST+ utilities makeblastdb and tblastn: for rotating and analyzing circular contigs.
8. Medaka: a tool for sequence alignment and correction (designed specifically for Oxford Nanopore long-read data).

3 Methods

3.1 DNA Extraction

Follow the kit protocol for DNA extraction from various plant materials to obtain high-quality, high molecular weight DNA (*see* **Notes 1** and **3**). We introduced a modification in the protocol by adding 1% w/v PVP40 to the CTAB-based buffer (P1) (*see* **Note 4**).

3.2 Microbial DNA Enrichment and Kit Selection

To enrich microbial DNA, we used a selective DNA enrichment method. Specifically, the NEBNext Microbiome DNA Enrichment Kit was chosen for its performance in our procedure. Alternative kits may be used if validated through appropriate quality checks.

3.2.1 DNA Preparation and Quantitation

Any method for the purification of protein-free genomic DNA can be used, including Proteinase K treatment followed by phenol/chloroform extraction and ethanol precipitation, lysozyme digestion, silica spin column purification for genomic DNA, or other methods. Sonication, nebulization, chaotropic salts, enzymatic fragmentation, rough handling, multiple freeze-thaws, or any other procedure that would cause DNA to shear should be avoided. Samples should be in DNase-free TE buffer (pH 8.0 or 7.5), be at least 15 kb or greater in size, and be free of small molecular weight fragments. If the DNA fragment size is <15 kb, enrichment will be less efficient. Determine DNA quality and quantity by agarose gel electrophoresis of the sample alongside a DNA marker (*see* **Note 5**).

3.2.2 Methyl-CpG Binding Domain-Mediated Enrichment

Use protein A-conjugated magnetic beads and binding and wash buffers for the subsequent steps. We used NEBNext Protein A Magnetic Beads and NEBNext Bind/Wash Buffer for the procedure. Beads and buffers from other suppliers might be equally suitable, but the protocol must be adapted accordingly.

1. Determine the volume of protein A prepared beads required for enriching microbial DNA from your sample. Use 1 μL of MBD2-Fc-bound magnetic beads for every 6.25 ng of input DNA (Υ = volume of MBD2-Fc-bound magnetic beads = quantity of initial DNA (ng)/6.25 ng/μL; for example, 160 μL of MBD2-Fc-bound magnetic beads per μg of input DNA).
2. Resuspend Protein A Magnetic Beads by gently pipetting up and down until the suspension is homogeneous. Alternatively, gently rotate the tube on a rotating mixer for 15 min at 4 °C. Do not vortex (*see* **Note 6**).
3. Prepare 1X Bind/wash Buffer on ice by diluting 1 part of Bind/wash Buffer (5X) with four parts DNase-free water. One individual reaction, from start to finish, will require 4 mL of 1X Bind/Wash Buffer (*see* **Note 7**).
4. Add ($0.1 \times \Upsilon$) μL of MBD2-Fc protein and (Υ) μL of Protein A magnetic beads, where Υ is the volume calculated initially and based on the amount of input DNA. Mix by pipetting up and down until the beads are completely homogeneous, at least 5–10 times (*see* **Note 8**).
5. Mix the bead–protein mixture by placing the tube in a rotating mixer for 10 min at room temperature.
6. Briefly spin the tube and place it on the magnetic rack for 2–5 min or until the beads have aggregated at the wall of the tube and the solution is clear.
7. Carefully remove the supernatant with a pipette without disturbing the beads.
8. Add 1 mL of 1X Bind/Wash Buffer (kept on ice) to the tube to wash the beads. Pipette up and down until the beads are homogeneously resuspended.
9. Mix the beads on a rotating mixer for 3 min at room temperature.

10. Briefly spin the tube and place it on the magnetic rack for 2–5 min or until the solution is clear.
11. Carefully remove the supernatant with a pipette without disturbing the beads.
12. Repeat **steps 8–11** (*see* **Note 9**).
13. Remove the tube from the rack and add (*Y*) μL (where *Y* is the volume calculated in **step 1**) of 1X Bind/Wash Buffer to resuspend the beads. Mix by pipetting up and down a few times. The MBD2-Fc-bound magnetic beads are stable for up to 7 days at 4 °C.
14. Add the input DNA to the tube containing the MBD2-Fc-bound magnetic beads.
15. Add undiluted Bind/wash Buffer (5X) for a final concentration of 1X. Pipette the sample up and down until the beads are completely resuspended.
16. Agitate the tube on a rotating mixer for 15 min at room temperature (*see* **Note 10**).
17. Briefly spin the tube and place it on the magnetic rack for 5 min or until the beads have aggregated along the side of the tube and the solution appears clear.
18. Carefully remove the supernatant with a pipette, without disturbing the beads, and transfer it to a clean microcentrifuge tube. This supernatant contains the target microbial DNA. Store this sample at −20 °C or proceed directly to purification.
19. Purify the samples by either Paramagnetic bead cleanup or ethanol precipitation.

3.2.3 Option A: Paramagnetic Bead Cleanup

1. Vortex paramagnetic beads to resuspend them homogeneously.
2. If the enriched sample volume from the previous step exceeds your tube volume by 2.8 times, split it into two tubes.
3. Add 1.8X volume of resuspended Paramagnetic beads to the sample. Mix well by pipetting up and down at least five times, expelling all the liquid off the tip during the last mix.
4. Incubate samples for at least 5 min at room temperature.
5. Place the tube on the magnetic stand to separate the beads from the supernatant. If necessary, quickly spin the sample to collect the liquid from the sides of the tube before placing it on the magnetic stand (*see* **Note 11**).
6. After 5 min (or when the solution is clear), carefully remove and discard the supernatant. Be careful not to disturb the beads that contain DNA targets.
7. Add 400 μL of freshly prepared 80% ethanol solution to the tube while in the magnetic stand. Incubate at room

temperature for 30 s, and then carefully remove and discard the supernatant. Be careful not to disturb the beads that contain DNA targets (*see* **Note 12**).

8. Repeat **step** 7 once and remove all visible liquid after the second wash. If necessary, briefly spin the tube, place it back on the magnetic stand, and remove traces of ethanol with a pipette (P10 or P20).
9. Air dry the beads for up to 5 min while the tube is on the magnetic stand with the lid open (*see* **Note 13**).
10. Remove the tube from the magnetic stand.
11. Elute the DNA target from the beads by adding 50 μL of 1X TE (use 25 μL for each tube if the sample was split in two). Mix well by pipetting up and down 10 times. Incubate for at least 2 min at room temperature.
12. Place the tube on the magnetic stand. If necessary, quickly spin the sample to collect the liquid from the sides of the tube before placing it back on the magnetic stand.
13. After 5 min (or when the solution is clear), transfer the eluate to a new microcentrifuge tube (combine the eluates if the sample was split in two). The sample can now be used for NGS library construction or other downstream analysis.

3.2.4 Option B: Ethanol Precipitation

1. Add 2.5 volumes of 100% ethanol to the sample and incubate for 10 min on ice.
2. Centrifuge the sample for 30 min at 16,000× *g*.
3. Remove the ethanol and allow the pellet to air dry.
4. Resuspend the pellet in a small quantity (50 μL) of TE buffer. The sample can now be used for NGS library construction or other downstream analysis. If the ethanol pellet contains any residual beads, the resuspended sample can be placed on the magnetic rack for 5 min to concentrate the beads on the inner wall of the tube, and the supernatant can be transferred to a fresh microcentrifuge tube.

3.2.5 NGS Library Construction

Enriched samples are compatible with NEBNext DNA Library Prep reagents described in Subheading 2.4, used for the ONT library preparation.

3.2.6 DNA Quantitation Using Agarose Gel Electrophoresis

1. Prepare the unenriched sample by diluting 0.1 μg of unenriched input DNA in TE Buffer for a final volume of 20 μL.
2. Aliquot 20 μL of the unenriched DNA, the supernatant fraction containing the enriched microbial DNA, and the eluted fraction containing the eukaryotic host DNA into separate tubes. Add a loading buffer to each sample.

3. Load 20 μL of each sample alongside an appropriate DNA marker on a 0.8% agarose gel. Since the supernatant fraction containing enriched microbial DNA will be depleted of host DNA, the total amount of DNA seen on the gel will be reduced compared to the input control (*see* **Note 14**).

3.2.7 Validation of Microbial Enrichment by Quantitative PCR (qPCR)

1. Prepare the unenriched input sample by diluting the same quantity as previously used for enrichment in TE Buffer in a final volume of 50 μL.
2. For each condition (input, enriched, and purified microbial DNA, aliquoted, and purified host (plant DNA)), aliquot the volume appropriate for your qPCR reaction in triplicate. Do this for all primer pairs used (16S rRNA gene and other host-specific primers).
3. Add the appropriate volume of qPCR primers for your qPCR kit. Dilute as needed. Typically, primers are supplied at 20 μM each.
4. Perform qPCR and analyze results as appropriate for your qPCR reagents and instrument, by comparing quantification cycle (Cq) values. In the enriched microbial fraction, minimal or no change in Cq values is expected for the phytoplasma-specific control primers compared to the input. In contrast, a marked increase in Cq values is expected for host-specific primers. For the host DNA fraction, a moderate increase in Cq values for host-specific primers should be observed between the input and enriched fraction with host-specific primers. A significant increase in Cq values should be observed for the phytoplasma-specific primers, indicating microbial DNA depletion.

3.3 Library Preparation and Sequencing Strategies

In the following, for the library preparation procedure, the use of specific, proprietary reagents, such as DNA repair enzyme mix and an end-repair and A-tailing enzyme mix, is described. Similar reagents from other suppliers might be equally suitable, but the method must be adapted accordingly. This procedure is based on the standard ONT protocol [25], with additional notes on modifications introduced during our experiments. Specifically, to minimize DNA shearing, all mixing steps were performed by flicking the tubes gently instead of pipetting.

3.3.1 DNA Repair and End-Prep

1. Thaw the DNA Control Sample (DCS) at room temperature, spin it down, mix by pipetting, and place it on ice.
2. Prepare all reagents according to the manufacturer's instructions and keep them on ice (*see* **Note 15**). After thawing all the reagents on ice, mix by flicking or inverting tubes. Spin down all tubes before first daily use.

If Ultra II End-Prep Buffer or FFPE DNA Repair Buffer shows precipitation, allow the mixture to reach room temperature, mix by pipetting up and down several times to break up the precipitate, followed by vortexing for 30 s to solubilize any precipitate (*see* **Note 16**). A yellow coloration of FFPE DNA Repair Buffer is acceptable, as it does not affect performance.

3. Transfer 1 μg (or 100–200 fmol) of input DNA to a 1.5 mL low-binding tube. Adjust the volume to 47 μL with nuclease-free water. Mix thoroughly by flicking the tube and briefly spin down in a microcentrifuge.
4. In a 0.2 mL thin-walled PCR tube, add in order: DNA from **step 14**, 1 μL DNA Control Sample (optional), 3.5 μL NEBNext FFPE DNA Repair Buffer, 2 μL NEBNext FFPE DNA Repair Mix, 3.5 μL Ultra II End-Prep Reaction Buffer, and 3 μL Ultra II End-Prep Enzyme Mix (total volume of 60 μL). Flick the tube gently after each addition.
5. Thoroughly mix the reaction by gently pipetting and briefly spinning down.
6. Using a thermal cycler, incubate at 20 °C for 5 min (*see* **Note 17**).
7. Resuspend the paramagnetic beads by vortexing.
8. Transfer the 60 μL reaction to a clean 1.5 mL low-binding tube.
9. Add 60 μL of resuspended Paramagnetic beads to the end-prep reaction and mix by flicking the tube.
10. Incubate on a rotator mixer for 5 min at room temperature (*see* **Note 18**).
11. Briefly spin down the tube and place the tube on a magnetic rack until the supernatant is clear. Remove and discard the supernatant.
12. Keep the tube on the magnet and wash the pellet with 200 μL freshly prepared 80% ethanol (*see* **Note 19**). Carefully remove the ethanol by pipetting.
13. Repeat the previous step.
14. Spin down and place the tube back on the magnet. Pipette off any residual ethanol. Allow to dry for ~30 s, but do not dry the pellet to the point of cracking.
15. Remove the tube from the magnetic rack and resuspend the pellet in 61 μL nuclease-free water. Incubate for 10 min at room temperature.
16. Pellet the beads on the magnetic rack until the eluate is clear and colorless, for at least 1 min (*see* **Note 20**).

17. Transfer the entire 61 μL eluate to a clean 1.5 mL low-binding tube (*see* **Note 21**).

3.3.2 Adapter Ligation and Clean-Up

1. Spin down the Ligation Adapter (LA) from the library kit and NEBNext Quick T4 DNA Ligase and place them on ice (*see* **Notes 22** and **23**).
2. Thaw the Ligation Buffer (LNB) at room temperature, spin down, and mix gently by pipetting. Due to viscosity, vortexing this buffer is ineffective. Place on ice immediately after thawing and mixing (*see* **Note 24**).
3. Thaw the Elution Buffer (EB) and Long Fragment Buffer (LFB) at room temperature, mix by vortexing, spin down, and place on ice (*see* **Note 25**).
4. Assemble the ligation reaction in a 1.5 mL DNA low-binding tube, flick the tube after each addition, in the following order: 60 μL DNA sample from the previous step, 25 μL Ligation Buffer (LNB), 10 μL NEBNext Quick T4 DNA Ligase, 5 μL Ligation Adapter (LA) (total reaction volume 100 μL).
5. Mix thoroughly by flicking and briefly spin down the tube.
6. Incubate at room temperature for 10 min (*see* **Note 26**).
7. Resuspend the paramagnetic beads by vortexing.
8. Add 40 μL of resuspended paramagnetic beads to the reaction and mix by flicking the tube.
9. Incubate at room temperature on a rotator mixer for 5 min (*see* **Note 27**).
10. Spin down the tube briefly and pellet the beads by placing the tube on a magnetic rack until the supernatant is clear and colorless, and remove and discard the supernatant.
11. Wash beads with 250 μL LFB: flick to resuspend, spin down briefly, pellet the beads as in the previous step, and remove and discard the supernatant.
12. Repeat the previous step (*see* **Note 28**). After washing, spin down the tube briefly, place the tube back on the magnetic rack, and carefully remove any residual supernatant.
13. Air-dry the beads for approximately 30 s, avoiding over-drying.
14. Remove the tube from the magnetic rack and resuspend the beads in 15 μL EB. Briefly spin down the tube and incubate for 10 min at room temperature (*see* **Note 29**). For high molecular weight DNA, incubating at 37 °C can improve the recovery of long fragments.
15. Place the tube on the magnetic rack until the eluate is clear and colorless, for at least 1 min (*see* **Note 30**).

16. Transfer 15 μL of eluate, containing the DNA library, to a clean 1.5 mL DNA low-binding tube.
17. Depending on your required data output, prepare your final library to 35–50 fmol for high-output of simplex data, or 10–20 fmol for duplex data, in 12 μL EB (*see* **Notes 31** and **32**).

3.3.3 Priming and Loading the MinION and GridION Flow Cell

1. Thaw the Sequencing Buffer (SB), Library Beads (LIB) or Library Solution (LIS, if applicable), Flow Cell Tether (FCT), and Flow Cell Flush (FCF) at room temperature (*see* **Note 33**). Thoroughly mix all reagents by vortexing, then briefly spin down and place on ice until use.
2. Prepare the flow cell priming mix by combining FCF and FCT (*see* **Note 34**). If using single-use tube format, add 30 μL FCT directly to the tube containing FCF and mix gently by pipetting at room temperature. If using a bottle, combine 1170 μL FCF with 30 μL FCT per flow cell and mix gently by pipetting. Ensure that the total priming mix volume equals 1205 μL per flow cell. Use the priming mix immediately after preparation.
3. Consult the current Oxford Nanopore Technologies (ONT) guidelines for MinION™ or P2-Solo™ flow cell loading and follow all procedural details and critical considerations [25] (*see* **Notes 35–37**).

3.4 Basecaller Training

3.4.1 Fast-Bonito

```
conda env create -f env_gpu.yml  # Setup environment using conda

conda env update -f env_gpu.yml # use env_gpu.yml for GPU server

source activate fast-bonito

git clone https://github.com/EIHealth-Lab/fast-bonito.git

cd fast-bonito # download fast-bonito

python setup.py install  # Install fast-bonito
python setup.py bdist_wheel; pip install dist/*.whl # As an alternative

INPUT_FAST5_DIR="input_fast5"  # input fast5 directory

OUTPUT_BASECALL_DIR="output"  # output directory

OUTPUT_NAME_BASECALL="test.fasta"  # output fasta name
```

```
python basecaller.py --reads_directory $INPUT_FAST5 --output \
/$OUTPUT_BASECALL/${OUTPUT_NAME_BASECALL} # running fast-bo-
nito
```

3.4.2 Docker Setup

```
docker build -f Dockerfile_fast_bonito -t fast-bonito:0.2.2
nvidia-docker run -it --rm --ipc host \

        -v ${INPUT_FAST5_DIR}:/input_data \

  -v ${OUTPUT_BASECALL_DIR}:/output \
           swr.cn-north-4.myhuaweicloud.com/fast-bonito/fast-
bonito:0.2.2 \

      sh -c "cd /workspace/fast-bonito/ && \
                  python basecaller.py \
                    --reads_directory $INPUT_FAST5_DIR \
                           --output /$OUTPUT_BASECALL_DIR/
${OUTPUT_NAME_BASECALL}" # build and run bonito
```

3.5 Genome Assembly

3.5.1 Host Removal and Assembly

```
makeblastdb -in host_genome.fa -dbtype nucl -out host_geno-
me_db # Create BLASTn database of the host reference genome to
align the raw reads on host genome

seqtk seq -A reads_R1.fastq > reads_R1.fasta # Converting
fastq file to fasta file as blastn input is fasta

blastn -query reads_R1.fasta -db host_genome_db -outfmt "6 std
qlen qcovs qcovshsp" -out blast_results.tab # map reads to the
host genome
```

Parse the BLASTn output reads with an Identity 90% (*see* **Note 38**) and Alignment length/query length equal or higher than 80% to filter reads. At the end of this step, you will have a text file containing the read IDs (readid.txt) that are not mapped to the host genome and can be used for the genome assembly.

```
seqtk subseq reads_R1.fastq readid.txt > NotMapReads.fastq
#depleted data set from the host
 flye --nano-raw NotMapReads.fastq -o assembly #doing assembly
```

A comparison of assembled reads with a phytoplasma genome database by any homology search tool identifies phytoplasma-related contigs.

An alternative approach involves preparing the sequencing setup in advance to enable real-time alignment. This is achieved by loading the reference genome of the target organism onto the sequencing device and configuring the basecalling software to align reads to this genome during sequencing. Reads that are identified as mapping to the target organism can then be directly used for assembly with Flye.

To ensure the accuracy and completeness of the assembly, all basecalled reads can subsequently be aligned to the host genome. This additional validation step helps confirm the effective removal of host sequences. The filtered reads can then be assembled with Flye, and the results can be compared to the initial assembly to detect any missed reads or discrepancies.

3.5.2 Assembly Evaluation

Assess the quality of the assembly using QUAST (which provides standard assembly statistics such as N50, total length, and number of contigs):

```
quast.py -o assembly_stats assembly/assembly.fasta
```

Evaluate the completeness of the genome by BUSCO to assess the presence of conserved single-copy orthologs in the assembly (*see* **Note 39**):

```
busco -i assembly/assembly.fasta -l mollicutes_odb10
 -o busco_results -m geno
```

3.5.3 Improving Genome Assembly

If the assembly remains of low quality after completing all the above steps—for instance, if it is very short or consists of many short contigs—a hybrid assembly that leverages the advantages of short and long reads for a comprehensive genomics reconstruction approach can significantly enhance the quality of the draft genome. Below are comments and instructions for conducting a hybrid assembly (*see* **Note 40**). Additionally, in cases where the raw long reads are of low quality, possibly due to the use of older MinION flow cells, the assembled genome may lack accuracy and completeness: in this case, utilize short reads to map and correct potential errors within the formed contigs, thus enabling more accurate representations as well (Hybrid error correction: leveraging short reads).

Install Unicycler and hybrid assembly:

```
git clone https://github.com/rrwick/Unicycler.git
cd Unicycler

 python3 setup.py install

unicycler -1 short_reads_1.fastq.gz -2 short_reads_2.fastq.gz
-o output_dir # Illumina-only assembly
```

```
unicycler -l long_reads.fastq.gz -o output_dir # Long-read-only assembly

unicycler -1 short_reads_1.fastq.gz -2 short_reads_2.fastq.gz -l long_reads.fastq.gz -o output_dir # Hybrid assembly
```

Hybrid error correction: leveraging short reads (previously explained) and assembly polishing by Medaka

```
conda create -n medaka -c conda-forge -c bioconda medaka #installation
conda activate medaka #activation

medaka_consensus -i reads_R1.fastq -d assembly.fasta -o medaka # assembly.fasta is the draft genome assembly and reads_R1.fastq is the raw reads
```

4 Notes

1. We used the Macherey-Nagel NucleoSpin Plant II Midi kit to extract DNA with sufficient purity and high molecular weight to achieve satisfactory sequencing yield and read length on the Oxford Nanopore platform. While this kit performed well in our hands, other DNA extraction kits may also be suitable. We recommend performing DNA quality control to ensure that the extracted material meets the requirements for long-read sequencing.
2. Paramagnetic beads should be resuspended carefully before use.
3. If excessive genomic DNA shearing is observed, minimize vortexing once DNA is released from cells; mix by flicking tubes and pipette gently to avoid further fragmentation.
4. Low DNA purity can lead to reduced read sequence yield. Adding PVP40 to the extraction buffer has been shown to improve the yield in ONT flow cells, potentially due to the removal of polyphenolic compounds. This improvement is supported by findings from the study [26].
5. It is also important to quantify the amount of DNA in the experimental sample by A260 measurement using a spectrophotometer.
6. Proper resuspension of magnetic beads is essential for uniform MBD2-Fc protein binding and efficient DNA capture. Beads should be resuspended by gently pipetting up and down or using a rotating mixer. Avoid vortexing, as it can damage the

beads and affect their magnetic properties. If beads appear clumpy, additional gentle mixing may be needed.

7. Improper preparation of the 1X Bind/Wash Buffer can significantly impact the efficiency of protein binding and washing steps. To ensure optimal performance, the 1X Bind/Wash Buffer should be freshly prepared on ice by diluting one part of the 5X Bind/Wash Buffer with four parts DNase-free water. It is essential to use the correct volume for each reaction to maintain consistency and reliability across experiments.
8. Incomplete mixing of MBD2-Fc protein with Protein A Magnetic Beads can lead to suboptimal protein binding. To achieve optimal binding, it is crucial to thoroughly mix the MBD2-Fc protein and beads by pipetting up and down at least five to ten times. Ensure that the beads are fully suspended and free of any clumps. For enhanced mixing, a rotating mixer can be used to guarantee complete binding.
9. Ineffective washing of beads can result in contamination or loss of target DNA. To prevent this, it is essential to perform the washing steps carefully, ensuring that unbound proteins and contaminants are effectively removed. The beads should be thoroughly mixed in the wash buffer, and it is important to remove the supernatant completely after each wash. Care should be taken not to disturb the beads during the removal of the supernatant.
10. Inadequate mixing of DNA with beads can result in poor capture efficiency. To achieve optimal binding, after adding the DNA and Bind/Wash Buffer, pipette the mixture up and down five to ten times to ensure thorough mixing of the beads and DNA. To further enhance binding efficiency, the mixture should be agitated on a rotating mixer.
11. Insufficient magnetic separation time can lead to incomplete collection of beads or loss of the sample. To ensure complete separation, place the tube on the magnetic rack for the full duration specified (2–5 min). If the solution remains turbid, extend the separation time slightly to ensure optimal bead collection.
12. Incomplete removal of ethanol can negatively impact DNA recovery. To ensure optimal DNA recovery, carefully remove ethanol after each wash without disturbing the beads.
13. Do not over-dry the beads. This may result in lower recovery of DNA. Elute the samples when the beads are still dark brown and glossy looking, but when all visible liquid has evaporated. If the beads turn brown and start to crack, they are overdried.
14. Inaccurate quantitation of DNA can compromise downstream applications, such as NGS. To ensure the sample is suitable for

further analysis, use appropriate quantitation methods, such as agarose gel electrophoresis or qPCR. It is important to follow recommended protocols for quantitation to achieve accurate and reliable results.

15. Do not vortex the FFPE DNA Repair Mix or Ultra II End Prep Enzyme Mix.
16. It is important that the buffers are mixed well by vortexing.
17. Extending the 20 °C incubation time by 15 min can lead to an increased yield.
18. The incubation time can be extended if necessary. Tests with incubation times of 30 and 45 min were conducted without observing any negative effects on DNA length during sequencing.
19. To clean the pellet, rotate the tube by 180 degrees so that the pellet is on the opposite side of the magnet, leaving it in the magnetic rack.
20. The solution must be clear without residuals of Paramagnetic beads.
21. Take forward the repaired and end-prepped DNA into the adapter ligation step. However, at this point, it is also possible to store the sample at 4 °C overnight.
22. At the end of this step, to ensure optimal sequencing performance, load 35–50 fmol of the final library onto the R10.4.1 flow cell for high output of simplex data. This loading amount ensures that a high pore occupancy of greater than 95% is achieved. For duplex data, load 10–20 fmol of the final library, as loading more than 20 fmol of DNA can reduce the rate of duplex read capture. To calculate pore occupancy, refer to the relevant guidelines.
23. Although the recommended third-party ligase is supplied with its buffer, the ligation efficiency of the Ligation Adapter (LA) is higher when using the Ligation Buffer (LNB) supplied in the Ligation Sequencing Kit.
24. The LNB buffer must be pipetted slowly to aspirate the right volume due to its high viscosity.
25. The clean-up step following adapter ligation is designed based on the type of wash buffer used. When using a Long Fragment Buffer (LFB), the clean-up step is optimized to enrich DNA fragments greater than 3 kb. Conversely, using a Short Fragment Buffer (SFB) ensures the purification of DNA fragments across all sizes, without preferentially enriching for longer fragments.
26. The ligation can be extended even to 30 min if necessary.
27. The incubation time can be extended to 30 min if necessary.

28. Wash well the beads, removing the tube from the magnetic rack and inverting or flicking the tube to completely resuspend the beads.
29. The Incubation time can be extended to 20 min if necessary.
30. The liquid must be completely clear. Avoid the carryover of beads that will negatively affect the sequencing.
31. If the library yields are below the input recommendations, load the entire library.
32. The prepared library should be stored on ice or at 4 °C until ready to load into the flow cell. For short-term storage or repeated use, such as reloading flow cells between washes, libraries should be stored in low-binding tubes at 4 °C. For single-use and long-term storage exceeding 3 months, it is recommended to store libraries at −80 °C in low-binding tubes.
33. This kit is compatible only with R10.4.1 flow cells (FLO-MIN114). It is recommended to watch the "Priming and Loading Your Flow Cell" video before performing their first run to ensure proper preparation and handling of the flow cell.
34. For optimal sequencing performance and improved output on MinION R10.4.1 flow cells (FLO-MIN114), it is recommended to add Bovine Serum Albumin (BSA) to the flow cell priming mix at a final concentration of 0.2 mg/mL. Using another type of albumin, such as recombinant human serum albumin, is not recommended.
35. Take care when drawing back buffer from the flow cell. Do not remove more than 20–30 μL, and make sure that the array of pores is covered by buffer at all times. Introducing air bubbles into the array can irreversibly damage the pores.
36. Low DNA purity is a common issue, often detected using microvolume spectrophotometric DNA quantification. Suboptimal OD 260/280 ratios (below 1.8) and OD 260/230 ratios (outside the ideal 2.0–2.2 range) may indicate the presence of contaminants. These issues often result from the DNA extraction method not achieving sufficient purity. To improve results, consider using an alternative extraction method that reduces contaminant carryover. Additionally, an extra Solid Phase Reversible Immobilization (SPRI) clean-up step can help enhance DNA quality.
37. Low recovery of DNA can also occur due to several factors. One common cause is DNA loss when the Paramagnetic beads-to-sample ratio is lower than intended, often because Paramagnetic beads settle quickly—ensure they are well resuspended before adding them. Using a beads-to-sample ratio lower than

0.4:1 results in the loss of DNA fragments of any size. Low recovery can also happen when DNA fragments are shorter than expected, as lower bead ratios more stringently select against short fragments; it is important to determine the input DNA length via agarose gel or other electrophoresis methods and adjust the bead amount accordingly. Finally, low recovery after end-prep can occur if the washing step is performed with an ethanol concentration lower than 70%, leading to premature elution of DNA from the beads—always ensure the correct ethanol percentage is used.

38. Adjust -evalue, word_size, and other parameters as needed for optimal sensitivity and specificity.
39. Check the available databases in BUSCO to choose your microorganism database (busco --list-datasets).
40. Long reads that are used in this step must contain reads related to your microorganism. Before using Unicycler, run all the steps related to removing the host genome as described.

Acknowledgement

Zahra Golestani-Hotkani is funded by the *Consorzio Vini del Trentino (Trento, Italy)*. We thank the *Consorzio Vini del Trentino* for their support.

References

1. Xue C, Zhang Y, Li H et al (2023) The genome of *Candidatus* Phytoplasma ziziphi provides insights into their biological characteristics. BMC Plant Biol 23:309. https://doi.org/10.1186/s12870-023-04243-6
2. Tan CM, Lin YC, Li JR et al (2021) Accelerating complete phytoplasma genome assembly by immunoprecipitation-based enrichment and MinION-based DNA sequencing for comparative analyses. Front Microbiol 12:766221. https://doi.org/10.3389/fmicb.2021.766221
3. Chen LL, Chung WC, Lin CP, Kuo CH (2012) Comparative analysis of gene content evolution in phytoplasmas and mycoplasmas. PLoS One 7:e34407. https://doi.org/10.1371/journal.pone.0034407
4. Melamed S, Tanne E, Ben-Haim R et al (2003) Identification and characterization of phytoplasmal genes, employing a novel method of isolating phytoplasmal genomic DNA. J Bacteriol 185:6513–6521. https://doi.org/10.1128/jb.185.22.6513-6521.2003
5. Kube M, Mitrovic J, Duduk B et al (2012) Current view on phytoplasma genomes and encoded metabolism. Sci World J 2012:185942. https://doi.org/10.1100/2012/185942
6. Rodrigues Jardim B, Tran-Nguyen LTT, Gambley C et al (2022) Iodixanol density gradients as an effective phytoplasma enrichment approach to improve genome sequencing. Front Microbiol 13:937648. https://doi.org/10.3389/fmicb.2022.937648
7. Prezelj N, Nikolić P, Gruden K et al (2013) Spatiotemporal distribution of flavescence dorée phytoplasma in grapevine. Plant Pathol 62:760–766. https://doi.org/10.1111/j.1365-3059.2012.02693.x
8. Nair S, Manimekalai R (2021) Phytoplasma diseases of plants: molecular diagnostics and way forward. World J Microbiol Biotechnol 37:36. https://doi.org/10.1007/s11274-021-03061-y
9. Nijo T, Iwabuchi N, Tokuda R et al (2021) Enrichment of phytoplasma genome DNA through a methyl-CpG binding domain-

mediated method for efficient genome sequencing. J Gen Plant Pathol 87:154–163. https://doi.org/10.1007/s10327-021-00993-z
10. Lohmaneeratana K, Gutiérrez G, Thamchaipenet A, Wellinger RE (2024) Phytoplasma DNA enrichment from sugarcane white leaves for shotgun sequencing improvement. Plants 13:3006. https://doi.org/10.3390/PLANTS13213006
11. Noone J, Wallace RG, Rochfort KD (2023) Immunoprecipitation: variations, considerations, and applications. Methods Mol Biol 2699:271–303. https://doi.org/10.1007/978-1-0716-3362-5_15
12. Heravi FS, Zakrzewski M, Vickery K, Hu H (2020) Host DNA depletion efficiency of microbiome DNA enrichment methods in infected tissue samples. J Microbiol Methods 170:105856. https://doi.org/10.1016/j.mimet.2020.105856
13. Deaton AM, Bird A (2011) CpG islands and the regulation of transcription. Genes Dev 25:1010–1022. https://doi.org/10.1101/gad.2037511
14. Ari Ş, Arikan M (2016) Next-generation sequencing: advantages, disadvantages, and future. In: Barh D, Azevedo V (eds) Plant omics: trends and applications. Springer, Cham, pp 109–135. https://doi.org/10.1007/978-3-319-31703-8_5
15. Oxford Nanopore Technologies (2024) Flow cell (R10.4.1). https://store.nanoporetech.com/eu/flow-cell-r10-4-1.html. Accessed 25 July 2024
16. Giordano F, Aigrain L, Quail MA et al (2017) De novo yeast genome assemblies from MinION, PacBio and MiSeq platforms. Sci Rep 7:3935. https://doi.org/10.1038/s41598-017-03996-z
17. Vereecke N, Bokma J, Haesebrouck F et al (2020) High quality genome assemblies of Mycoplasma bovis using a taxon-specific Bonito basecaller for MinION and Flongle long-read nanopore sequencing. BMC Bioinformatics 21:517. https://doi.org/10.1186/s12859-020-03856-0
18. Ferguson S, McLay T, Andrew RL et al (2022) Species-specific basecallers improve actual accuracy of nanopore sequencing in plants. Plant Methods 18:121. https://doi.org/10.1186/s13007-022-00971-2
19. Espinosa E, Bautista R, Larrosa R, Plata O (2024) Advancements in long-read genome sequencing technologies and algorithms. Genomics 116:110842. https://doi.org/10.1016/j.ygeno.2024.110842
20. Zhang H, Jain C, Aluru S (2020) A comprehensive evaluation of long read error correction methods. BMC Genomics 21:889. https://doi.org/10.1186/s12864-020-07227-0
21. Kolmogorov M, Yuan J, Lin Y, Pevzner PA (2019) Assembly of long, error-prone reads using repeat graphs. Nat Biotechnol 37:540–546. https://doi.org/10.1038/s41587-019-0072-8
22. Walker BJ, Abeel T, Shea T et al (2014) Pilon: an integrated tool for comprehensive microbial variant detection and genome assembly improvement. PLoS One 9:e112963. https://doi.org/10.1371/journal.pone.0112963
23. Vaser R, Sović I, Nagarajan N, Šikić M (2017) Fast and accurate de novo genome assembly from long uncorrected reads. Genome Res 27:737–746. https://doi.org/10.1101/gr.214270.116
24. Oxford Nanopore Technologies (2025) Medaka: sequence correction. https://github.com/nanoporetech/medaka. Accessed 10 Mar 2025
25. Oxford Nanopore Technologies (2022) Ligation sequencing DNA V14 (SQK-LSK114). https://nanoporetech.com/document/genomic-dna-by-ligation-sqk-lsk114. Accessed 13 Sept 2024
26. Pearman WS, Arranz V, Carvajal JI et al (2024) A cry for kelp: evidence for polyphenolic inhibition of Oxford Nanopore sequencing of brown algae. bioRxiv. https://doi.org/10.1101/2024.06.26.600698

Chapter 19

Complete Genome Sequence of Phytoplasma Obtained by Nanopore Sequencing Using High Molecular Weight DNA

Yi-Ching Chiu, Helen Mae Mejia, Chih-Horng Kuo, and Jun-Yi Yang

Abstract

Nanopore sequencing can obtain long reads, which help to overcome the challenges posed by AT- and repeat-rich genomes, such as those found in phytoplasmas. Additionally, removing short DNA fragments improves the read length distribution before sequencing. Here, we present a methodology for extracting genomic DNA from phytoplasma-infected plants, the removal of short DNA fragments under 10 kb, and utilizing nanopore sequencing to generate lengthy reads. Furthermore, we outline procedures for assembling a circular genome by removing the plant sequence and refining it using short reads generated by Illumina sequencing to correct sequencing errors.

Key words Phytoplasma, Nanopore, SRE, Illumina

1 Introduction

Advances in sequencing technologies have increasingly made whole-genome sequencing accessible for biological research. Traditional short-read sequencing technologies, such as Illumina sequencing, are widely used for initial genome assembly due to their high accuracy and high throughput [1, 2]. However, these methods often fall short in resolving repetitive sequences and large structural variations, especially in AT- and repeat-rich phytoplasma genomes [3–6]. To overcome these limitations, long-read sequencing technologies such as Oxford Nanopore Technologies (ONT) have emerged as valuable complements. ONT generates longer reads that improve genome continuity and integrity [2, 7].

This protocol introduces a two-stage method for *de novo* genome assembly by integrating Illumina and ONT reads [8, 9]. For the first stage, draft assembly is performed using Velvet with only the Illumina short reads. Putative phytoplasma contigs

Katrin Janik and Mattia Tabarelli (eds.), *Phytoplasma: Methods and Protocols*, Methods in Molecular Biology, vol. 3008, https://doi.org/10.1007/978-1-0716-5104-9_19,

are identified by BLASTX search against the NCBI non-redundant protein database. Those having a significant hit to phytoplasma proteins are further screened by BLASTN search against the NCBI standard nucleotide collection database and an inspection of sequencing depth. For the second stage, the Illumina and ONT reads are mapped to those phytoplasma contigs from the first stage using Burrows-Wheeler Aligner (BWA) and Minimap2, respectively [10, 11]. The mapped reads are extracted for hybrid assembly using Unicycler [12]. For validation, all raw reads are mapped to the contigs produced by Unicycler for an iterative process of validation and correction using SAMTOOLS and IGV. This iterative process is continued until a complete genome assembly is achieved, supported by raw read mapping results.

This approach combines the strengths of different sequencing technologies, ensuring high accuracy and completeness through multiple validation and correction steps. It also includes detailed gene prediction and annotation processes and an exemplary workflow on how to identify potential phytoplasma effector proteins, providing a solid foundation for future functional genomics research.

2 Materials

2.1 Chemicals and Reagents

The products mentioned as examples in this section are indicative. Similar products from other companies or suppliers may be equally suitable for the method described.

1. Liquid nitrogen.
2. PBS buffer (0.137 M NaCl, 2.7 mM KCl, 10 mM Na_2HPO_4, 1.8 mM KH_2PO_4, pH 6.8) (*see* **Note 1**).
3. 70% ethanol.
4. ddH_2O (nuclease-free).
5. Protein A-coated magnetic beads (e.g., Dynabeads™ Protein A, Invitrogen).
6. Polyclonal anti-Imp antibody.
7. Silica-matrix-based DNA purification kit (e.g., DNeasy Plant Mini Kit, Qiagen).
8. RNase A.
9. Magnetic bead-based system (e.g., KAPA HyperPure Beads kit, Roche).
10. Ligation-based library preparation kit (e.g., KAPA HyperPrep Kit, Roche, in combination with the Ligation Sequencing Kit, Oxford Nanopore Technologies).

11. Fragment depletion kit (e.g., Short Read Eliminator Kit, PacBio).
12. Priming kit (e.g., Flow Cell Priming Kit, Oxford Nanopore Technologies).

2.2 General Equipment

1. 1.5 mL microcentrifuge tubes.
2. PCR tubes.
3. Mortar and pestle.
4. 100 μm cell strainer.
5. Magnetic stand.
6. Fluorometer for nucleic acid quantification.
7. Flow cell for sequencing.
8. MinION R9.4.1 flow cell.
9. Water bath.
10. Centrifuge (for 13,000× *g*).
11. Thermocycler.

2.3 Bioinformatic Resources

1. Guppy v3.3.3 for base-calling [13].
2. Velvet v1.2.10 [14].
3. BLASTX v2.10.0 [15].
4. BLASTN v2.10.0 [15].
5. BWA v0.7.17 [10].
6. Minimap2 v2.15 [11].
7. Unicycler v0.4.9b [12].
8. SAMTOOLS v1.9 [16].
9. IGV v2.5.0 [17].
10. RNAmmer v1.2 [18].
11. tRNAscan-SE v1.3.1 [19].
12. Prodigal v2.6.3 [20].
13. BLASTP v2.10.0 [15].
14. OrthoMCL v1.3 [21].
15. GenBank [22].
16. KEGG [23].
17. COG [24].
18. SignalP v5.0 [25].
19. TMHMM v2.0 [26].

3 Methods

In the Methods section, specific reagents or materials from particular suppliers are mentioned for illustrative purposes. These should not be viewed as exclusive options. Equivalent alternatives from other suppliers can be used, as long as the protocols are appropriately adjusted according to the manufacturer's guidelines.

3.1 Tissue Processing and Immunoprecipitation-Based Enrichment

1. Gently homogenize tissues with liquid nitrogen using a mortar and pestle.
2. Suspend 3 g of ground tissues in 6 mL of PBS buffer.
3. Filter the suspension using a 100 μm cell strainer to remove unfragmented tissues.
4. Incubate antiserum containing polyclonal anti-Imp antibody with protein A-coated magnetic beads for 30 min at room temperature.
5. Place the tube on the magnetic stand and remove the supernatant.
6. Incubate the filtered cell suspension with 3 mg of the antibody-coated beads for 30 min at 4 °C.
7. Use a magnetic stand to separate the bead-bound phytoplasma cells from the remaining supernatant to remove unwanted substances.
8. Wash the remaining tissue material containing the phytoplasma cells twice with 1 mL PBS buffer.
9. Collect the washed material containing the enriched phytoplasma cells for DNA extraction.

3.2 Genomic DNA Extraction and Purification

1. Grind material containing enriched phytoplasma cells with liquid nitrogen and transfer to a 1.5 mL microcentrifuge tube (*see* **Note 2**).
2. Use a silica-matrix-based DNA purification kit for DNA extraction (e.g., the DNeasy Plant Mini Kit, Qiagen, *see* **Note 6**). In the following, the buffers from the DNasy Plant Mini Kit are mentioned; however, when using a different kit, use the corresponding buffers and refer to the respective instructions.
3. For every 100 mg ground sample, add 400 μL lysis buffer and 4 μL RNase A. Vortex briefly and incubate in a water bath at 65 °C for 10 min, inverting the tube every few minutes.
4. Add 130 μL precipitation buffer, mix, and incubate for 5 min on ice. Centrifuge at 13,000× *g* for 5 min.
5. Transfer the lysate into a shredder spin column and centrifuge at 13,000× *g* for 2 min.

6. Add 1.5× volumes of Buffer AW1 to the supernatant, mix by pipetting or vortexing.
7. Place a silica spin column in a 2 mL collection tube. Load the mixture onto the column, centrifuge at 13,000× *g* for 1 min, discard the flow-through, and repeat with the remaining sample (*see* **Note 3**).
8. Add 500 μL wash buffer to the column, centrifuge at 13,000× *g* for 1 min, discard the flow-through, and repeat once.
9. Dry the column membrane by centrifuging at 13,000× *g* for 3 min. Place the spin column in a new 1.5 mL microcentrifuge tube, add 50–100 μL of ddH_2O, incubate at room temperature for 5 min, and centrifuge at 13,000× *g* to elute the DNA (*see* **Note 4**).
10. Store the eluted DNA at −20 °C for long-term storage or at 4 °C for immediate use.
11. Ensure that the DNA sample is eluted in ddH_2O, adjust the volume to 100 μL if necessary.
12. Use a magnetic-bead-based system for DNA purification as follows.
13. Resuspend magnetic beads by vortexing (*see* **Note 5**), add 0.4 × volume of beads to the DNA sample, mix thoroughly, and incubate at room temperature for 5 min.
14. Place the tube on a magnetic stand for 2 min, transfer the supernatant (containing fragments <1 kb) to a new tube.
15. Remove the tube from the stand, add 30 μL ddH_2O to the beads, mix thoroughly, and incubate at room temperature for 2 min.
16. Place the tube back on the magnetic stand for 2 min or until the solution is clear. Carefully transfer the supernatant, which contains the purified DNA fragments ≥1 kb, to a new tube (Fig. 1, *see* **Note 4**).
17. Store the eluted DNA at −20 °C for long-term storage.

3.3 Small Fragment Depletion

1. Adjust the DNA sample to 60 μL with a DNA concentration of 50–150 ng/μL. Add 60 μL Buffer SRE from a fragment depletion kit, such as the Short Read Eliminator (SRE) Kit (*see* **Notes 6** and 7), mix thoroughly. Incubate the tube at room temperature overnight.
2. Centrifuge at 13,000× *g* for 30 min. Carefully remove the supernatant without disturbing the DNA pellet.
3. Wash with 200 μL 70% EtOH, centrifuge at 13,000× *g* for 2 min. Carefully remove the wash solution without disturbing the pellet. Repeat this step.

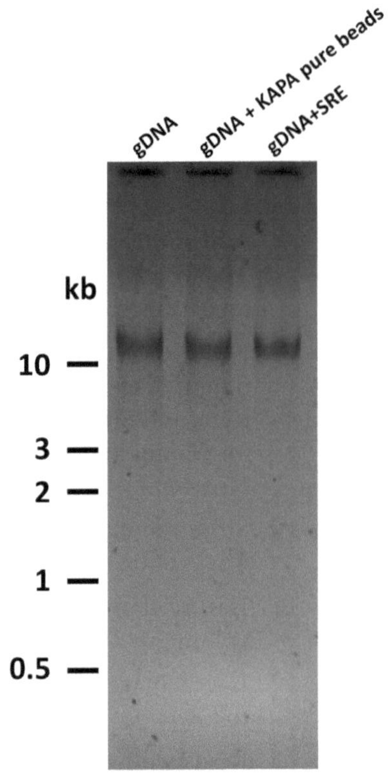

Fig. 1 DNA electrophoresis on a 1% agarose gel for the size selection of extracted genomic DNA. Lane 1 shows the extracted genomic DNA without size selection. Lane 2 shows the removal of <1 kb fragments using beads. Lane 3 shows the removal of <10 kb fragments using SRE. Although the DNA size distribution might appear similar across all three lanes, it is noteworthy that the lower molecular weight smear present in the first (gDNA) and the second (gDNA+KAPA pure beads) lanes is absent in the third lane (gDNA+SRE). Therefore, SRE treatment is recommended to enrich for long DNA prior to nanopore sequencing

4. Add 50–100 μL Buffer EB from the SRE kit, incubate at room temperature for 20 min. Gently tap the tube to ensure DNA is properly resuspended and mixed. This solution contains the purified DNA fragments ≥10 kb (Fig. 1).
5. Analyze DNA recovery and purity using the fluorometer for nucleic acid quantification.

3.4 Nanopore Library Preparation

1. Combine 49 μL genomic DNA (100 fmol), 7 μL End Repair & A-Tailing Buffer, 3 μL End Repair and A-Tailing Enzyme Mix, and 1 μL DNA Control Strand (DCS) as provided in the KAPA HyperPrep Kit or another suitable ligation-based library preparation kit. (*see* **Notes 6** and **8**).
2. Prewarm the thermocycler to 20 °C. Incubate at 20 °C for 30 min, then at 65 °C for 30 min, hold at 4 °C.

3. To the 60 μL end-repair and A-tailing reaction product from **step 2**, add 5 μL AMX, 5 μL nuclease-free ddH_2O, 30 μL Ligation Buffer, and 10 μL DNA Ligase, as provided in the Ligation Sequencing Kit (Oxford Nanopore Technologies) or another suitable ligation-based library preparation kit (*see* **Note 6**). Incubate at 20 °C for 60 min in a thermocycler.
4. Add 44 μL magnetic beads and mix for 5 min. Place on a magnetic stand for 2 min. Discard supernatant.
5. Wash the beads twice with 250 μL LFB (Long fragment buffer), as provided in the Ligation Sequencing Kit (Oxford Nanopore Technologies) or another suitable ligation-based library preparation kit (*see* **Notes 6** and **9**). Elute the DNA with 14 μL Elution Buffer at room temperature for 10 min.
6. Quantify the eluted DNA using a fluorometer for nucleic acid quantification.
7. Prepare the priming mix by adding 30 μL FLT to a tube of FB as provided in the Flow Cell Priming Kit (Oxford Nanopore Technologies) or another suitable priming solution (*see* **Note 6**).
8. Carefully load the mixture into the flow cell and wait for 5 min to activate the flow cell (*see* **Note 10**).
9. Prepare the loading mix by combining 12 μL adaptor-ligated DNA library (corresponding to 50 fmol), 37.5 μL Sequencing Buffer (SQB), and 25.5 μL Loading Beads (LB), as provided in the Ligation Sequencing Kit (Oxford Nanopore Technologies) or another suitable ligation-based library preparation kit (*see* **Note 2**). Gently load the mixture into the activated flow cell.
10. Perform Nanopore sequencing. Base-call reads using Guppy v3.3.3 [13] (*see* **Note 11**).

3.5 De Novo Genome Assembly (Fig. 2, See *Note 11)*

3.5.1 Initial Assembly Using Illumina Reads

1. Assemble Illumina reads using Velvet v1.2.10 [14]. Set k-mer length to 151 and minimum contig length to 2000 bp (*see* **Note 12**).
2. Compare all contigs against the NCBI non-redundant protein database [21] using BLASTX v2.10.0 [21]. Select contigs with hits to phytoplasma proteins (e-value <1e-15) for manual inspection (*see* **Note 12**).
3. Identify false positive contigs derived from plant chloroplast and mitochondrial genomes based on differences in sequencing depth (*see* **Note 13**).
4. Verify these contigs using BLASTN v2.10.0 [15] against the NCBI standard nucleotide collection database.

5. Use BWA v0.7.17 [10] to extract phytoplasma reads from the Illumina data set with an alignment score cutoff of 30 (*see* **Note 12**).
6. Use Minimap2 v2.15 [11] to extract phytoplasma reads from the ONT data set with an alignment score cutoff of 1000 (*see* **Note 12**).

3.5.2 Hybrid Assembly

1. Perform de novo hybrid assembly using Unicycler v0.4.9b [12] (Fig. 2, *see* **Note 11**).
2. Validate contigs as belonging to the phytoplasma genome using BLASTN v2.10.0 searches as in Stage 1.
3. Map all Illumina and ONT raw reads to the draft assembly using the methods described.
4. Use the "mpileup" function of SAMTOOLS v1.9 [16] and IGV v2.5.0 [17] for programmatic and manual inspection of mapping results.
5. Identify and correct assembly errors based on the continuity of ONT long reads and validate using mapping results in subsequent iterations.
6. During early iterations, use ONT read mapping results for scaffolding information and to validate the organization of the circular chromosome, focusing on junctions between repetitive and unique regions.
7. During later iterations, use Illumina read mapping results to validate base-pair scale indels and sequencing errors introduced by ONT reads.
8. Repeat the process until a complete genome assembly is obtained, with all regions supported by raw read mapping results.
9. Use the "depth" function of SAMTOOLS [16] to calculate sequencing coverage.

3.6 Gene Prediction and Annotation

1. Predict rRNA genes using RNAmmer v1.2 [18], tRNA genes using tRNAscan-SE v1.3.1 [19], and protein-coding genes using Prodigal v2.6.3 [20] (Fig. 2, *see* **Note 11**).
2. Use BLASTP v2.10.0 [15] and OrthoMCL v1.3 [21] to identify homologous gene clusters present in representative phytoplasma genomes (e.g., AYWB, OYM).
3. Manually curate the identified genes based on information from GenBank, KEGG, and COG databases [21–23].

3.7 Potential Effectors Identification

1. Use SignalP v5.0 [24] based on the Gram-positive bacteria model to predict putative secreted proteins (Fig. 2, *see* **Note 11**).

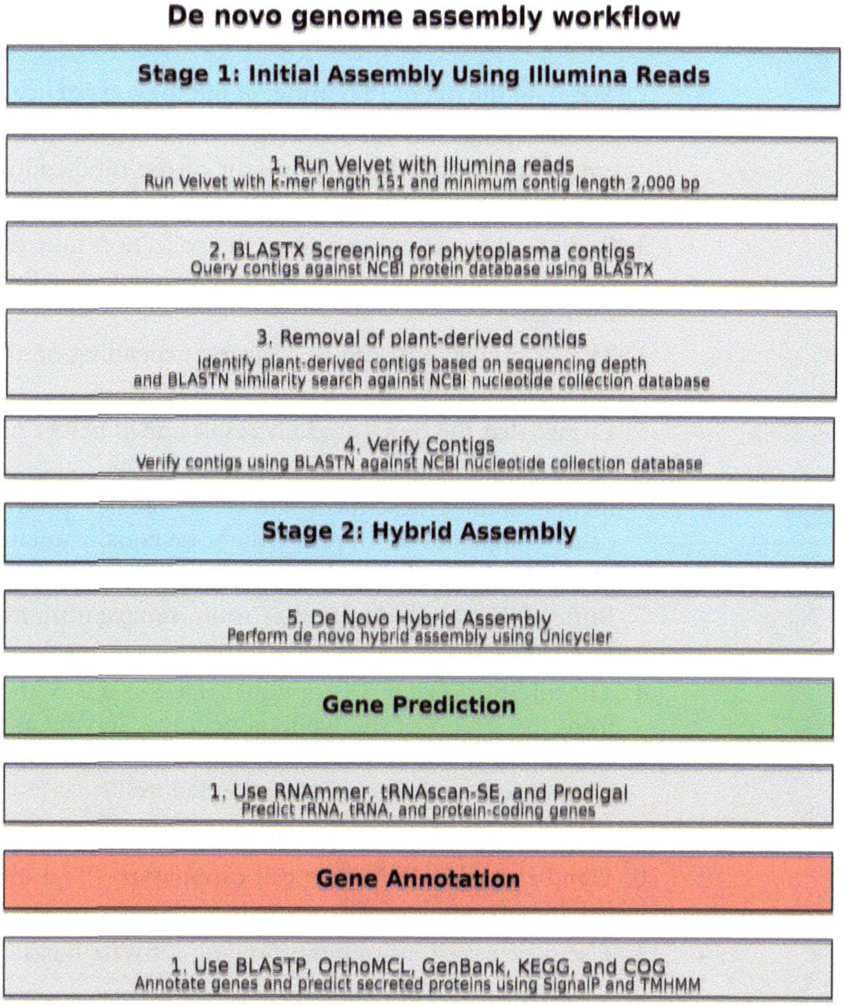

Fig. 2 Pipeline for *de novo* assembly of phytoplasma genomes. The *de novo* genome assembly (blue) involved two stages. The first stage used only Illumina reads. The second stage combined both Illumina and ONT data sets. Gene prediction is performed using RNAmmer, tRNAscan-SE, and Prodigal (green). The annotation is based on the homologous gene clusters present in other phytoplasma genomes identified by BLASTP and OrthoMCL and then manually curated based on the information obtained from GenBank, KEGG, and COG databases (pink). Prediction of putative secreted proteins using SignalP based on the Gram-positive bacteria model. Those candidates with transmembrane domains recognized by TMHMM were deleted

2. Use TMHMM v2.0 [25] to identify and remove candidates with transmembrane domains.
3. Ensure the remaining candidates have a signal peptide length in the range of 21–52 amino acids.

4 Notes

1. Maintain the pH of PBS buffer at 6.8 for consistency in results.
2. To determine the sample weight, subtract the weight of the empty tube from the total weight of the tube containing the sample tube.
3. Remove the spin column from the collection tube carefully so that the column does not come into contact with the flow-through.
4. The elution volume can be adjusted depending on the desired concentration of the DNA.
5. Ensure that the beads are fully resuspended before use.
6. The product mentioned in this step worked well in the experimental setting described. However, similar products from other companies or suppliers might be equally suitable for the method described.
7. Buffer SRE should be used at room temperature for optimal DNA recovery.
8. The addition of DCS is optional. DCS is a 3.6 kb amplicon from the 3′ end of the Lambda genome, serving as a positive control for quality assurance. It helps users to distinguish sample failure from library preparation failure.
9. Use LFB to enrich DNA fragments of 3 kb or longer.
10. Handle the MinION flow cell carefully to avoid introducing bubbles during loading.
11. Use newer versions of bioinformatic software if available.
12. Given the inherent variability in phytoplasma genome sequencing projects, optimal parameter settings for each bioinformatic tool may depend on factors such as the exact species analyzed, proportion of phytoplasma DNA in samples, read length distribution, and sequencing depth. Thus, the recommended parameter settings should be considered practical starting points rather than rigid prescriptions. Users are encouraged to critically evaluate and adjust these parameters according to the specific characteristics of their datasets.
13. It is challenging to provide a universally applicable guideline due to extensive variability in phytoplasma titer among samples. Nevertheless, host organelle genomes typically exhibit much higher sequencing depth compared to phytoplasma-derived sequences, making these differences generally easy to distinguish in practice.

Acknowledgement

This work was financially supported (in part) by grants-in-aid from the National Science and Technology Council (111-2313-B-005 -014 -MY3), the Advanced Plant and Food Crop Biotechnology Center from The Featured Areas Research Center Program within the framework of the Higher Education Sprout Project by the Ministry of Education (MOE) in Taiwan, and Academia Sinica.

References

1. Cheng C, Fei Z, Xiao P (2023) Methods to improve the accuracy of next-generation sequencing. Front Bioeng Biotechnol 11: 982111. https://doi.org/10.3389/fbioe.2023.982111
2. Slatko BE, Gardner AF, Ausubel FM (2018) Overview of next-generation sequencing technologies. Curr Protoc Mol Biol 122:e59. https://doi.org/10.1002/cpmb.59
3. Andersen MT, Liefting LW, Havukkala I et al (2013) Comparison of the complete genome sequence of two closely related isolates of '*Candidatus* Phytoplasma australiense' reveals genome plasticity. BMC Genomics 14:1–15. https://doi.org/10.1186/1471-2164-14-529
4. Kube M, Schneider B, Kuhl H et al (2008) The linear chromosome of the plant-pathogenic mycoplasma '*Candidatus* Phytoplasma mali'. BMC Genomics 9:1–14. https://doi.org/10.1186/1471-2164-9-306
5. Marsili L, Duque KR, Bode RL et al (2022) Uncovering essential tremor genetics: the promise of long-read sequencing. Front Neurol 13:821189. https://doi.org/10.3389/fneur.2022.821189
6. Music MS, Samarzija I, Hogenhout SA et al (2019) The genome of '*Candidatus* Phytoplasma solani' strain SA-1 is highly dynamic and prone to adopting foreign sequences. Syst Appl Microbiol 42:117–127. https://doi.org/10.1016/j.syapm.2018.10.008
7. Udaondo Z, Sittikankaew K, Uengwetwanit T et al (2021) Comparative analysis of PacBio and Oxford Nanopore sequencing technologies for transcriptomic landscape identification of *Penaeus monodon*. Life 11:862. https://doi.org/10.3390/life11080862
8. Huang C-T, Cho S-T, Lin Y-C et al (2022) Comparative genome analysis of '*Candidatus* Phytoplasma luffae' reveals the influential roles of potential mobile units in phytoplasma evolution. Front Microbiol 13:773608. https://doi.org/10.3389/fmicb.2022.773608
9. Tan CM, Lin Y-C, Li J-R et al (2021) Accelerating complete phytoplasma genome assembly by immunoprecipitation-based enrichment and MinION-based DNA sequencing for comparative analyses. Front Microbiol 12:766221. https://doi.org/10.3389/fmicb.2021.766221
10. Li H, Durbin R (2009) Fast and accurate short read alignment with Burrows–Wheeler transform. Bioinformatics 25:1754–1760. https://doi.org/10.1093/bioinformatics/btp324
11. Li H (2018) Minimap2: pairwise alignment for nucleotide sequences. Bioinformatics 34: 3094–3100. https://doi.org/10.1093/bioinformatics/bty191
12. Wick RR, Judd LM, Gorrie CL et al (2017) Unicycler: resolving bacterial genome assemblies from short and long sequencing reads. PLoS Comput Biol 13(6):e1005595. https://doi.org/10.1371/journal.pcbi.1005595
13. Wick RR, Judd LM, Holt KE (2019) Performance of neural network basecalling tools for Oxford Nanopore sequencing. Genome Biol 20:129. https://doi.org/10.1186/s13059-019-1727-y
14. Zerbino DR, Birney E (2008) Velvet: algorithms for de novo short read assembly using de Bruijn graphs. Genome Res 18:821–829. https://doi.org/10.1101/gr.074492.107
15. Camacho C, Coulouris G, Avagyan V et al (2009) BLAST+: architecture and applications. BMC Bioinformatics 10:1–9. https://doi.org/10.1186/1471-2105-10-421
16. Li H, Handsaker B, Wysoker A et al (2009) The sequence alignment/map format and SAMtools. Bioinformatics 25:2078–2079. https://doi.org/10.1093/bioinformatics/btp352
17. Robinson JT, Thorvaldsdóttir H, Winckler W et al (2011) Integrative genomics viewer. Nat Biotechnol 29:24–26. https://doi.org/10.1038/nbt.1754
18. Lagesen K, Hallin P, Rødland EA et al (2007) RNAmmer: consistent and rapid annotation of

ribosomal RNA genes. Nucleic Acids Res 35: 3100–3108. https://doi.org/10.1093/nar/gkm160

19. Lowe TM, Eddy SR (1997) tRNAscan-SE: a program for improved detection of transfer RNA genes in genomic sequence. Nucleic Acids Res 25:955–964. https://doi.org/10.1093/nar/25.5.955
20. Hyatt D, Chen G-L, LoCascio PF et al (2010) Prodigal: prokaryotic gene recognition and translation initiation site identification. BMC Bioinformatics 11:1–11. https://doi.org/10.1186/1471-2105-11-119
21. Li L, Stoeckert CJ Jr, Roos DS (2003) OrthoMCL: identification of ortholog groups for eukaryotic genomes. Genome Res 13: 2178–2189. https://doi.org/10.1101/gr.1224503
22. Benson DA, Cavanaugh M, Clark K et al (2018) GenBank. Nucleic Acids Res 46:D41–D47. https://doi.org/10.1093/nar/gkx1094
23. Kanehisa M, Goto S, Furumichi M et al (2010) KEGG for representation and analysis of molecular networks involving diseases and drugs. Nucleic Acids Res 38:D355–D360. https://doi.org/10.1093/nar/gkp896
24. Tatusov RL, Fedorova ND, Jackson JD et al (2003) The COG database: an updated version includes eukaryotes. BMC Bioinformatics 4:1–14. https://doi.org/10.1186/1471-2105-4-41
25. Almagro Armenteros JJ, Tsirigos KD, Sønderby CK et al (2019) SignalP 5.0 improves signal peptide predictions using deep neural networks. Nat Biotechnol 37:420–423. https://doi.org/10.1038/s41587-019-0036-z
26. Krogh A, Larsson B, Von Heijne G et al (2001) Predicting transmembrane protein topology with a hidden Markov model: application to complete genomes. J Mol Biol 305:567–580. https://doi.org/10.1006/jmbi.2000.4315

Index

Katrin Janik and Mattia Tabarelli (eds.), *Phytoplasma: Methods and Protocols*, Methods in Molecular Biology, vol. 3008, https://doi.org/10.1007/978-1-0716-5104-9,

R

S

T

U

V

MIX
Papier aus verantwortungsvollen Quellen
Paper from responsible sources
FSC® C105338

If you have any concerns about our products,
you can contact us on
ProductSafety@springernature.com

In case Publisher is established outside the EU,
the EU authorized representative is:
Springer Nature Customer Service Center GmbH
Europaplatz 3, 69115 Heidelberg, Germany

Printed by Libri Plureos GmbH
in Hamburg, Germany